INSECT-BORNE DISEASES IN THE 21ST CENTURY

INSECT-BORNE DISEASES IN THE 21ST CENTURY

MARCELLO NICOLETTI

Academic Press is an imprint of Elsevier
125 London Wall, London EC2Y 5AS, United Kingdom
525 B Street, Suite 1650, San Diego, CA 92101, United States
50 Hampshire Street, 5th Floor, Cambridge, MA 02139, United States
The Boulevard, Langford Lane, Kidlington, Oxford OX5 1GB, United Kingdom

Notices
Knowledge and best practice in this field are constantly changing. As new research and experience broaden our understanding, changes in research methods, professional practices, or medical treatment may become necessary.

Practitioners and researchers must always rely on their own experience and knowledge in evaluating and using any information, methods, compounds, or experiments described herein. In using such information or methods they should be mindful of their own safety and the safety of others, including parties for whom they have a professional responsibility.

To the fullest extent of the law, neither the Publisher nor the authors, contributors, or editors, assume any liability for any injury and/or damage to persons or property as a matter of products liability, negligence or otherwise, or from any use or operation of any methods, products, instructions, or ideas contained in the material herein.

Library of Congress Cataloging-in-Publication Data
A catalog record for this book is available from the Library of Congress

British Library Cataloguing-in-Publication Data
A catalogue record for this book is available from the British Library

ISBN: 978-0-12-818706-7

For information on all Academic Press publications
visit our website at https://www.elsevier.com/books-and-journals

Publisher: Charlotte Cockle
Acquisitions Editor: Anna Valutkevich
Editorial Project Manager: Mona Zahir
Production Project Manager: Omer Mukthar
Cover Designer: Mark Rogers

Typeset by SPi Global, India

Contents

About the author

Marcello Nicoletti is full professor at the University Sapienza of Rome, Italy. He graduated at the same university in Chemistry (1974) and Industrial Chemistry (1976). Later, he went on to complete a postdoctorate at the University of Cambridge (England), focusing on biochemistry of aminoacids. Nowadays, he is full-time Professor of Pharmaceutical Biology at the Environmental Biology Department of the University Sapienza of Rome, teaching in the Courses of the Faculty of Pharmacy and Medicine, in particular Pharmaceutical Biology and Pharmacognosy.

His research activity was first dedicated to the study of the plants utilized in local traditional medicine. He has spent several years in many countries of South America and Africa, focusing initially on curare alkaloids, and later on antimalarial alkaloids and glycosides to treat benign prostatic hypertrophy. He specialized in the chemical, nutritional, and biological study of plant species, including the isolation and identification of active constituents and chemical markers useful for quality control. He developed integrated analytical techniques for the determination of qualitative and quantitative profile control of medicinal and food plants. This scientific interest involved the utilization of advanced analytic techniques in solving biological problems, including environmental impact factors and consequences of climate changes on diseases of global interest, like malaria and Zika virus.

Professor Nicoletti is the author of more than 300 papers in international scientific journals and several books. During the last years, more than 100 of his papers and book chapters have been dedicated to the control of insect-borne diseases by the utilization of new natural substances and plant-derived products. Among the utilization of natural products in control of insect-borne diseases, he has dedicated attention to nanotechnology applications, developed by the collaboration with several international research groups. He is co-editor of *Natural Products Research*. Recently, he has also taken an interest in meristemotherapy, a medical approach based on utilization of juvenile plant tissues to treat several pathologies, leading to the publication of the book *Gemmotherapy, Scientific Foundations of a Modern Meristemotherapy*, co-authored with Dr. F. Piterà, which has been translated into several languages. He lives in the countryside of Rome, spending his free time growing food for himself and his family.

Foreword

The book *Insect-Borne Diseases in the 21st Century* by Professor Marcello Nicoletti offers a unique multidisciplinary view about hot issues on insect-borne diseases on a worldwide scale. This is not a classic book about insect science. Indeed, the book takes advantage of the author's solid background in organic chemistry and environmental biology. Starting by citing the famous book *Silent Spring* by R. Carson, Professor Nicoletti moves forward to present novel concepts and new perspectives to face the overuse of synthetic pesticides in the attempt to tackle current insect-borne issues of economic importance. Recent emergencies about the spread of important vectors such as the Asian tiger mosquito, *Aedes albopictus* (Diptera: Culicidae), the spittlebug vectors of *Xylella fastidiosa*, and microbial pathogens have been considered.

The author considers insects attacking plants, animals, and humans, and summarizes recent advances in the field of "green insecticide development" to manage several of these issues. From a personal point of view, I very much admire the strong motivation and multidisciplinary preparation of Professor Nicoletti, who has worked hard in many countries worldwide on an outstanding number of natural products of interest for life science research, including curare alkaloids, antimalarial quinoline alkaloids, and insecticidal neem limonoids, to cite just some key examples.

Overall, the book covers an impressive knowledge about chemistry, biology, and pharmacology, outlining many connections between solid published research and our everyday lives. For these reasons, I believe that the present book will be of interest to many readers worldwide, from different research fields within life science.

Pisa, December 6, 2019.

Dr. Giovanni Benelli, Ph.D.
Department of Agriculture, Food and Environment, University of Pisa, Pisa, Italy

Preface

This book deals with diseases vectored by arthropods, with focus on insects, and caused by microorganisms, like bacteria, protists, and viruses. In other words, the arguments are mainly related to viral and bacterial illnesses derived from insect (bug) bites, like mosquitoes, sand flies, ticks, and fleas, but in this category we can also include plague, which involves a bacterium and a flea as well as rats. It is complicated matter, involving the role of several actors and targets, not forgetting the environment, the main factor dominating the life of any organism.

A book should fill a gap in the knowledge of many people about a selected argument or be able to turn on the fantasy of some reader. In the second case, limits are expected to be overcame, whereas in the first one, limits should be respected as the boundaries of the concerned argument. In a scientific sector, the book must be a bridge of knowledge, using what is already known to predict the future, by expanding the present. Books written without this belief risk consisting of boring repetition or being a useless exercise.

In this case, there were several important reasons to write an innovative book at this time and on this subject. It is necessary to explain what is going on in terms of the alerts concerning insect-borne diseases, involving human health as well as strategic agricultural and livestock production, and why they are part of a general phenomenon of global relevance. Information about insect-borne diseases risks being contaminated by interests of different types, which may be influenced by political factors or deviated by attempts to alert or sedate public opinion. All of this has happened before, is happening, and will happen again; the only difference is in the entity, and this depends mainly by us. Insect-borne diseases are natural occurrences; they are part of the natural equilibrium and become epidemics in particular situations. These situations are often caused by humans.

The book is dedicated mainly to scholars and specialists, but also to ordinary people who may be interested in understanding what is going on in the environment from a different point of view and the possible consequences in any sectors of daily life. The entomologic aspects and examples are considered as opportunities to reflect and introduce exercises to ponder the reality around us, challenging the dominant axioms and expanding arguments in several directions.

This book was inspired by the possibility of finding new ideas about a central argument for our future, even our survival. Its roots are in scientific experiences, which generated the main body with branches in several possible applications, and finally the shoots for possible new solutions. In many cases, the problem is a consequence of a rapid change, the crisis of a dominant paradigm, or the emergence of a deep disconnection inside a consolidated scenario. The actuality of the matter is evident, but the argument can be used as the key to interpret several events that have been occurring over the years. We can focus on migration of insects, but the interpretation of the same phenomenon can be found in human migrations. We can argue about pathogens and vectors, but the reasons of the arising of a disease is essential to understand impacts of the radical changes in act and our role inside the current phenomenon of spread and impact of the infection. We can investigate the present scenario to compare with the past and imagine the future, and find solutions.

We are focusing both on general considerations and some selected samples to be used as general representatives, using insect-borne diseases as the focus on human affairs. Previous facts about insect-borne diseases could provide important inputs and sparks for interpretation. Knowledge and interpretation of previous episodes are essential to understand what is going on now and in the near future. Several times mankind has erroneously considered insect-borne diseases to be forever defeated or passed away. The resurgence of such diseases must be considered a natural event and the methods previously successful will surely become inadequate.

That in essence is my introduction, but I would like to add some more information about the genealogy of this book and myself as its author. In several parts of the book, attention will be focused on the importance of innovations, which are usually the simple result of pure intelligence. The tendency to consider science as a collection of experimental data is retracted by many episodes wherein the fundamental advances were produced by pure mental work. We can perform any kind of experiment and obtain data as we want; the important moments are when we are thinking.

The ancient Greeks were able to perform incredible explorations of reality. To achieve these extreme results, they used only the potentiality of the mind, arriving at summits probably never achieved by humanity before or since. Imagine yourself, alone on a little island in the Mediterranean sea, sitting under the shadow of an olive tree. In front is the brilliant blue sea, around you are rooks and rare vegetation, and above is a shining sun animating all the colors of the landscape. Near you, a wizard is resting on a stone, looking for a hole to go inside. Then, we can start thinking about the possibility of an inner linkage of everything around. You have not special experience, no experiments or data to refer to. Only you and your brain. That's like mathematics and physics moved their initial paths.

First, you consider that everything—the water of the sea, the sun in the sky, the wizard, the stone—albeit so different in form, must have a common constitution, and this can be found by breaking up any object into smaller and smaller pieces. This process results in a unique unit that cannot be separated or reduced any further: this is an atom. These atoms are generated somewhere and they move from this common origin following parallel trajectories, until some of them deviate and during their new paths, they meet another atoms. From these impacts, new atoms or solid mixtures of atoms arise.

Of course, we now know that any atom is composed of many subparticles and it is possible to violate its integrity, but the model of Bohr's atom is still successfully used by chemistry, and the theory of the Big Bang and the subsequent inflation more or less tells the same story of Democritus' atomism.

Nowadays, however, we collect evidence of the reality around us, forgetting that Nature's mystery can tell us all. A simple collection of data and experiments, without considering the importance of the interpretation of any event and trying to obtain inspirations through the conscience, is a sterile exercise. In science, there is a current persistent tendency of collecting data in order to publish them. Most papers only report numbers as the results of one experiment. These experiments are obsessively repeated by changing just a few details of the investigation. The result is a terrible fragmentation of scientific knowledge and a loss of fantasy. The possibility of advancing hypotheses and personal interpretations is considered practically abnormal and left to extemporary popularizers, leading to terrible results and bad information. In a given sector of scientific investigation, published papers are at least very similar in the format of reference. A paper is considered acceptable if it is adequate to the conceptual form of the journal and the tables are well presented. Concern about very specialized aspects of the matter is dominant and often data are used to confirm previous axioms. However, a general view is necessary and interesting. This is also supported by my personal experience, considering my data as recorded by Research Gate Agency; my ordinary published papers register several reads and few citations, but papers concerning general arguments (in particular about environmental care and insect control including nanotechnology) have received thousands of reads and a relevant number of citations.

There are several instances of evidence of this tendency. In the last 20 years, the number of scientific journals has increased exponentially, and even the legendary *Nature* is now divided into several subcategories.

In several cases, the exaggerated cost made the publication a business. All this is very good for the curriculum and its impact factor, and is the necessary fuel for an academic career, but science needs integration of different knowledge and not only the exasperate specialization, because its prevalent aim should be solutions of problems of ordinary people. I imagine the embarrassment of many scientists, encountering people in the street who ask the scientists to explain in simple words the results of their research, and in particular if the results have changed in some way the quotidian life of ordinary people.

These considerations can be considered the background of this book and first emerge from an exigence of syncretism between different scientific inputs, like biology, chemistry, and ecology, but even culture, history, and arts, due to the multidisciplinary approach that insect-borne diseases require. In any case, this need is adherent to my cultural formation. Like the atoms of Democritus, my scientific life was deviated by crucial crosses, in details when I meet important personages (Fig. 1).

After 5 years of classical studies including Greek, Latin, and Philosophy at one of the historic Lyceums of Rome, the "Giulio Cesare," going to the University La Sapienza of Rome (now simply changed in Sapienza), whose name is dedicated to the goddess Athena, my idea was to change completely in favor of the exact sciences, in particular to Chemistry. The statue of Athena stands triumphantly at the center of the university's campus, and students are very proud of this symbol—with the exception of examination

Fig. 1 The main square of Sapienza University with the statue of the goddess Athena.

Fig. 2 Prof. Marini-Bettolo with the author, attending a congress.

days, when they believe it is better not to look at the statue. After 5 years of molecules and reaction, and a second degree in Industrial Chemistry, I was looking for a job, at which point Professor Giovanni Battista Marini-Bettolo Marconi (related to Giovanni Marconi) entered my life (Fig. 2).

At that time, the Professor was an eminent figure in the national and international scientific panorama. In 1950 he was President of the Italian Academy of Science and President of the Pontificia Academia of Sciences in Vatican, and teacher at the University of Rome and at the Catholic University "Sacro Cuore" of Rome. First, he decided a restructuring of my knowledge was necessary, and sent me to the University of Cambridge (England) for postdoctoral study in Biochemistry under the direction of Sir Alan A.R. Battersby. On returning from England, I worked in Rome at the Istituto Superiore di Sanità. In that period, Prof. Marini-Bettolo was the director and he was working with the Nobel Prize-winning Daniel Bovet on the chemical characterization of the curare alkaloids. I had the opportunity to participate in this very important research and therefore understand the importance and the pharmaceutical power of the natural products. From a mixture of plants in the middle of the intricate forest of Amazonia, a population of Indios without any scientific knowledge was able to select from the environment and correctly utilize substances with enormous biological activity. The study of curares evidenced the Amazonian bisindole alkaloids' capacity to block the nervous system and opened the way to modern medical surgery (Figs. 3–6).

Later, the professor sent me to Latin America, where most of the directors of Organic Chemistry in different countries were his disciples, to work on natural products from plants utilized in local traditional medicines. I spent 3 years in Brazil, several in Chile, and varying periods in other countries. My research included also important collaborations in Africa: Madagascar, Somalia, Cameroon, and Nigeria. In the years after the Second World War, the world had been divided in two blocs, USA versus USSR, without any possibility of interchange and collaboration. Using his position in the neutral zone of Vatican, the Professor was able to set up meetings between American and Russian scientists and to break down to some degree the dangerous competition between the two superpowers. He was very significant in the liberation of Sacharov, and I still have the photographs of the Pravda of his meetings with the Soviet Union President Breznew. The name of

Fig. 3 The blowpipe used by Amazonian Indios.

Fig. 4 The arrows with the curare on the tip of the arrow.

Fig. 5 The container of the curare.

Professor Marini-Bettolo is not written in the books of history, but his hidden work was important in making the world as now it is.

At the beginning of the 21st century, I was in trouble in my laboratory at the university; I was in a deep crisis about the results of my research. I had studied many plants, published many papers and some books; I had a brilliant academic career becoming rapidly full professor—but only one product in commerce in the pharmacies was a result of all these studies. I then received an unexpected visit. It was a delegation from Ferrero (manufacture of

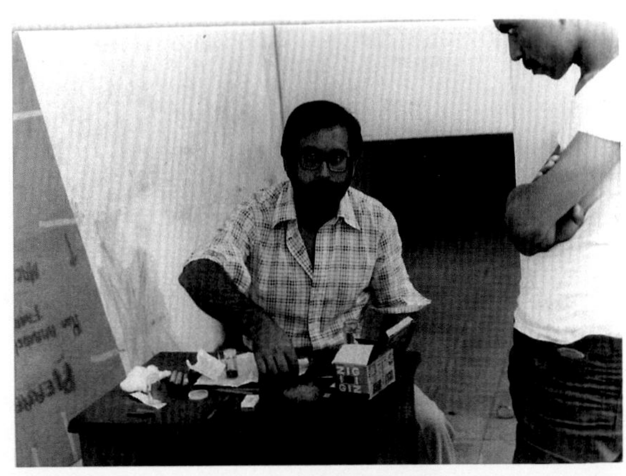

Fig. 6 The author examining the curare and extracting a sample for chemical investigation.

Nutella and other products) asking for my collaboration with a new product where they wanted to introduce extracts of digestive medicinal plants. In this way, I was able to learn the logic of the food industrial production and how to insert science into commerce. I also had occasion to meet Mr. Michele Ferrero, the leader of a global super-industry that employed 30,000 workers, and I learned why science should enter supermarkets in order to avoid terrible errors in alimentation. The linkage with the industrial world never stopped, and later I had occasion to work on products for pet care, fertilizers from algae, medicines from plant buds, and nutraceuticals, always searching for novel and rational utilization of natural sources. To support these products, it was necessary to invent a new analytical approach adapted to the control of quality of the complex natural products' mixtures.

During my search on the useful utilizations of natural products, I received an invitation to participate to a research about the new alien species *Aedes albopictus* (see Chapter 2) by my friend and colleague, Dr. Susanna Mariani. Susanna was working at ENEA, an institution created in 1980 to supervise the utilization of nuclear power. When, by national referendum, the utilization of central of nuclear power was banned in Italy, ENEA became an environmental research center. We started a collaboration to find the way to control the pest with the extracts of the neem tree, immediately obtaining important results (see Chapter 5). Therefore, most of my scientific interest was focused on insect pest control, and in particular reinforced by collaboration with Prof. Kadarkarai Murugan, Vice-Chancellor of Thiruvalluvar University, Tamil Nadu, India and his large group of researchers.

The knowledge of the Indian situation pushed me to focus my interest on this argument and the consciousness of its key importance for humanity. About the consequences of this further deviation, an enormous quantity of researchers made by groups worldwide converging on developing new approaches in pest control and a consequent high quantity of published papers. Special recognition must go to Dr. Giovanni Benelli, University of Pisa. Thanks to his special work on insect behavior, it was possible to substantiate most of the ideas and experiments developing insecticidal formulations to be used in integrated vector and pest management. Dr. Paola Del Serrone, at the Centro di Ricerca Zootecnia e Acquacoltura CREA ZA, Rome, was in charge of the microbiologic researches and the field experiments on livestock with neem. Other collaborations can be found inside the book and are reported in the references, including with Prof. Galeffi's team at the Istituto Superiore di Sanità and with Prof. Rasoanaivo of Madagascar, on the research of alkaloids for treatment of malaria.

Progressively, these arguments took the center of the experimental activities, with more than 100 published papers on insect-borne diseases and resistance in insects and microorganisms. As a collateral effect of these events, I received an invitation from Elsevier to write and publish this book. All collaborators at Elsevier were fundamental in supporting my efforts to give adequate form to the galaxy of my convictions on this subject. Another collateral effect is the possibility for you, reader, to consider the arguments in this book and perhaps contribute positively to solutions to these problems.

At the end of each chapter, there is a list of references concerning the treated arguments. It is a crude selection since the literature about insect-borne diseases is immense. For example, just for the term "Plasmodium falciparum," more than 60,000 references can be found; selecting the information to include was thus challenging. Most of the reported research concerns the last 10 years and this is significant, as it indicates that scientists are hardly working on this matter. However, the conversion of such efforts, for internal reasons like the fragmentation, and external condition like the utilization of produced data, so far gave risible results. Let us hope that this is a necessary preliminary gestation step, and in due course the research will be converted into advancements for life and health conditions.

CHAPTER ONE

Past, present, and future of insect-borne diseases

Image courtesy of Shutterstock

Insect-Borne Diseases in the 21st Century
https://doi.org/10.1016/B978-0-12-818706-7.00001-2

Listening to the silence

We start with two books, which contain seeds of the further arguments. The first one is *The Silent Spring* by Rachel Carson (Carlson, 1962). The second one is *The Origin of Species by Means of Natural Selection* by Charles R. Darwin (Darwin, 2003). Both had deep cultural and scientific influences, for different reasons, and nowadays they must be considered as the starting points of theories about habitats needing careful consideration.

In 1962, Rachel Carson published *The Silent Spring*, intending to document the effects on the environment of the indiscriminate use of synthetic pesticides, starting from the absence of the usual songs of birds in spring. The book explicitly identifies DDT and other pesticides as responsible for enormous damage to the environment, causing in general threat to wildlife and a series of negative effects like the increase of cancer cases in humans. Carson directly evidences the responsibility of the chemical industry not only in its excessive use of chemicals, but also by covering its activity through spreading misleading information about the consequences, in collaboration with public officials. Chemical companies reacted by masking and justifying their actions, but the book had a great impact on the American public, acting as a seminal event for the environmental movement and generating a strong debate about habitat care. The policy on the utilization of chemical insecticides was in part affected and another consequence was the creation of national agencies in defense of environmental equilibria. The argument is still totally open, since the effects of neonicotinoids on honeybees and birds are nowadays under consideration. However, the most interesting point is arguably that until *The Silent Spring* was published, all the above issues were considered acceptable and even normal.

Probably, no ideas were subjected more to misleading and manipulation than the results presented by Charles Darwin about animal social behaviors. A clear example of a diverted utilization of scientific theories by people, who never had read one word of his books. Some interpretations of Darwinism gave the impression of a giant omnipresent struggle among organisms for surviving and winning the competition for resources and reproduction, obtaining sexual advantages. However, we know that there are innumerable examples of cooperation, mutual help, positive coexistence, and symbiosis between very different organisms. However, there are also example of slavery by an organism for another to satisfy personal needs, and examples of clear sloth (as reported also by Darwin). This is probably the key to understand several factors affecting the future of the environment and humankind, from the gut microbiome to the Angiosperms reproduction.

In several examples, selecting those related to insect-borne diseases, the integration and incorporation between very different organisms—i.e., bacteria, insects, plants, and fungi—gives rise to a complete network, working as a natural mechanism, wherein each organism has a precise and defined role.

We are in the midst of epochal planetary changes, whose consequences are becoming more and more evident in the developing scenarios. The powerful weapons utilized by humans to control insect-borne diseases, consisting of chemical-made antibiotics and insecticides, are becoming useless. For a long time, the pathogen organisms were easily killed and controlled. Insects and microorganisms, thanks also to other allied organisms, after a long period of passivity are finally reacting properly to the lethal continuous attacks by humans, whose minimum goal was the complete extermination of these insects and microorganisms. So far, the counteractions by microorganism and insects have been mainly defensive and limited, but they are ready to become brutal and offensive, and perhaps decisive.

The counteraction by target organisms is the most obvious in any war: nullify the enemy's weapons by resistance, and fight back. This resistance is already being worked on, and is ready to become very effective on a large scale. The incoming front of the resistance's phenomenon is based also on more efficient methods of diffusion, including organisms so far latent and in a revision of strategies to survive and diffuse. Insect-borne diseases are evolving in this scenario and therefore they are ready to play again their central role in the never-ending fight for survival (Mehlhorn, 2015a,b; Mahmud et al., 2017).

Entomology will need to play an important role. However, it is necessary to reconsider its goals, which have too often focused on taxonomic problems, and consider the needs for new and original approaches able to face novel challenges. It is time to revise several dominating axioms on the light of the occurrence of a series of important phenomenons, which are acting as current motors of radical changes. In this book, we will introduce the key concepts of superorganisms, system biology, and bionetwork, and present some examples to verify these approaches. Examples must consider the several aspects involved, including target organisms or selectivity effects. It is necessary to understand what is going on and the role played by each organism. Several examples will be presented and their related solutions, based on current or recent episodes of public concern, including health and production of food. Therefore, the philosophy and the strategies reported will generally find their evidence and concreteness in selected cases.

The main goal of the book consists in encouraging readers to consider the possibility of thinking in another way, without accepting the dominant paradigms. The book will also ask each reader to contribute where possible to another possible style of living, considering surrounding organisms, such as insects or bacteria, to be not annoyances to be removed, but potential allies to play a daily and fascinating scenario with alternative costs and new hopes.

In this chapter, the two main actors of an insect-borne disease, the microparasite and the vector, will be examined, considering in particular their current evolution and consequent effects on the occurrence of diseases. Let us start with the parasite, in consideration of its key responsibility in the disease.

Several signals of changes are converging to create a new environmental scenario. The 21st century announced its advent with radical planetary events disclosing enormous impacts. Among the main influencing factors, the enormous advances obtained by technology are changing any aspect of our life. The instantaneous planetary connection is opening the door to a planet globalization of the information, but not only the news are travelling everywhere. Therefore, continuous innovations in ordinary life are fueled by a progressive dependence on the artificial intelligence, allowing the possibility to exchange everything can be moved, like ideas and materials of any kind, including pathogens and parasited. Changes are rapidly affecting quality of life and health, including deep evolvements in social organization, evident in the crisis of the tribal and family models. Continuous and rapid challenges of dominant paradigms inside the global network are actively changing the planet, but with different effects in each part. Most people consider these changes to be simple collateral effects of scientific advancements, whereas everything is still moved by the usual eternal motivations: the possibility of surviving and growing in the best environmental conditions, the research of habitat sources to be utilized efficiently in the best way, in total indifference of the consequences necessary to achieve the expected goal. Now, as ever, climate changes generate migrations of humans and animals, moved by their usual needs. In some cases, organisms move to conquer territories previously closed to them, disrupting previous equilibria. Migration is a natural phenomenon, and always has consequences (Bezirtzoglou et al., 2011; Lamb, 1995; Cook, 1992; Wigley et al., 1981).

Environmental changes can offer new possibilities, not only damages. Survival needs, or simply homeostasis rules and imperatives, push organisms to find better territories or more favorable living conditions. It is a thermodynamic contest, like water moving freely from two containers or the equilibration of the temperatures between two adjoining rooms. Continuously, the brave vanguard of any organism try out the boundary of their territory in search of opportunities. When movements are successful, they become massive and overflow, finding resistance from previous native inhabitants, but defensive damage is largely counterbalanced by the absence of the usual natural enemies remaining in the old territory. The fight for natural sources is open. Therefore, the alien species enlarges its distribution as soon as possible, whereas autochthonous species experience difficulties and the whole environment is highly affected.

It is important to focus on the mechanism of the migration. During the first steps of the migration, some epigenetic changes can occur, generating more aggressive populations, and these are more motivated to move. Several genotypes of the alien species can move in sequence and the strongest one takes supremacy during the starting step. When the rooting is completed and the migrant alien population is integrated and favorable, the second step, consisting of diffusion, can start with increasing efficiency, causing the dramatic diffusion, like an expansion of the oil stain can start with increasing efficiency, causing the dramatic diffusion, like an expansion of the oil stain. Therefore, the epidemic stage, so large, abundant, and evident at the height of its manifestation, is the result of the action of a super-selected vanguard. The success of the initial step is crucial, and explains why for a long time the invasion was not possible. The route of the invading species in the whole phenomenon, from the starting initial territory to the final one, can be visualized in the form of an hourglass. Over time, the sand in the upper glass tube (the starting population of the species in its territory) will decrease in favor of the other one (the new habitat), but it is necessary to satisfy the initial condition of flowing.

The environmental changes influence our ordinary life, including the possible advent of great threats, which will affect previous situations already ripening over a short or long time. The changes are just the development of previous situations already ripening over a short or long time. Several signals are announcing the incoming future, but correct interpretations and, in particular, necessary counteractions are largely lacking. Scientists are modern haruspices, like Cassandras dedicated to the interpretation of signals from the habitat, and as in ancient times they may not be listened to by powerful people, who are more concerned with maintaining their power. Therefore, the only hope is in the ordinary people. The rise of a general and capillary consciousness is the necessary key to face the new challenges, influencing the behaviors of everyone and forcing solutions that will benefit all mankind.

Let us recall the aforementioned concepts, using the key words: signals, interpretation, and counteractions. Lack of knowledge of this sequence increases the chances of something unexpected happening, such as a sudden catastrophe or real emergency. The natural consequences of the planet's movements and migrations of its inhabitants are considered unexpected and unusual. It is a return to times of ignorance. Once the hubris of gods was considered as responsible of outbreaks, and now we live in times when that ignorance is simply substituted by other kinds of fear, counterbalanced by frantic manifestation of man omnipotence. There are many examples of this aptitude. For these reasons, the themes of this book are fundamental and crucial to the pathway toward our secure future. A general consciousness of the peril of the current pathway is vital. Some causes of the changes can be attributed to the planet, others to human influences, but in any case the lack of counteractions in the right direction will be our fault. Selected cases will be exposed to evidence the ongoing trends and speculate on the coming years.

Possible scenarios are in conflict

Current changes in planet climate are going to be fundamental as never before for the success of every human activity and enterprise, from agriculture to trade. Temperature rise and desertification are generating massive migrations from rural areas to urban ones, and from the global south to the global north, remodeling animal and human distribution (Reiter, 2001; Mouchet and Carnevale, 1997; Córdoba-Aguilar, 2018). Technology is a key actor in this changing scenario. Resources availability and food production are highly dependent on access to high technology, causing a new form of colonization and continuous migration toward advanced countries for humans and constant consequent movement of the myriad creatures associated with human activities. Several factors are remodeling everyone's concept of life and welfare. Current changes are likely to be connected mainly with the increase in life expectancy, including the emerging of new pathologies and health disorders, and with the revolution in the nutritional environment, due to radical changes in food. If we want to imagine our near future, considering the effects of this "evolution sap" derived from these concomitant factors, we must consider the alternative utilizations of natural resources, facing the challenge between sustainability and overconsumption. In this moment, mankind are not alone, deciding their destiny.

In every moment of human history, we have had to face direct or indirect discrete formidable attacks from invisible enemies, able to threaten our lives. The COVID-19 outbreak is only a further example of the potentiality of parasite of changing economy and lifestyle. Although in decline, pathogen microorganisms and parasites remain among the main causes of deaths globally, and their virulence is far from being dominated. Each episode of this eternal fight is different and needs careful interpretation. In the last period, much attention was focused in production and use of insecticides useful against vector-borne diseases, with the aim of eradicating their presence and therefore save as many lives as possible. Epidemic emergencies can occur everywhere. Insects are vectors of important diseases involving non-human targets, causing important effects on plants and animals. Recently, some of these diseases rapidly increased in profile and generated great alarm about the potential consequences of their diffusion. The economic negative effects are enormous and damage to the local living system is dramatic. The global incidence of insect-borne diseases is relevant considering the population at risk and the number of reported cases, but the percentage of death is around 0.2% compared to 15% for tuberculosis and 0.3% for the similar disease influenza. These data indicate an endemic presence of these diseases, whose effects must be mainly considered from social and production points of view (Reiter, 2001; Mouchet and Carnevale, 1997).

Besides the insect-borne diseases concerning human beings, recent cases of widespread insect-borne diseases not directly endangering human health will also be reported. In these cases, no successful strategies or pesticides are available, but many new proposals have been presented (Lounibos, 2002; Wright and Sutherland, 2007; Khater et al., 2017; Benelli, 2015, 2019, Nicoletti et al., 2016; Rogers and Randolph, 2000; Tanwar et al., 2014; Benelli and Mehlhorn, 2018; Willcox et al., 2005).

Super agents, supervectors, and superbugs

Microorganisms are dominant in the planet's biomass and affect any organic equilibrium. One of the main roles of pathogen microorganisms in habitats is the turnover of organic matter. Nobody knows how, but they are able to feel the absence of life, and, as soon as possible, immediately after the death of an organism, a plethora of "wreckers," mainly virus, bacteria, and microfungi, assault the body to obtain short and available substances, useful as their food. A side effect is the cleaning and scavenger action on the habitat. Otherwise, we should be covered by residues of organic matter, as happens with plastic. The main reason is that molecules must be exchanged, and rapidly, to ensure new organisms can replace the old ones. This can be achieved by acceleration of the catabolism up to the point of death of the attacked target. Infection is the first step, often consisting of a small vanguard, which are usually destroyed by the natural defenses of the target, but in some cases not totally, causing a rapid increase of the infecting population. The start is the crucial step and needs first the introduction of the pathogen in the body of the target organism, but there is a preliminary act. Initially, the microorganism must be present near to the target—near enough to be able to obtain the inoculation. As a result, microorganisms need to solve the problem of their dimensions. The world around them is at least 10^6 bigger, and any movement is therefore virtually irrelevant. The difficulties of the microorganism reaching the target are usually insurmountable, preventing it from performing its mission. Furthermore, the utilization of abiotic agents, like wind or rain, have low probabilities of success. The transfer must be efficient and performed to keep the pathogen alive and efficient.

This problem is related to the next step and is focused on the diffusion of the infection, meaning reproduction of the agent and consequent propagation of the infected organisms. We have already established the key role of the vanguard, which is to test possibilities. In an epidemic scenario there is a starting point—an insignificant place on the map—wherein a more potent and efficient population unexpectedly appears, as a consequence of a genome change or a migration caused by environmental situations. The starting epidemic area is called the "plague focus'" or "plague reservoir." The Ebola epidemic started in one small village in Guinea and was able to generate a rapid spread in Central Africa. This minuscule key starting point is normal in biology. According to genetic data, the 7 billion *Homo sapiens* currently on the planet can be phylogenetically related to the population of a village in some part of Eastern Africa, later widespread in every part of the world. In the epidemic model, the diffusion initially develops slowly, but at a certain point it increases dramatically to reach the exponential curve, until it reaches a plateau due to the shortage of nutritional input in comparison with the quantity of the population (Fig. 1.1). However, the shape of the resulting curve, meaning the time to wait until the end of the epidemic, cannot be easily predicted, as is evident in many cases. It is often the same when we have a fever, being sure that in the next two days everything will be solved; we should probably be more patient, avoiding the

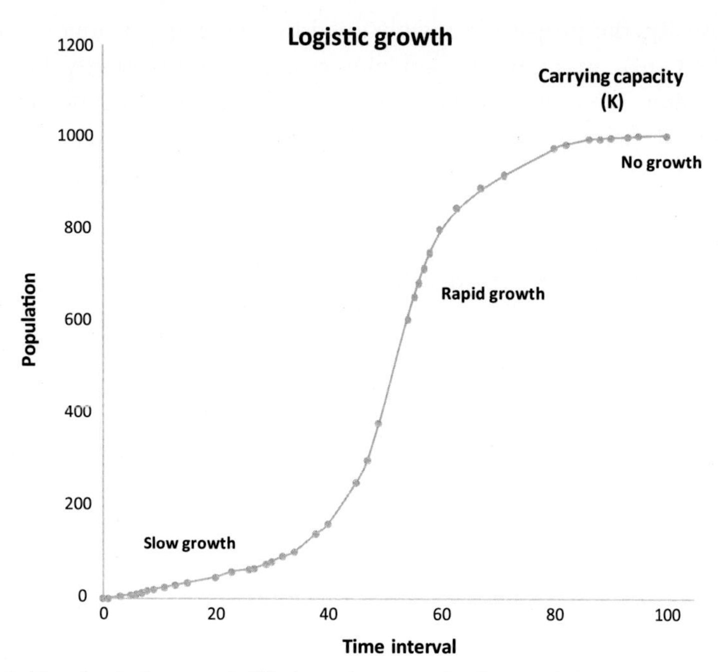

Fig. 1.1 The classical curve of diffusion of an organism's population.

use of drugs including antibiotics. Another complication is that the virulence of the insect-borne disease can differ in accordance with the agent, perhaps a virus or a bacterium or a protozoan, but the initial symptoms are very similar, and not very different from ordinary infections. Furthermore, medical doctors often do not have the necessary experience to make a diagnosis that relates to tropical diseases.

In other words, the detection of the origin of the disease and its nature are fundamental to act positively in the first steps, and also when choosing the therapy. In other words, more the patient is able to live, more are the possibilities to survive. After vaccination, giving the immunity system time to react is the most efficient therapy to fight and defeat the agent of the infection. However, vaccination is not always available, in particular for emerging diseases. A new insect-borne disease can contain several elements of novelty. Even the re-emergence of an already known disease will contain new aspects, and the approach of control must be reconsidered. This is evident in particular for viremias, which are predominant in insect-borne diseases.

Usually, the propagation is obtained by utilizing the subjects already attacked, like in influenza and pneumonia, not necessarily being the main target. Another efficient and smart method, essential in the first steps, consists of transportation, a sort of lift, by an efficient and rapid agent. In our cases, the adjutant is called a vector and is a flying insect or another arthropod. The choice of an insect, in particular a biting one, has excellent advantages. In this way, the disease can be diffused rapidly and the infection of a large territory achieved without any loss of efficiency or energy by the microorganism, which is hosted and protected. The insect is also in charge of the introduction of the microorganism inside the target. However, the main target is an organism very different from the vector, requiring important morphological changes by the pathogen. The result is an astonishing protean performing capacity, which is completely absent in organisms considered much more advanced and specialized, like us. Imagine waking up in a completely different form, like in Kafka's *The Metamorphosis*.

In Greek mythology, Proteus was a prophetic old sea god, subject to Neptune. Proteus knew all things—past, present, and future—but he disliked divulging what he knew. To consult him, you first had to surprise and bind him during his noonday slumber. However, even when caught, he would try to escape by assuming all sorts of shapes, from an elephant to a mouse, but this was possible only for a short period. If his captor held him fast and for long enough (like Heracles did), the god at last returned to his proper shape, gave the wished-for answer, and then plunged into the sea. The word *protean*, one meaning of which is "changeable in shape or form," is derived from Proteus. Proteus was able to assume whatever shape he pleased, and therefore, he can be regarded as a symbol of the original matter (Gea) from which the world was created. In our contest, it could be considered the metaphor of the ancestral bacterium, the first type of Life, still able to convert easily itself, thanks to an astonishing changing capacity, whereas we are forced to be coherent to our unique form. However, it is noteworthy that insect vectors, although in a limited way, are also able to perform metamorphosis.

However, I believe that the legend of Proteus can be considered as a metaphor of the research in biology. The researcher's aim is to understand what is going on in an organism under examination, but meanwhile the biological continuum is in action, able to change the results and the data already obtained, under the influence of a pressure due to a series of variables, acting together and generating confusion during the experiment and the interpretation of data. The researcher must be absolutely patient and consistent throughout their experiments, repeating everything many times, until the correct explanation discloses the shadow of the inherent complexity of the phenomenon, to obtain the coherent explanation.

Once inside its target, a microorganism is subjected to a series of changes to survive in its new environment and to multiply efficiently. In many cases, the consequence is a disease or even the death of the target. This appears to be an incongruence. Why should the etiological agent attack the host, causing debilitation and damage to its health? This is an energetic problem, i.e., transfer of negative entropy: the microorganism is a parasite subtracting the energy of the host (the transient habitat) for its homeostasis, i.e., to survive and reproduce.

Living systems, like any type of organic organisms, work by subtracting negative entropy from the environment. A similar phenomenon can be observed in other systems, like crystals, but life is based on a specialized order, constantly fighting against the environmental chaos, which is asking back its subtracted energy. Thus, insect-borne diseases, as well as resistance, are only consequences of these energy transfers, occurring to ensure survival and maintain the adequate favorable state. Once they have utilized the negative entropy of other living systems, they react in opposite manner, defending themselves and counteracting. The environment seems to have gained back its energy and the organism will desperately fight to survive until death, when its energy and its molecules will be recycled. Biologists are attempting to understand and reveal the mechanisms through which the negative entropy's transfer is obtained.

However, why should the microorganism kill the host, which is necessary for its reproduction and survival? After all the procedures to infect the host and the necessary metamorphoses? The reason for this apparent suicidal behavior is in the interest of the species versus that of the individual. The best way to propagate an infection is a dead target body: the cadaver is a perfect medium for the reservoir of the pathogen and its subsequent expansion, without the opposition of the immunity system. This is why corpses were often burned in ancient times, even though nothing was known about the strategies of the infective agents. Nowadays, we have a lot of information and science, but it is necessary to remember that if this is the general scenario, every infection is different and the strategy must be tailored carefully for any single case.

First act: The attacked microorganism reacts

We must consider the entity of the subject in question. Fear, terror, dear, and desperation are ancestral words associated with major pathogen diseases. Microorganism pathogens were from early on in history the most dangerous enemy to mankind. As already reported, in principle they are appointed with the important mission to clean habitats of degenerated organic matter and to accelerate the turnover of molecules by causing the death and decomposition of living organisms. The work of micropathogens is therefore continuous and necessary, but unpleasant when they cause diseases. In terms of the latter, micropathogens are able to cause death on a large scale. The history of humanity contains plenty of epidemic episodes that can be attributed to infections of various types. Let us consider, for instance, the ten plagues in the Bible and the situation at the opening of the *Iliad*, based on a medical emergency due to an epidemic causing the death of Achaean warriors and their animals.

The greatest catastrophe in the history was not a war, but a bubonic epidemic plague, known as the Black Death or Great Plague. The Black Death epidemic killed 30%–50% of the entire population of Europe, affecting between 75 million and 200 million people within a few years. The disease started in 1348 when the plague reached the harbor of London, where the city was extremely dirty and overcrowded, but no one knew what caused this dreadful pestilence. Only a few years ago from now, scientists confirmed by DNA analysis that it was caused by the bacterium *Yersinia pestis* and first appeared in wild rodents in places where they lived in great numbers and density. The plague reached humans when the black rats—rodents very common in human habitations—became infected.

The pandemic moved fast, spreading everywhere with terrifying speed and staggering mortality. Whole villages died within a few weeks, and fear spread even faster than the infectious agent. Some towns barricaded themselves in the futile hope of saving themselves via isolation. Mothers abandoned husbands and children—and vice versa—for fear of catching the contagion. Otherwise, fighting the contagion by fire was considered the final, unique solution. Houses and villages were burned to the ground—with the inhabitants inside, if they were known to be ill. Disease can spread easily, causing a new supply of victims, and every time the efforts to contain such pandemics become far more difficult. Ordinary parish burial grounds were insufficient to hold the massive numbers of dead, and new plague cemeteries were opened, often consisting of a mass grave roughly dug. The social and economic havoc created by a plague is almost beyond imagining, and the impacts of pandemics are still stamped in the minds of humans, such as when a new epidemic jeopardizes humanity or an old one reappears, like recent Ebola or Zika epidemics. It is noteworthy that the mass graves were the best source of information for recent studies based on molecular biology, which were able to rewrite all the story of human plagues.

Until recent times, the causes were totally unknown, but the reports are sufficiently clear. At that time, the physical responses were simple and linear: burning of the cadavers and performing any tentative to limit the area of diffusion of the outbreak by separation of the bodies. The spiritual sphere was involved with prayers and offers to divinities, the plague being considered by some a punishment for some unknown sin. For a long time such diseases remained a mystery, until a fundamental episode, consisting of the discovery of the secret world of microorganisms through the invention of the microscope, became evident in the classic case of ergotism. However, this was not a solution to the diseases.

The situation changed radically in our favor with the introduction of antibiotics. In 1929, penicillin was reported as an antibacterial agent by Alexander Fleming, produced by the mold *Penicillium*. In 1938, Howard Florey, Professor of Pathology at Oxford University, began research on the use of penicillin as a medical drug. Doctor Florey started his treatment due to the potential consequences of the war with Germany and the possible invasion of Britain. Therefore, he focused on cultivating the most productive mold and purifying penicillin. Microorganisms live in a very difficult situation, where most competitors are other microorganisms, like bacteria and fungi, searching for exactly the same opportunities to feed and grow. This microwar is not physical, but based on production of secondary metabolites, synthetized to damage the proliferation of competitors and called antibiotics by us.

In 1941, a police constable called Albert Alexander was the first patient to be clinically treated with penicillin as an antibacterial drug. Constable Alexander was a human volunteer, being in a terminal condition due to an infection accidentally achieved from a rose scratch 2 months earlier. Twenty-four hours after intravenous infusion of 160 mg of penicillin, the infection had begun to heal. After 4 days of treatment, Alexander was well on the way to recovery, but the stock of penicillin ran out. He died a month later. Therefore, the treatment was switched to sick children, who required smaller quantities of the drug, demonstrating the efficacy of this "miracle drug." In 1944, penicillin was followed by streptomycin, chloramphenicol (1947), cephalosporin (1948), etc., and selection of abnormal strains able to produce enormous quantity of antibiotics. The Age of Antibiotics was born.

Since then, antibiotics have been used successfully to cure a series of infections, such as septicemia, meningitis, pneumonia, and infections of sinuses, joints, and bone, with effects absolutely never experienced by human populations, but also with social consequences.

In advanced countries, life expectancies rose over the centuries in accordance with the increase of civilization and availability of food. During the Egyptians' age, life expectancy was about until 20 years. At the times of Jesus Christ and Alexander the Great, the life expectancy was 33 years, exactly the length of their lives, until an arrow or a pestilence or an accident or simply a deficiency of food or water might bring life to an end. Life expectancy then increased slowly until the advent of antibiotics, and within a few decades the estimated lifetime advanced from 50 to 65 years, at least in advanced countries. The introduction of antimicrobial drugs was fundamental to save lives from simple infections and insect-borne diseases. Nowadays, life expectancy in advanced countries is 77–78 for men and 81–82 for women. This difference occurs because during a woman's fertile years, she is protected by the production of special hormones. After this time, Nature's special care ends and the speed of aging is exactly the same for both sexes. Considering a future when cancer and cardiovascular diseases may be defeated, and aging may be slowed down due to better living conditions, life expectancy in advanced countries is considered to reach an average of 112 years. The rest of the world is still fighting an ongoing war against hunger, famine, and diseases, including old and new vector-borne diseases.

The use of antibiotics was extended to any kind of animal of interest and even in agriculture. The confidence in the value and efficacy of antibiotics was without any shadow. As a consequence, widespread abuse of antibiotics occurred. However, microorganisms are trained to react to environmental changes. In the exact moment that antibiotics were first used, some bacteria resulted that were resistant to these drugs. In other words, when a population of sensitive bacteria are exposed to antibiotics, they will mainly die except the few resistant bacteria, which are already present in the population or created by mutation. These bacteria can continue to grow due to the absence of competitors. Continuing the use of the antibiotics, they will be favored until they are the only dominant ones, causing the inefficiency of the old antibiotic.

Insecticide resistance

In 1946, a year after the use of penicillin became widespread, some Staphylococcus aureus strains had already become resistant to it. During the next decades, the cases of resistance raised exponentially, including strains of the most common bacteria, starting the phenomenon of the multidrug resistance. Resistance means that in a population of organisms, some of them develop the capacity to render harmful the substances or drugs currently used (Semmler et al., 2009; Mehlhorn, 2015a,b; Gale et al., 1981; Natham and Cars, 2014; Dondorp et al., 2009; Trdan, 2016; Karaagac, 2011; Naqqash et al., 2016). Multidrug resistance is the result of appearance of bacterial strains that could survive exposure to several different classes of antibiotics, and on the other side allied insect vectors became themselves resistant to insecticides. The eternal fight against our most dangerous enemies could be lost in the near future, or at least we may be defeated in the current battle. Several studies are predicting the end of the Antibiotic Era, when most antibacterial drugs will have no effect against microorganism attacks. This situation was already clear to some scientists, but it became an emergency when very important institutions alerted their populations about the incoming problems. Measures adopted so far are too late or insufficient. Currently, use of antibiotics is banned in agriculture and should be for farm animals only, in cases of real necessity, but the real issue concerns inefficacy due to increasing multi-resistance. Once used, the antibiotic remains in the processed food and is accumulated by the consumer. Considering the difficulty of producing new active molecules in the pharmaceutical industry using the established model, it is time to explore new solutions, like the use of natural substances, novel mechanisms of action, and multi-component drugs.

The eternal invisible and devious enemy is coming back. Nowadays in advanced countries the principal causes of death these days are cancer, cardiovascular diseases, and diabetes. In contrast, in the global south, the situation is practically unchanged, with malaria and other insect-borne diseases still dominant. The Antibiotic Age is now in decline. It is possible a more democratic equalized future, at least in terms of causes of death.

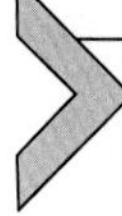

Timeline of DDT (dichlorodiphenyltrichloroethane)'s rise and fall (U.S. EPA, 1975)

1900–1935: Most insecticides are constituted of inorganic ingredients, and a few organic compounds, such as nicotine, pyrethrin, and rotenone.

1914: First records on resistance to inorganic insecticides.
1939: DDT's insecticidal action was discovered by the Swiss chemist Paul Hermann Müller (Fig. 1.2).
1940: DDT was introduced as an insecticide, becoming rapidly the principal actor of the period known as the "pesticide revolution," responsible of the wide utilization of pesticides everywhere. In this year the use of DDT became dominant in pest control. DDT was mainly employed with the aim of eliminating or controlling the density of undesired insect populations, but it also affects other insects. DDT was used latterly in World War II to control malaria and typhus among civilians and troops.
1945: In October, DDT was made available for public sale in the United States. Its use was promoted by the government and industry as a safe and efficient agricultural and household pesticide. Once DDT became available, it played a key role in the eradication of malaria in Europe and North America.
1947: Report on occurrence of DDT resistance in houseflies, followed by many other reports in next years. Among others, Dr. Bradbury Robinson, a physician and nutritionist practicing in St. Louis, USA, warned of the negative effects of DDT in agriculture.
1948: Müller was awarded the Nobel Prize in Physiology or Medicine "for his discovery of the high efficiency of DDT as a contact poison against several arthropods."
1955–1965: Relying largely on DDT utilization for mosquito control, the World Health Organization (WHO) started a worldwide program to eradicate malaria in countries with low to moderate transmission rates. The program was able to eliminate the disease in North America, Europe, and the former Soviet Union, and to reduce mortality in several countries. Therefore, it seemed possible to eradicate malaria forever; however, in practice the program was only really effective in areas with

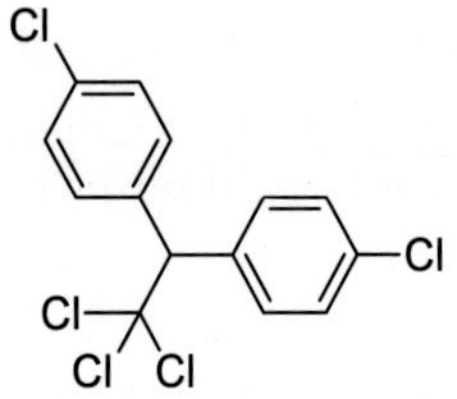

Cl
Cl
Cl
Cl Cl

Structure of DDT (dichlorophenyltrichloroethane)

Fig. 1.2 The structure of DDT.

"high socio-economic status, well-organized healthcare system and relatively less intensive or seasonal malaria transmission," as later reported by the same WHO. On the other hand, the failure to sustain the program everywhere resulted in an increasing mosquito tolerance to DDT and a parallel parasite tolerance, leading to a progressive resurgence of the disease. In many areas, early successes were partially or completely reversed, and in some cases rates of transmission increased.

1962: Rachel Carson published the book *The Silent Spring*, focusing on the concerns about massive use of DDT from the beginning of its utilization and denouncing the negative effects on habitats, including birds. The book had an increasing impact on public concern and generated a large public outcry about the environmental damage from widespread use of DDT and other pesticides, in particular in terms of harm to beneficial insects.

1972: Spraying programs (especially using DDT) were curtailed due to concerns over safety and environmental effects (accumulation of insecticide in the soil and in beneficial organisms), as well as problems regarding administrative, managerial, and financial implementation. Utilization of DDT was reduced and its agricultural use was finally banned in the United States.

2001–2004: A worldwide ban on DDT's agricultural use was formalized under the Stockholm Convention on Persistent Organic Pollutants, but its limited and still-controversial use in disease vector control continued in several parts of the world. Attempts at eradication were abandoned and attention was instead focused on controlling and treating the disease. Efforts shifted from spraying to the use of bednets impregnated with insecticides and other interventions.

2014: At least 590 species of insects were reported as resistant to insecticides as registered to one or more insects.

Second act: The post-antibiotic era

Meanwhile, although micropathogens were considered defeated, they were preparing a great return. Resistance can be extended to the entire repertoire of available therapeutic agents. Emergence of resistance to multiple antimicrobial agents in pathogenic bacteria has become a significant public health threat as there are fewer (or even sometimes no) effective antimicrobial agents available for infections caused by these bacteria. Gram-positive and Gram-negative bacteria are both affected by the emergence and rise of antimicrobial resistance. The problem of increasing antimicrobial resistance is even more threatening when considering the very limited number of new antimicrobial agents that are in development.

The economic costs of antimicrobial resistance are dramatic. For example, the yearly cost to the US health system alone has been estimated at US $21–34 billion dollars, accompanied by more than 8 million additional days in hospital. The European Centre for Disease Prevention and Control estimated that 25,000 deaths per year were caused by antimicrobial resistant organisms and a cost of approximately 1.5 billion euros per year and 2.5 million additional days in hospital, as reported by the WHO in 2014. In 1970, at least 440,000 cases of multi-drug resistance tuberculosis were detected in 69 countries, resulting in around 150,000 deaths. In 2011, around 25,000 deaths a year in the EU were caused by multidrug resistant infections, with the paradox that two-thirds of these were caught by hospital in-patients. Resistant bacteria from hospitals are also causing more "community-acquired" infections. Difficulty in treating infections with effective antibiotics has increased, because some resistant bacteria have also acquired toxins that make them more virulent, like leukocitin, which causes necrotic lesions that can kill patients in 72 h.

So far, most antibiotics have been used not for therapeutic purposes, but for prevention of infectious diseases in livestock and to promote animal growth in intensive livestock production, amplifying their diffusion behind the abuse of medicinal drugs. Therefore, European farmers are moving to alternative measures such as improved husbandry, increased biosecurity, and nutrition, as well as selective vaccination programs. As a consequence, more than 80% of Organisation for Economic Co-operation and Development (OECD) countries have banned the use of antibiotics for growth promotion, but across developed and developing countries they are widely used to prevent disease, and often when one animal becomes sick the whole herd is treated.

After a long period of prevalence thanks to antibiotics, bacteria are getting out of control. What will happen when there are no more antibiotics left to treat infections? This possibility is not so distant as we might think.

In 2011, Margaret Chan, General Director of the WHO, choosing the theme "Combat Drug Resistance," reported:

"We are now on the brink of losing this precious arsenal of medicines. The use and misuse of antimicrobials in human medicine and animal husbandry over the past 70 years have increased the number and types of microorganisms resistant to these medicines, causing deaths, greater suffering and disability. If this phenomenon continues unchecked, many infection diseases risk becoming uncontrollable. In the absence of urgent corrective and protective action, the world is heading towards a post-antibiotic era, in which many common infections will no longer have a cure."

Her words evidence a change in the antibiotic story, when the inadequacy of medicine enters into the problem. The infinite trust in the power of therapy is cracked.

Let's reconsider the resistance phenomenon and the possibility of avoiding its insurgence. To be effective, an antibiotic or an insecticide should be lethal to the great majority of individuals in a normal population. The treatment can lose its efficacy if many populations, or many individuals in a population, develop resistance to the toxic effects. Let us focus on this key point, considering that it is only a further example of the consequences of the human tendency to overexploit natural resources, in order to obtain the maximum effects and not considering the consequences. The problem is inherent: resistance is related to a massive and persistent use of chemicals. Many species may have numerous resistant populations, which can resist one or many treatments. As a result of the chemical treatment, some individuals in a among the population become resistant. Individual genomic differences are inherent to biology. Sensitive microorganisms exposed to the chemical will die, except for the few resistant ones, which can continue to develop and proliferate. Continuing in this way, they will be favored. More use of chemicals fuels the dominance of the resistant part of a population. The consequence of the mechanism is that sooner or later, medicines or insecticides that were once effective are not sufficient to control microorganisms or insects, respectively. However, recalcitrance is not solely caused by resistance but also implies peculiar cells, named persistents, which are drug-tolerant. Tolerance of persistents is not genetically manifested, since they are as susceptible as their parent strains. Stress responses, as in the case of antimicrobial use, may act as general activators of persistents formation.

Nowadays, the situation concerning the future efficacy of antibiotics is wondering, but the key argument is that current resistance could be only the tip of the iceberg. Resistance could be associated with changes at the genome level, giving rise to more virulent organisms. In this case, the scenario changes radically and dangers arise. Bacteria can resist the action of antimicrobial agents by several mechanisms: target modification, target over-expression (e.g., folate inhibitors pathway), antibiotic inactivation (e.g., beta-lactams and aminoglycosides), and modifications of the outer membrane permeability by reducing the expression of outer membrane-proteins (OMPs) or by increasing the expression of multidrug transporters.

Resistance to chemotherapeutic agents can be the consequence of horizontal or vertical transfer of resistance genes and/or intrinsic resistance arising by adaptive response to drugs exposure.

In nature, i.e., in normal situations, the mechanism of multi-resistance is possible, but very unlikely, and challenging to develop. A massive introduction of human-derived products in the environment enables exceptional effects to occur, like speed genomic changes, outside of the Darwinian natural selection laws. Newer organisms are forced to spread out as a consequence of this unnatural treatment. Therefore, resistance could be considered a natural phenomenon, which can be simply considered as a retort of the micropopulation to abnormal environmental conditions. The meaning of the world abnormal, in this case must be identified with massive utilization of antibiotics or insecticides.

Antibiotics nowadays in use, just as is the case for insecticides, all act with three or four mechanisms, and in a short time multi-resistance is a reality. Multi-resistance means that all the substances in use lose any effect, independently by the structure and the action mechanism. Multi-resistance is an important phenomenon first studied in medicinal drugs, like those used against cancer. It has generated several novel concepts in treated complex pathologies.

Third act: The antibiotic emergency

More than 150 antibiotics belonging to at least 17 different classes are now potentially available. They are used mainly for medical treatments and for farm animals and pets. Each antibiotic operates on a specific target or site within the bacterial cell. On the other side, the microorganism has a defense to counteract the effect of the drug.

The range of antibiotics' mechanisms of action is large. Very common antibiotics attack the cell wall, inhibiting the wall synthesis. This class includes the beta-lactams, i.e., glycopeptides, cephalosporins, carbapenems, monobactams, and glycopeptides and cyclic lipopetides, including daptomycin. The response of the bacterium is the enzymatic cleavage of the beta-lactame ring. Other antibiotics act on cell membranes (like polymyxins) or at the metabolic level inhibiting synthesis of proteins (e.g., aminoglycosides, chloramphenical and tetracycline), nucleic acids (e.g., fluoroquinones, rifamycins), or target particular biochemical pathways (e.g., methotrexate, sulfonamides) or cross-link to cysteines on enzymes (e.g., metronidazole). The response of antibiotics includes alteration of the target site, bypassing an inhibited reaction by alteration of the metabolic pathway, reduced drug accumulation by decreasing the drug permeability, expulsion of the drug, and dilution of the drug's concentration inside the target cell.

Among the issues at the beginning of the 21st century, we must consider a sort of rapid swinging in political tendencies. Any speculation about the future is complicated by this and other factors causing rapid and dramatic changes of counteractions. Further steps in the struggle against multi-resistance are not clear. It is necessary that any counteraction should be based on the knowledge of the peculiarity of the phenomenon and its effects, but other influences can interact and confuse the context. The same debate is taking place about consequences of climate changes, even if a real change is going on. Although scientists are almost in total accordance about the situation and its strong impact, as well the need of rapid responses, the related decisions are out of their hands. What is going on in the US administration is a clear example, with the radical changes after the election of Donald Trump, whose administration stopped former important acts in the right direction. The administration of former US President Barack Obama stepped up its efforts to combat the rising problem of antibiotic resistance. In 2014, Obama's policy started a series of acts to face the multi-resistance phenomenon. First, he signed an executive order establishing a new inter-agency task force charged with developing a national strategy to combat antibiotic-resistant bacteria.

Dr. John Holdren, Director of the White House Office of Science and Technology Policy and Assistant to the President, said the problem is a serious challenge to public health and national security: "We are clearly in a fight against ... bacteria where no permanent treatment is possible." The order also established a Presidential Advisory Council made up of non-governmental experts, who would provide advice and recommendations to strengthen surveillance of infections, research new treatments, and develop alternatives to antibiotics for use in agriculture. The administration released the "National Strategy on Combating Antibiotic-Resistant Bacteria," a 5-year plan to prevent and contain outbreaks and develop the next generation of tests, antibiotics, and vaccines. The President's Council of Advisers on Science and Technology (PCAST) reported on future scenarios and also released their opinions on combating antibiotic resistance. There are three main components to the report: (a) improve surveillance of antibiotic-resistant bacteria and stop outbreaks; (b) increase the shelf-life of current antibiotics and develop new ones, as well as promote research accelerating clinical trials; and (c) finally increase economic incentives to develop new antibiotics. A $20 million prize was set up to be given to spur development of tests that health care professionals can use to identify highly resistant bacterial infections. The task force, responsible for counteractions, was to be co-chaired by the secretaries of Health & Human Services, the Department of Defense, and the Department of Agriculture. The task force submitted its national action plan to the President by February 15, 2015. As a consequence, in March 2015 President Obama declared the fight against multi-resistance and the decision to ban definitely every use of antibiotics for farm animals.

Obama's resolution is the result of an incredible situation. Most antibiotics used in the US and UK are given to animals, and not for therapeutic purposes. In 2001, the Union of Concerned Scientists estimated that more that 70% of the antibiotics used in the US were given to animals reared for food (chickens, pigs, and cattle) also in the absence of diseases. The idea is that antibiotics at low doses not only furnish a shield against microorganisms, but also promote health and therefore growth in farm animals, although there is no scientific evidence of these effects. The situation in the EU is not very different despite more restrictive regulations. The situation is similar for human use. In most countries, antibiotics are available from pharmacies without medical prescription, or this restriction can be easily overcome. This means that anyone who feels sick with an upset stomach or diarrhea or who feels any symptom of influenza can directly buy whatever antibiotic treatment is available, whether it is appropriate or not. Treated or untreated fecal matter is a source of contamination for water supplies. The potential for rapid spread of antibiotics in the environment appears to be greater in emerging countries, as well as in Europe and North America, but we must consider also the greater availability and the major possibility of expanding the sanitary service and reacting in case of emergency, epidemics, or pandemia, in richer countries. Obama's resolution was a crucial sign, but it was subjected to political influences and economic interest. It ended a long strike with farm industries that previously had been successfully blocked by the food and pharmaceutical industries.

Is it time to face the possible end of Antibiotics Era? Considering the consequences of the absence of these products and the necessity of their presence, we have only one answer: declare the end of the first period of antibiotics, and continue directly with the second period. We must hurry up, since we have only 30 years to solve the problem. The way to solve the problem is based on the exploration in other directions with research and helping mankind to progress the lifestyle against any dystopia. Once again, if this is the scenario proposed by scientists and confirmed by common experience, the consequent decisions can be incoherent. Furthermore, in these arguments there is always a sort of tunnel vision, focusing on human health and not considering side environmental effects.

Of course, there are already signs of reactions. The WHO called for "push" incentives to encourage certain classes of antibiotics. The US Senate introduced the Generating Antibiotic Incentives Now Act in 2011, to "spur development of new antibiotics to combat the spread of antibiotic resistant bacteria." However, President Trump is now denying any global environmental changes and effects of antibiotics, just as previous president Ronald Reagan did, who focused his electoral campaign on this argument.

Nowadays, the Trump administration is resisting the WHO's effort to limit sharply use of antibiotics in farm animals, a move intended to preserve these drugs' effectiveness. Instead, the US is helping to draft an alternative approach that appears more favorable to agribusiness, since antibiotics are mainly used as growth promoters, although this use should be not allowed. Therefore, on name of the independence of any country in matter of developing, the U.S. Agriculture Department termed the "WHO position as shoddy science," causing the leave of WHO representatives out of the agreement to avoid potential conflicts. This is a clear example of a conservative position. Even if one considers the scientific considerations about resistance to be exaggerated, research on this field should be reinforced, just in case.

In recent years, the spread of antibiotic resistance among bacteria has reached worldwide proportions. In ordinary consideration, antibiotics have realized a utopian role as a panacea against all kinds of invisible attacks to health. All the policies and measures intended to contain or slow the development of antibiotic resistance have become inadequate and immediate actions are vital to understand the phenomenon of "antibiotic resistance pollution." The use of antibiotics in a wide range of applications, from health care to agriculture, is key to the evolution of antibiotic-resistant organisms following the phenomenon defined as "use it and lose it." This wide utilization of antibiotics means that the limit of use in medical drugs is only part of the problem, as a consequence of the food production chain. However, lacking of other solutions and drugs, in this moment we cannot renounce to antibiotics.

So far, as matter of fact, the pendulum between outbreak and control of pathogens resulted in a stand by and, as for the environmental changes, no real solutions or changes came from the rituality of global conferences and consequent declarations. Therefore, the key word is "new." To face new challenges, we need new solutions. To obtain new substances we must explore new possibilities, and to do it we must be brave and intelligent. Most pharmaceutical companies are not able to be free in this exploration.

We must learn from our enemies. In the struggle, they are able to adapt quickly and even sacrifice their precious past. These organisms, so simple and microscopic, have ruled Earth from the beginning and now, again, they try to claim absolute dominion.

To face an epidemic alert, attention is usually focused on the present, in search of efficient responses to halt the disease as soon as possible. The expectation is that usually the disease disappears, but this could be only for a while, and then the disease may return, stronger and more virulent. Activities should thus be concentrated also on the reasons and the events related to the origin of the disease. It is worth focusing on point zero at time zero—the start of the disease, when the pathogen changes and becomes a powerful danger. In the past, the organism was subjected to control by antibiotics or insecticide. Resistance modified the scenario radically. What are the origins of resistance? How does the resistant strain rise in the population? What factors activate and influence this event? What was the deflagrating factor, which was able to cause a significant change inside a species that suddenly became so destructive after a long history of normal disease? Let us reconsider the nature of the pathogen in the light of actual knowledge and some recent important inputs.

In clinical practice, an organism is resistant to a therapeutic agent if treatment with that agent results in clinical failure at the in vivo concentration achieved. As we have seen, resistance is the key phenomenon to understand the present and imagine the future of the therapeutic treatment. Resistance involves both protagonists, insect vector and microorganisms, in the first place, and is a consequence of human activity. Most resistance factors can be intrinsic or acquired. The acquisition of antibiotic resistance genes is possible through the acquisition of genetic mobile elements such as plasmids, transposons, and gene cassettes. In bacteria, horizontal gene transfer (HGT) plays an important role in the spread of antibiotic resistance genes and virulence factors also among phylogenetically unrelated organisms.

Bacteria can resist the action of each antimicrobial agent type by several mechanisms:

- target modification (e.g., fluoroquinolones);
- target over-expression (e.g., folate inhibitors pathway);

- antibiotic inactivation (e.g., beta-lactams and aminoglycosides);
- modifications of the outer membrane permeability by reducing the expression of outer membrane-proteins (OMPs); and
- increasing the expression of multidrug transporters.

Genetic lessons

An evolutionary change should consist of a genetic modification, resulting in the acquisition of characters in better accordance with the habitat changes (Barlow and Hall, 2002; Andersson and Levin, 1999; Miller Jr, 2013). The correlation between environmental factors and related genome changes is not easy to verify. A character may be the result of the co-occurrence of several genes and casual breeding. A gene can consist of thousands of bases, and the change of a single base in a key position can generate the production of an amino acid or something else, but this does not result automatically in a phenotypic change, as is evident in genetic studies. Most attempts at correlation between the presence of a certain allele and the phenotypic character of a human population were wasting time and misunderstanding, although in several cases the occurrence of an interested utilization of data can be usually used to denigrate or morally criticize certain targeted groups. Like the phoenix rising from the ashes, eugenics reappears with its promises of improving genetic quality by excluding. Neanderthals and *Homo sapiens* were considered a priori incompatible, until the DNA sequences evidenced exactly the contrary. However, science needs models and formulas, based on the study of simplified situations.

The lesson of genetic starts with Mendel's laws, although they are not really useful to clarify our genetic origins, and all the results reported by the agencies to reveal the pedigrees of confident citizens are arguably a smart method to accumulate money, flying the flag of genome sequencing reliability. The reality is that in very few cases the human characters were demonstrated as a consequence of an adaptation response to the environmental pressure in defined regions, although everybody knows the example of melanin production. Interestingly, another successful example is related to insect-borne diseases. Among some populations subjected to malaria caused by *Plasmodium falciparum* and *Plasmodium vivax*, owing to a mutation of gene of hemoglobin beta, the shape and structure of red cells are altered. The erythrocytes changes their usual classic discoidal round form into a stretched shape, and therefore acquire partial protection against malaria. In sickle cell anemia, the red blood cells become rigid and sticky and are shaped like sickles or crescent moons. However, there are two variants of the mutation. People with only a single changed copy of the gene will have a sufficient number of regular blood cells and few symptoms of the disease. People with two modified genes will develop pains, infection, and ictus, because irregularly shaped cells can get stuck in small blood vessels, which can slow or block blood flow and oxygen to parts of the body. However, they will also develop a resistance against malaria as a collateral effect, since the *Plasmodium* is used to living inside normal blood cells. The shift in blood composition toward the prevalence of modified red cells is shown in Fig. 1.3.

The development of antibiotic resistance by bacteria is considered by evolutionists to be a demonstration of evolutionary change. Bacteria can be used as an appropriate model for studying evolution steps in a homogenous population selecting the environmental factors and their influences. Therefore, the so-called "evolution in a Petri dish" is based on some aspects very useful in the study: rapid rate of replication, easy of analysis and detection of changes, wide range of conditions generated in the laboratory, and recently molecular analysis of the bacterial genome, in such a way that the responses of various strains can be compared. The bacterial genome is a powerful tool to understand trends, since it is possible to use it to write the events of thousands of years of evolution.

The development of antibiotic resistance was referred to by Miller as the consequence of evolution's "creative force" and by Barlow and Hall as "the unique opportunity to observe evolutionary process over the course of a few decades of the several millennia that are generally required for these processes to occur." In this way, an evolutionary change is the result of a so-called adaptation, which is a consequence of "beneficial" changes in the genotype. The consequent descent with modification from the prototype should require "evolutionary" acquisition of characters obtained by a vertical sequence of generations, which means a consequence of mutations capable of the adequate genetic changes.

Fig. 1.3 The comparison between normal blood cells and sickle cell anemia.

The study of resistance mechanisms, although forced by the necessity to find a therapeutic solution, has afforded important information about the biology of microorganisms, including real discoveries. Molecular biology studies on bacteria have shown that these organisms, so simple and microscopic, are able to become resistant by different and unexpected mechanisms, some already discovered and others under examination.

This is a classic interpretation of the genomic response in case of the resistance phenomenon. We can imagine the situation as a tensor acting in vertical and determining the normal sequence of the generations in the species and another tensor operating as a disturbance able to generate changes in the above sequence.

Resistance to chemotherapeutic agents can be the consequence of horizontal or vertical transfer of resistance genes and/or intrinsic resistance arising by adaptive response to antimicrobial exposure. In Gram-negative bacteria, many of these genes are associated with mobile genetic elements such as plasmids, transposons, and gene cassettes with their integrons.

In fact, this is an efficient mechanism quite common in resistant bacterial strains, but it can account only for the rapid spread since resistance genes must be present already in the bacterial world. The only origin of real evolutionary changes must be in mutations that must be regarded as beneficial when they increase the survival chances of bacteria in the presence of antibiotics, by a homeostatic mechanism. However, resistance can occur in the reverse direction to the ordinary sense, i.e., subtracting and not adding. This mechanism is often underestimated or even considered absent in multicellular organisms, but it has been clearly evidenced in unicellular populations. This approach could be a key explanation of many evolutionary changes for many organisms, including mammalians. The general interpretation of an evolutionary step consists of an advance of the complexity and possibilities of an organism, but this is not correct. In these cases, modifications consist in reduction or loss of cell functions previously occurring, such as lack of membrane selective transport by proteins or porins and protein binding affinities, a decrease or block of enzyme activity, and modification of proton motive force and other regulatory control systems. In such cases, introduction of a cell drug can be efficiently limited or the drug easily extracted, disarming the drug by limiting its concentration inside the cell.

Another hypothesis of mechanisms concerning resistance is based on genetic mobile elements, which are part of a huge "intrinsic resistome" in bacteria. The resistome is composed of genes of varied phylogenetic origin that act as resistance genes only in the presence of the ultimate drug, as a form of survival by the target organism. The success of resistant organisms contributes to the constant accumulation of a genetic platform and vehicles able to recruit and spread novel resistance genes efficiently. This phenomenon is called "genetic capitalism."

Antibiotics and synthetic insecticides produce effects that should be analyzed under the light of the multilevel selection theory. All biological-genetic elements at any level of hierarchy should become targets for intervention against the antibiotic-resistant phenomenon. Such a perspective indicates the need and possibility of drugs acting not necessarily to cure the individual, but to cure specific environments from resistance and to prevent or weaken the evolutionary possibilities of the biological elements involved in this phenomenon. This approach is referred to as the "ecological and evolutionary" approach.

Classes of antibiotics

Antibiotics belong to several classes that are classified by their mode of action (Table 1.1). As consequence, the organism target can generate several ways to counteract the pathogen. The general idea is that in the next 20–30 years, bacteria could develop resistance to any kind of antibacterial drug, independently of the mode of action. This situation is called multi-resistance and means that it is necessary to investigate several areas to understand the what, where, and when of the insurgence and success of the resistance. Let us start from the current situation.

It is important to stress again the importance of the resistance phenomenon in microbial pathogens, always considering that in all this argumentation, most of the considerations with some obvious differences can be translated to insecticides. However, microorganisms could be considered a better model to observe, study, and understand the phenomenon.

Research in the field of antimicrobial agents has reached a dead end. The production of antibiotics was and still is totally in private hands. Large pharmaceutical companies, for economic reasons, are no more interested in developing new antibiotics. The costs of research and the risk of antibiotic failure caused by the almost simultaneous appearance of resistance are unsustainable. Based on clinical outcomes, the costs of new antibiotics would too high to be comparable with the older ones. This is only one aspect, probably the most searing of the general problem in producing new medicinal drugs because of the enormous cost of the clinical trials required. Research centers, mainly within universities, are working to explore new solutions, but the shortage of public grants is a great limit.

Table 1.1 Summary of antibacterial classes and mode of action.

Antibacterial classes	Mode of action
Penicillins, cephalosporins, carbapenems, monobactems, glycopeptides, polypeptides	Cell-wall construction inhibitors of peptidoglycan synthesis or cross-linking functions resulting in osmotic lysis
Lipopeptides, polypeptides	Cell-membrane disruption altering the structure and function of the cell membrane, thus causing cellular leakage
Aminoglycosides, tetracyclines	Protein synthesis inhibitors binding to the 30S ribosomal subunit, thus preventing translation initiation and tRNA binding
Macrolides, oxazolidinones, streptogramins, phenicols	Protein synthesis inhibitors binding to the 50S ribosomal subunit, thus disrupting translocation and peptidyl transferase activity
Rifampin	RNA synthesis inhibitors preventing the synthesis of mRNA by binding to DNA-directed RNA polymerase
Quinolones	DNA synthesis inhibitors prevent DNA replication by binding to topoisomerase IV or DNA gyrase
Trimethoprim	Folic acid metabolism inhibitors preventing the synthesis of nucleotide bases by blocking the synthesis of tetrahydrafolate
Sulfonamides	Folic acid metabolism inhibitors inhibiting nucleic acid synthesis by preventing the synthesis of folate

Toward a new antibiotic age, or the end of a fundamental health tool?

The beginning or the end? Among the new approaches, we must consider the increasing consideration assigned to the whole human microbiome, in relation to homeostasis and performances of the human body. Remember that, until the human genome changes in terms of acquiring resistance to bacteria, we are dependent on antibiotics. The possible speed of change is about 0.1% in 10,000 years, although we can see that something has already moved under the disease pressure. Use of antibiotics in the last 70 years has transformed human health. We can survive bacterial infections that routinely killed our ancestors, but without new antibiotics, we may soon be exposed once more to terrible epidemics. In any case, we need a solution, urgently.

Recent studies on our microbiota are changing radically our point of view about bacteria. Usually, people are concerned about microorganisms only in the case of infection. Attention is focused on effects of the presence of pathogens and people are only interested in killing the bad bacteria as soon as possible to reach the previous status of health again. The problem is that the consequence of antibiotics treatments are not only beneficial. We have more bacterial cells (around 10^{14}, accounting for 1–2.5 kg of our body weight) than eukaryotic cells (around 10^{13}). The bacterial biodiversity is far more abundant, considering that in our gut, more than 500 different species have been found, albeit 10 are predominant, and in our mouth we have hundreds of other species different from those of the gut; the human skin carries several hundred more. Introducing huge numbers of fast-growing and virulent bacteria, any antibiotic resistant genes that appear in response to selection pressure could become established and continue to spread. However, the most important cause for alarm concerns not their misuse as medical drugs, but the environmental aspect.

The widespread antibiotic resistance among pathogenic bacteria affects not only the treatment of infectious diseases, but also many other medical practices such as surgery and immunosuppression in transplants. The spread of antibiotic-resistant bacteria in hospitals "means that commonplace medical procedures once previously taken for granted could be conceivably consigned to medical limbo. The repercussions are almost unimaginable" (WHO "Overcoming antibiotic resistance report," 2009). Hospitals, instead of being helpful, constitute one of the main reservoirs of antibiotic-resistant microorganisms. Patients with resistant bacterial infections are in close proximity with other patients whose vulnerable states make them susceptible to acquire such nosocomial infections.

The studies of our microbiota have completely changed the paradigm of the role of the invisible enemies. Currently, we know that without this microscopic symbiotic help, we could not be able to live and that even our feelings are probably influenced by our microbiota. Therefore, we know that there are good and bad bacteria. In addition, a good bacterium can change and become aggressive and dangerous. The microbe world is continuously subjected to change according to its environment.

Furthermore, it is clear that our microbiota are different and change in parts or organs of our body. Therefore, probiotic and prebiotic products must be tailored on this consideration. Products must be also tailored for ambient situation. For instance, a computer keyboard is usually a preferred area for some aggressive bacteria, and several bacteria may be transferred by the use of the same computer by different persons. Therefore, in particular epidemic situations, some places must be monitored and cleaned. The cleaner must consider the type of bacteria usually present in this case. A place can appear clean, but might not be disinfected. Nowadays there are very simple and low-cost kits to detect the presence of dangerous bacteria, but few people know them and very few are using these cell sensors. An alternative solution is the use of gloves or/and other protection, but workers do not usually like this solution, especially for long periods of time. These aspects were not usually considered as part of people's experience until an emergence generates a general need to face the attack of a new pathogen, more dangerous than those already known. It was so far strange, until the Coronavirus pandemia, for Europeans to see Asiatic tourists wearing face masks in public places, but this is familiar in other countries, and several epidemics have had their origins in the Orient. In other words, the need for familiar products to maintain hygiene and prevent diffusion of microscopic parasites, from viruses to microorganisms, is fundamental, being the first frontline for control of insect-borne diseases, as well as for other outbreaks.

The microbial communities of humans are characteristic and complex mixtures of microorganisms that have co-evolved with their human hosts, for better and for worse. Humans and their bacteria share the same evolutionary fate in which mutualistic interactions are essential for human health. For instance, several diseases are the result of perturbation of this equilibrium caused by changes occurring in the ecology or genetic of the bacterial world. Taking ecology and evolution into account might provide new strategies for restoring and maintaining human health. However, the dominant aspect of the economic cost must always be considered. The "win-win" approach, based on sustainability and respect for the environment joined with economic benefit, is now probably a utopia, but it is a new perspective for the future, probably the unique able to work to maintain the perspective of a sustainable state.

A current definition of resistance

This section is dedicated to a partial revision of the arguments already exposed, in consideration of the current debate about the resistance and the consequent counteractions. As a consequence of the multiple attacks, bacteria have evolved a large range of protection to deactivate, remove, or otherwise circumvent the toxicity of antibacterial compounds, thereby leading to today's multidrug-resistant organisms. As a result of this evolutionary ping-pong between attack and defenses, a bacterium is able to decrease the concentration of the drug in its cell, enable the effects, and/or interfere in the mechanism of action. In such cases, the medical response is to change the drug to another considered more active or able to surprise the pathogen. The final consequence is an incredible sequel of antibiotic drugs, exploring all the possible chemical derivatives of the leading molecules, which evidenced some kind of efficacy (Table 1.2). Although the possibilities of chemical variations are practically infinite, several routes have already been exhausted.

A table similar to Table 1.2 can be obtained also in case of insecticides resistance (Table 1.3). The sequence of emergence of resistance for the antibiotic vancomycin is reported in Fig. 1.4. However, the scenario evidenced in Table 1.3 does not totally explain the resistance to antibiotics. In many cases, such as cephalosporins and tetracyclines, several generations of related drugs have been developed and introduced in therapy with the hope of obtaining a better result. This strategy is still being used, but so far the problem has not been overcome. As can be deduced from the data reported in Table 1.3, it is possible that the time for the appearance of resistance is going to be progressively reduced, in accordance with the multidrug phenomenon being the result of a wider utilization of antibiotics in any field. The use of antibiotics in a wide range of applications, from health care to agriculture, is

Table 1.2 Timelines of the introduction of the antibacterial drug in chemotherapy and first appearance of resistance for most common antibiotics.

Antibiotic	Introduction of antibiotic drug in therapy	First resistance report
Sulfonamides	1935	1945
Penicillin	1940	1945
Chloramphenicol	1945	1957
Tetracycline	1948	1953
Streptomycin	1952	1986
Erythromycin	1957	1985
Vancomycin	1960	1987
Methicillin	1962	1964
Ampicillin	1963	1973
Linezolid	2000	2004
Daptomycin	2004	2005
Tigecycline	2005	–

Table 1.3 Examples of utilized insecticides and insurgence of resistance.

Type of insecticide	Year of introduction to the market	Year of insurgence of resistance
Modulator of sodium channels		
▪ DDT	1940	1947 (banned in 1950s
▪ Pyrethroids	1960	and 1960s) 2018
Acetylcholinesterases (AChE)		
▪ Organophosphates (parathion and malathion)	1938	1961 (banned in 1970s)
Chlorine channel antagonist regulated by GABA		
▪ Synthetic phenylpyrazolones	1930	2010
Juvenile hormones		
▪ Pheromones	1960	1989
Chitin synthesis inhibitors		
▪ Diflubenzuron	1970	1978

1944 Utilization of penicillin against *Staphylococcus aureus* → Resistance to penicillin

1962 Methicillin → Resistance to methicillin by *S. aureus*

1990 Vancomycin → Resistance to vancomycin by *Euterococcus*

2002 Vancomycin derivatives → Multidrug resistance by *S. aureus*

Fig. 1.4 The sequence of emergence of resistance for vancomycin.

the key of the evolution of antibiotic-resistant organisms following the phenomenon defined as "use it and lose it." Accordingly to the WHO definition, "antimicrobial resistance (AMR) is the ability of a microorganism (like bacteria, viruses, and some parasites) to stop an antimicrobial (such as antibiotics, antivirals and antimalarials) from working against it. As a result, standard treatments become ineffective, infections persist and may spread to others." Consulting other current definitions, resistance can be defined as "a heritable change in the sensitivity of a pest population that is reflected in the repeated failure of a product to achieve the expected level of control when used according to the label recommendation for that pest species," as reported by IRAC (Insecticide Resistance Action Committee) of the University of Nebraska. Here, we will consider also a further approach, wherein resistance will be considered as "the inherited ability of an organism to become tolerant and/or resistant to a dosage of the chemical that would be lethal to a definite species." In fact, resistance, being related to genome, appears to be a no-turn phenomenon, meaning that we have to face its consequences in mankind's future.

Antibiotics are the natural result of the usual war between microorganisms to dominate a common territory. In addition, infected hosts are involved in the production of such compounds. The consequence is that when an antibiotic drug is no longer effective, the usual response involves a shift to a new drug. This measure is always wasting time, or at least delaying, since in a short time the new drug will become ineffective. The manifestation of the antibiotic resistance is very rapid, as is well evidenced by the sequence in Fig. 1.4. This is in accordance with the presence of resistant strains already inside the targeted population of microorganisms. How is this possible? Most antibiotics are obtained or derived from natural products, whose structures are already in the memory of the attacked microorganism. After a short gap for adaptation, the resistance resurges.

However, it is possible that the natural reservoir of biocides natural products is not totally explored, in particular in plants, considering the lower use of antibiotics in agriculture and the important case of pyrethroids. Pyrethroids are a group of a synthetic pesticides, whose structures are deduced from the natural pesticide substances, pyrethrins and related terpenes, contained in the flowers of the perennial plant pyrethrum (*Chrysanthemum cynerariifolium*, Compositae) (Fig. 1.5). The use of seeds and flowers as insecticide dates from thousands of years, since in China and in Iran, chrysanthemums were crushed and used as insecticide powder as early as 1000 BCE. This so-called Persian Powder was widely used for centuries, as an insecticide in household use and as a repellent for mosquitos, fleas, and body lice, such as by French soldiers in the Napoleonic Wars. Flower of this plant are

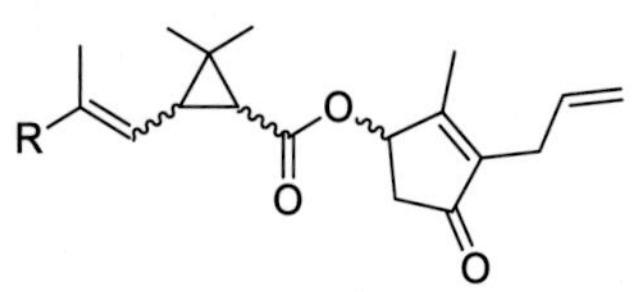

Fig. 1.5 The general structure of pyrethroids.

similar to those of the common daisy, but they are bigger and all yellow. Although more than 1000 pyrethroids have been made, only a few have been selected for this use, mainly as a domestic insecticide. The strange structure of these compounds is characterized by the presence of one rare cyclopropane unit, essential for the insecticide activity a pentacyclic lactone alpha, beta unsaturated and several chiral centers. In particular, pyrethrum is the extract from the chrysanthemum plant, containing pyrethrins. Therefore, pyrethroids are the man-made version of the natural pyrethrins, but while pyrethrum extract is composed of six esters which are insecticidal, a synthetic pyrethroid is usually composed of only one chemically active compound, in accordance with the kind of activity typical of natural products, consisting of the effects of a mixture of constituents.

Resistance is an emerging phenomenon, demonstrated in interesting several cases of the interaction of humans with the environment, including microorganisms' resistance to antibiotics and that of insects to insecticides. Resistance is an increasing problem, whose solution may be crucial for any future scenario for mankind, dramatically involving not only directly human health, but also future feed and food. Furthermore, it is necessary to consider that resistance is a widespread phenomenon. Antibiotics are everywhere, and bacteria that are resistant to chemically modified and synthetized antibiotics are present in any environment.

Multidrug resistance

Multidrug resistance (MDR) is among the most important causes of infections in nosocomial and community settings. Emergence of resistance to multiple antimicrobial agents in pathogenic bacteria has become a significant public health threat as there are fewer (or even sometimes no) effective antimicrobial agents available for infections caused by these parasites.

The same definition of MDR is actually inadequate to describe the phenomenon. To date, the adjectives "extensively drug resistant" (XDR) and "pandrug resistant" (PDR) have been introduced to describe the degree of resistance to a determined number of different classes of antibiotics. For instance, PDR organisms show resistance to all available antimicrobials. The idea is describe the different levels of the phenomenon, and this distinction is useful to identify its mechanism.

As already reported, to be effective, an antibiotic should be lethal to the great majority of the individuals in a normal population. However, a population is a mix of individuals, each having similar but different metabolisms, and therefore having personalized responses to the antibiotic attack. The antibiotic can lose its efficacy if many populations, or many individuals in a population, develop resistance to the toxic effects. Let us focus on this key point, considering that is only a further example of the consequences of the human tendency to overexploit natural resources, in order to obtain the maximum effects and not considering the consequences. The problem is inherent: resistance is related to a massive and persistent use of chemicals. More use of chemicals fuels the dominance of the resistant part of a population. The consequence of the mechanism is that chemicals that were once effective become insufficient to control insects. However, antibiotic recalcitrance is not solely caused by resistance, but also implies peculiar cells, named persistents, which are drug-tolerant. Antibiotic tolerance of persistents is not genetically manifested, since persistents are as susceptible as their parent strains. Stress responses, as in the case of antimicrobial use, may act as general activators of persistents formation.

The key argument is that the current resistance occurrence could be only the tip of the iceberg. This resistance could be associated with genome changes, giving rise to more virulent organisms. In this case, the scenario changes radically and dangers arise. The dominance of resilient organisms have consequences not limited to the target population, but all the environment is engaged and affected.

Multi-resistance is the final boomerang step, related to an intense and massive use of insecticides, exactly like antibiotics in microorganisms: many species have numerous resistant populations, as the normal range of possible reactions to the habitat changes. In nature, i.e., in normal situations, the mechanism is possible, but very unlikely and challenging to develop, and subjected to randomness. Newer organisms are forced to spread out as a consequence of this unnatural treatment. However, the environmental conditions are very important and even decisive, as we already have seen, for the emergence and spread of the epidemic phases.

References

Andersson, D.J., Levin, B.R., 1999. The biological cost of antibiotic resistance. Curr. Opin. Microbiol. 2, 489–493.

Barlow, M., Hall, B.G., 2002. Phylogenetic analysis shows that the OXA beta-lactamase genes have been on plasmids for millions of years. J. Mol. Evol. 55, 314–321.

Benelli, G., 2015. Research in mosquito control: current challenges for a brighter future. Parasitol. Res. 114 (8), 2801–2805.
Benelli, G., 2019. Managing mosquitoes and ticks in a rapidly changing world—facts and trends. Saudi J. Biol. Sci. 26 (5), 921–929.
Benelli, G., Mehlhorn, H., 2018. Mosquito-Borne Diseases, Implications for Public Health. Springer, Switzerland.
Bezirtzoglou, C., Dekas, K., Charvalos, E., 2011. Climate changes, environment and infection: facts, scenarios and growing awareness from the public health community within Europe. Anaerobe 17, 337–340.
Carlson, R., 1962. Silent Spring. Farwcett World Library, USA.
Cook, G.C., 1992. Effect of global warming on the distribution of parasitic and other infectious diseases: a review. J. R. Soc. Med. 85, 688–691.
Córdoba-Aguilar, A. (Ed.), 2018. Insect Behaviour. From Mechanisms to Ecological and Evolutionary Consequences. Oxford University Press, Oxford (UK).
Darwin, C., 2003. Origin of Species. Mass Market Paperback, New York (2003).
Dondorp, A.M., et al., 2009. Artemisinin resistance in *Plasmodium falciparum* malaria. N. Engl. J. Med. 361, 455–467.
Gale, E.F., et al., 1981. The Molecular Basis of Antibiotic Action. John Wiley & Sons, Inc., New York, NY.
Karaagac, S.U., 2011. Insecticide resistance. Chapter 21 in insecticides. In: Perveen, F. (Ed.), Advances in integrated Pest management. InTech, Croatia, Rijeka.
Khater, H., Govindarajan, M., Benelli, G. (Eds.), 2017. Natural Remedies in the Fight Against Parasites. InTechopen, Rijeka, Croatia.
Lamb, H.H., 1995. Climate, History and the Modern World. Routledge, London.
Lounibos, L.P., 2002. Invasions by insect vectors of human disease. Annu. Rev. Entomol. 47, 233–266.
Mahmud, R., Ai Lian Lim, Y., Amir, A., 2017. Medical Parasitology. A Textbook. Springer, Switzerland.
Mehlhorn, H., 2015a. Host Manipulations by Parasites and Viruses. Springer, Switzerland.
Mehlhorn, H. (Ed.), 2015b. Encyclopedia of Parasitology, fourth ed. Springer, New York.
Miller Jr., W.B., 2013. The Microcosm Within. Evolution and Extinction in the Hologenome. Universal-Publisher, Boca Raton, FL. 2013.
Mouchet, J., Carnevale, P., 1997. Impact of changes in the environment on vector-transmitted diseases. Sante 7, 263–269.
Naqqash, M.N., et al., 2016. Insecticide resistance and its molecular basis in urban insect pests. Parasitol. Res. 115, 1363–1373.
Natham, C., Cars, O., 2014. Antibiotics resistance—problems, progress, and prospects. N. Engl. J. Med. 371, 1761–1763.
Nicoletti, M., Murugan, M., Benelli, G., 2016. Emerging insect-borne diseases of agricultural, medical and veterinary importance. In: Trdan, S. (Ed.), Insecticide Resistance. InTech, Rijeka, Croatia.
Reiter, P., 2001. Climate change and mosquito-borne disease. Environ. Health Perspect. 109 (Suppl. 1), 141–161.
Rogers, D.J., Randolph, S.E., 2000. The global spread of malaria in a future warmer world. Science 289, 1697–1968.
Semmler, M., et al., 2009. Nature help: from research to products against blood-sucking artropods. Parasitol. Res. 105, 1483–1487.
Tanwar, J., Das, S., Fatima, Z., Hameed, S., 2014. Multidrug resistance: an emerging crisis. Interdiscip. Perspect. Infect. Dis. 541340. Available at https://doi.org/10.1155/2014/541340.
Trdan, S. (Ed.), 2016. Insecticide Resistance. InTech, Rijeka, Croatia.

U.S. EPA, 1975. DDT Legulatory History: A Brief Survey (to 1975).
Wigley, T.M.L., Ingram, M.J., Farmer, G. (Eds.), 1981. Climate and History. 1981. Cambridge University Press, Cambridge.
Willcox, M., Bodeker, G., Rasoanaivo, P., 2005. Medicinal Plants and Malaria. CRC Press; Taylor & Francis, Boca Raton, FL.
Wright, G.D., Sutherland, A.D., 2007. New strategies for combating multidrug-resistant bacteria. Trends Mol. Med. 13, 260–267.

CHAPTER TWO

New scenarios arising from radical changes in diseases

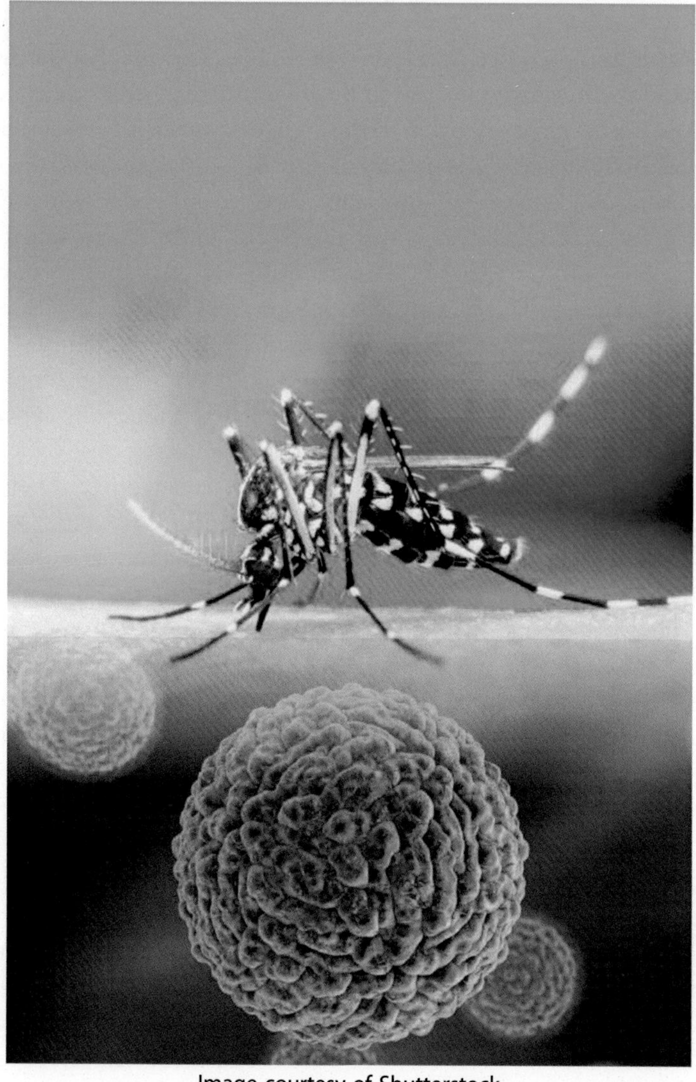
Image courtesy of Shutterstock

Insect-Borne Diseases in the 21st Century
https://doi.org/10.1016/B978-0-12-818706-7.00002-4

Facing antimicrobial resistance

Very little is known about where resistance originates and how resistant gene move through the food system. We have much more knowledge about some foodborne pathogens, and some of these have important resistance phenotypes, but there is almost non published research regarding resistance genes in the larger microbial community.

Dr. Paul Morley, Colorado State University (2015).

The study of antimicrobial resistance represents an important opportunity to understand the mechanisms exerted by environmental stress on the organism. How the bacterial cell reacts to the antibiotic attack is considered a reliable model to understand the process of resistance insurgence and the related mechanisms concerning genome, but evident in phenotype behavior (Sadashiv and Kaliwal, 2016; Hayes and Wolf, 1990; Davies and Davies, 2010; Natham and Cars, 2014). The intensive use of antibiotics in clinical therapy operates a Darwinian selective pressure, inducing bacteria to improve their ability to survive in new environmental conditions: a phenomenon that we name adaptation. However, it is nowadays simplistic to think that the mere presence of antibiotics is able to induce the appearance of resistant phenotypes. In other words, the resistance response by a microorganism must be the activation of a mechanism already present before the advent of antibiotics.

Any population of living organisms, including man as later evidenced, constantly harbors inside a subpopulation. In case of microorganisms or insects, the subpopulation is refractory to the drug treatment, meaning that the cure has no effect on the subpopulation. The size of this tolerant subpopulation increases in adverse growth conditions such as depletion of nutrients. This condition can be commonly retrieved during infections; in particular, during the early stages, the low availability of nutrients induces bacteria to a stationary phase. In this approach, the development of resistance is considered a complex phenomenon, which can be parceled out in a sequence of steps. The experimental procedure is based on intermittent antibiotic exposure, and consequent controls during the antibiotic treatment, in order to detect and clarify the process leading to the insurgence of antibiotic resistance. The experimental data were interpreted as an evolution of tolerance, consisting in the ability to survive under treatment without developing resistance. That is, tolerance precedes resistance and boosts the chances for resistant mutations to spread in the population. Tolerant mutations can be considered the first genetic reaction, paving the way for the rapid subsequent evolution of resistance. Therefore, through the evolution of tolerance it could be possible to develop a new strategy for delaying the emergence of resistance. The physiological basis of this phenomenon, termed antibiotic tolerance, is still poorly understood. This argument will be reconsidered later in the proposed strategy for controlling parasites of malaria by the coordinated utilization of natural substances or/and acting on basic behavior, including shortage of energy.

Another strategy is based on the study of the mechanisms of the genetic response to environmental pressure. Exposure to stress, consisting of any environmental perturbation that decreases organism homeostasis, affecting growth rate or survival, induces the expression of stress-response pathways that can result in profound changes in gene expression and physiology. Some details of such pathway have been explored in the study of insurgence of resistance in bacteria populations. In order to survive in various environmental conditions, cells can rely on a pool of genes that they can choose to express or silence according to their needs. The mechanism is called jump-starting DNA repair, or the SOS response, since it consists of an inducible DNA repair system that allows bacteria to survive sudden increases in DNA damage. The term "SOS response" was introduced in 1975 by Miroslav Radman, who first reported on the existence of an inducible DNA repair network in *Escherichia coli* (Radman, 1974). The SOS response and its regulatory network are now underscored and considered widely present in bacteria, reflecting the need for all living cells to maintain the integrity of their genome. During the past 35 years, the SOS response has become a paradigm for the field of DNA repair. The study of the SOS response is considered as a classic approach to the interpretation of bacterial response, since bacterial cultures are treated by a DNA-damaging agent and further analyses must determine the genes fused to an SOS promoter, or the involved proteins network by immunoblotting. However, several researches have addressed different explanations concerning the real functionality of the SOS genes and mechanisms that fine-tune their regulation. In fact, although a repertoire of 40 genes was shown to be under SOS control, most of them are DNA repair genes, but there are several genes that still have no known function (Little and Mount, 1982; Celina, 2008; Michel, 2005; Bellio et al., 2014a,b).

In this perspective, exposures to environmental agents interfering with DNA replication or which are able to cause DNA damage are involved in the expression of the set of genes of the SOS response. Most SOS-related genes are involved in DNA repair, recombination, replication, biofilm formation, and cell division. One of the most active inhibitors seems to be the Indian yellow spice curcumin, a polyphenolic compound isolated from *Curcuma longa*, whose antiinflammatory, anticarcinogenic, and antioxidant activities have been well-characterized (Bellio et al., 2014a,b).

The more we study bacterial strategies to adapt, the more we discover incredible facts about the plasticity of their genome. How it is possible that resistance, once started, can be spread rapidly throughout a microorganism population? The SOS response is related to specific bacterial agents. The regulation of this response is related to the functionality of two proteins: a repressor named LexA and an inducer, the RecA filament. The RecA filament is a nucleoprotein complex that induces the LexA cleavage reaction. RecA is a highly common protein, present in nearly all bacteria and conserved in all organisms, including humans. It specifically binds single-stranded DNA (ssDNA), forming a nucleoprotein filament that has two functions. SOS genes are repressed to different degrees under normal growth conditions. During normal growth, the LexA repressor binding to a specific sequence of the SOS box, present in the promoter region of SOS genes, prevents their expression. This effect depends on the exact sequence of their SOS box, i.e., the region of a promoter that is recognized by LexA, on the position in the promoter region and the strength of the promoter. Therefore, when a signal of increasing DNA damage alerts the cell, the LexA repressor undergoes a self-cleavage reaction and the SOS genes are de-repressed.

In conclusion, the SOS response can be considered as the expression of mutagenic bacterial stress-response pathways with evolutionary benefits. Starting from this consideration, we can speculate if the SOS response provides a short-term advantage or if mutagenesis provides a long-term advantage by accelerating the evolution of resistance. However, stress-response is often associated with the expression of error-prone, translesion synthesis DNA-polymerase enzymes that elevate the mutation rate under stress. This evidence can suggest a pathway in two steps: the primary benefits of the SOS response, increasing bacterial competitive ability, further activating a stress-induced mutagenesis.

Another aspect is related to the exponential curve of the epidemic episode. Once resistance appears in a population, what factors influence the spread? We are focusing in the first part of the exponential rise of the curve, when the mechanism of diffusion is not fueled by the high number of infected subjects. This is a delicate step, where environmental factors play a central role, as we will see in insect-borne diseases involving bacteria, such as plague, and virus, such as Zika. We are talking of transfer of genetic material between resistant bacteria to others of the same generation, as part of the antibiotic resistance spread among a bacterial population. The gene transfer can be horizontal or lateral genes, involving also homologous recombination, transposition, transformation, conjugation, and transduction. A successful horizontal transfer relies not only on the introduction of DNA into a recipient cell's cytoplasm, but also on the heritability of the transferred sequences in the recipient microorganism. Integrons play an important role in genetic mechanisms that allow bacteria to adapt and evolve rapidly through the stockpiling and expression of new genes.

Integrons are mobile DNA elements and were first described at the end of the 1980s. They consist of an integrase gene and a combination site. Integrons are part of the genetic mechanisms that allow bacteria to adapt and evolve rapidly through the expression of new genes. These integron genes are embedded in a specific genetic structure called a gene cassette, named an integron cassette in this case, containing genes that are almost exclusively related to antibiotic resistance. Gene cassettes are small, discrete mobile elements, which generally comprise a single gene and a 59-base element, which acts as a recombination site. The mobile gene cassettes can be directionally inserted into the nearby recombinant site. Alternatively, the enzyme integrase excises the cassettes by a site-specific recombination mechanism. Although most integrons were initially described in human clinical isolates, they have now been identified in many nonclinical environments, such as water and soil. Integrons are present in $\approx$10% of the sequenced bacterial genomes and are frequently linked to mobile genetic elements (MGEs), particularly the class 1 integrons.

Genetic linkage to a diverse set of MGEs facilitates horizontal transfer of class 1 integrons within and between bacterial populations and species. The mechanistic aspects limiting transfer of MGEs will therefore limit the transfer of class 1 integrons. However, horizontal movement due to genes provided in trans and homologous recombination can result in class 1 integron dynamics independent of MGEs. A key determinant for continued dissemination of class 1 integrons is the probability that transferred MGEs will be vertically inherited in the recipient bacterial population. Heritability depends on both genetic stability and the fitness costs conferred on the host.

Integrons are capable of incorporating exogenous and promoterless open reading frames (ORF), referred to as gene cassettes, by a specific recombination site called *attC*. Transferable genes encoding antibiotic resistance to major antibiotics such as beta-lactams and aminoglycosides are often carried by class I integrons in Gram-negative pathogens. Integrons are a class of site-specific recombination elements which insert and excise mobile antibiotic resistance gene cassettes. This process is mediated by the action of integrase, a member of the tyrosine recombinase family. Gene cassette expression is driven by a promoter located in the integron platform upstream of the primary site of cassette integration, referred as an *attI* site. In the case of the class 1 integron, the most common in bacteria, the gene-cassette promoter is inside the *intI1* gene, which encodes the cassette recombinase. In this peculiar organization of the integron, the closer a gene cassette is located to the *attI* site, the higher its expression is. The expression of the integron integrases is controlled by the SOS response.

Fig. 2.1 reports generations of individuals in comparison under an environmental pressure due the utilization of antibiotics or insecticides. The normal individuals of a population of microorganisms are in green and the resistant ones are in red. With some minor changes, the sequence can be also used for any pest resistance. The idea is that resistance is essentially a process of reinforcing of some special individuals inside a population and in such way they acquire the capacity to survive to the antibiotics, and in this way they become dominant whereas the other individuals are victims of the treatment. In the first line, the consequence of the use of the antibiotic is reported.

The meaning of the two sequences is that the dominance of the resistant individuals is practically a direct consequence of the massive use of the product, although the potentiality of the resistance was already present, but so far practically unexpressed. Attention is totally focused on the red organisms, expecting their decreasing and obtaining exactly the opposite.

Most of the antibiotics currently used are not sufficiently selective, therefore beneficial normal individuals are also killed, and in general the overall population of microbiota is highly affected. Continuing the use, the red

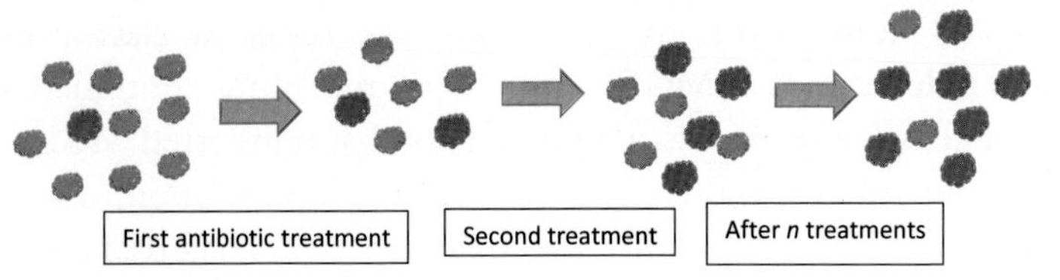

Fig. 2.1 The insurgence of resistance in a population.

individuals will be favored in restoring the new situation against the normal ones. Usually, the concomitant utilization of probiotics is not totally useful to counteract the loss rapidly. This means that all the population is completely affected in the equilibrium with difficulties affecting all life of the host, whose health mainly depends on its microbiota.

The problem is that so far attention has been focused on the red individuals, being the natural target as responsible for the disease, not considering the equilibrium of the total community. So far, the dominant paradigm of good antibiotics has been essentially to find the best weapon to kill the bad microorganisms in the most rapid and efficient way. The green individuals were considered unimportant, and in any case a natural restoration of the normal situation after some time was expected. Consideration of the importance of our helpful prokaryotes, also in the resistance phenomenon, gave rise to a completely different point of view, pushing research for more selective antibiotics and the preservation of the present useful ones, which is exactly how immunity works.

It is possible to work on the development of a new paradigm. The new antibiotic should effective against the bad population, but should not totally affect the microbiota; it should be used at lower concentrations, even if the effects could be less rapid and evident. The minor effectiveness should be compensated by the activity of other substances, maintaining the good population and immunity responses of the host.

Now, we can complete the revision of the previous paradigm, changing the word "resistance" to "reaction," evidencing more the natural tendency of living organisms to adapt and react to adverse situations, in particular when superactive molecules are introduced into their habitats. Furthermore, the term most evidencing the character of the new pesticides or antibiotics is "persistence," considering the necessity of maintaining the equilibrium in the microbiota, that first the attack of the pathogen and then the action of the antibiotic were able to compromise and destabilize. On the basis of new scientific approaches, the concept of persistence contains the characters of sustainability and environmental care, but we must never forget that first we need the agent to kill the pathogen using the necessary drug.

Such general consciousness if the fil rouge of this approach, leading to the production of fueling ideas, which can be crystallized in marketed products. In accordance with the idea that it is necessary to consider all the steps of the pathway leading to the solution of the antibiotic resistance challenge, the control of the pest can be divided into four aspects concerning research and the related production of products: feed, food, drug, and environment.

As far as possible, the new antibiotic should be selective and friendly to the microbiota. A similar approach has already been studied in the fight against pest insects, known as the Integrated Management System; in this case, the environmental factor is more evident and comprehensible compared to the microorganisms one.

The general consideration of this chapter is that an insect-borne disease is a complex system involving several interacting factors. These factors should be monitored and followed in their evolution. Considering an organism's tendency to maintain its homeostasis, responses to environmental pressure must be first considered along with their different actions, including unexpected accelerations. Any environmental perturbation that causes stress can affect growth rate, inducing profound changes in gene expression and physiology. Resistance study revealed that bacteria are key to understand the survival strategies of any cell of any living creature. This consideration involves also the side of the etiological agent, which is the causative agent of a disease condition, and it often consists in a microorganism. On the vector's site, it would be important to define: the effective vector's abundance and its behavior; the number of bites, uninfected or infected, on the focal host as a result of the vector's feeding rates; the probability that the bite transfers efficiently the disease, meaning host susceptibility and vector infectiveness; the turnover of the vector's population in the observed period; the vector's longevity and incubation time; and the capacity of the vector's survival and its reaction to the insecticide's treatment. In addition, these factors are subjected to variation by environmental pressure and utilization of pesticides, but they do not consider the diffusion of the disease by other pathways than the insect's ones. Other similar approaches should be considered for the host, and the data inserted into a database and subjected to the natural algorithm, which we use to call habitat.

Novel antibiotic pathways

The investigation of the potential antimicrobial activity of natural molecules from the secondary metabolism of living wild organisms could provide us, as in the past, with new templates for the synthesis of more efficient and potent antimicrobial agents.

In my search of active substance from organisms living in extreme habitats, I was involved in the study of secondary metabolites from lichens. Lichens, thanks to their dual nature, algal and fungal, are able to live in very difficult environmental conditions where other organisms cannot survive, like glass. However, there are many difficulties in studying lichens. Their taxonomy is very complex and needs a dedicated specialist. Often they are present in small populations associated with other organisms and their collection is complicated. In this case, the lichens studied came from Antarctic zones, in particular the Chilean part. Lichens derive from an integration between an alga (a microalga) and a special fungus. Usually, the two organisms can live alone, but their association generates a mutual advantage, since the alga can survive out of the water thanks to the fungal supply and the fungus can satisfy its heterotrophic needs and offers protection and water supply. The mutualistic association is very intimate, since the hypha of the fungus can penetrate inside the algal cell. The consequence is the production of a special metabolism, which cannot be overimposed to the algal or the fungal ones. In fact, considering lichens as a core symbiosis between a mycobiont (fungal partner) and a photobiont (photoautotrophic partner), the consequent assemblages of microorganisms are able to produce metabolites that do not bring them direct benefit but are useful to the lichen as a whole. There is evidence to suggest that the unique nature of the symbiosis has played a substantial role in shaping many aspects of lichen chemistry. We must now consider that often the involved alga is a cyanobacterium, meaning a prokaryote is connected with our interest in antibiotic resistance. However, it is possible to find some novelties since lichens present a production of special metabolites, which cannot be considered as related to the algal or the fungal metabolism, but are produced by one organism. The majority of secondary lichen products are aromatic phenolic polyketides, and a number of them have exhibited marked biological activity, including polycyclic substances, named depsides and depsidones, but secondary metabolites biosynthetically derived from the acetyl polymalonyl, mevalonic, and shikimate pathways are also present (Vu et al., 2015).

The substances isolated from lichens, collected in several southern regions of Chile (including Antarctica), were evaluated against methicillin-resistant clinically isolated strains of *Staphylococcus aureus*, *S. haemolyticus*, and *S. warneri*. These compounds were able to interact synergistically with therapeutically available antibiotics, whose efficacy against multidrug resistant is compromised by various mechanisms of resistance (Segatore et al., 2012; Bellio et al., 2015; Celenza et al., 2013). In particular, the synergistic interaction between gentamicin and lichen compounds usnic acid (Fig. 2.2) and pannarin was evidenced (Celenza et al., 2012). Usnic acid, first isolated from *Usnea barbata*, is a well-known antibiotic agent and it is a lichen compound utilized in medicine. Specifically, these compounds are able to reverse resistance dramatically

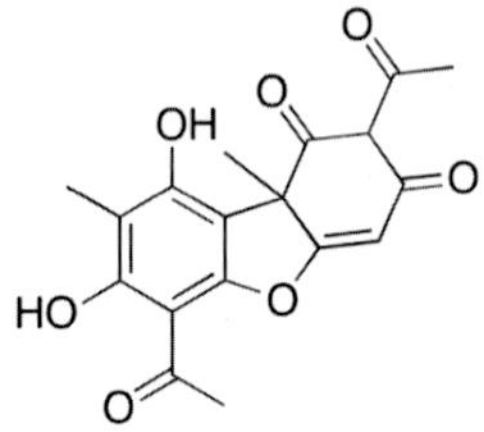

Fig. 2.2 Structure of usnic acid.

against gentamicin. Special attention was dedicated to the in vitro antimicrobial activities of pannarin, a depsidone isolated from lichens. Fig. 2.3 shows the structures of two depsidones. Their structures are derived from two polysubstituted benzenes joined by a ester linkage, whose hydrolysis generates the volatile compounds responsible for the typical smell in most parts of wood and that are largely utilized by perfumery. Pannarin activity was evaluated alone and in combination with five therapeutically available antibiotics, using a checkerboard microdilution assay against methicillin-resistant clinically isolated strains of *S. aureus*, by MIC(90) and MIC(50), as well as MBC(90) and MBC(50). A moderate synergistic action was observed in combination with gentamicin, while antagonism was observed in combination with levofloxacin. Treatment with pannarin produced bactericidal activity without significant calcein release, with a lack of even significant structural damage to the cytoplasmic membrane, including lysis and disruption. Furthermore, pannarin showed low hemolytic activity and a moderate cytotoxic effect on peripheral blood mononuclear cells. These findings suggest that pannarin, and in general lichen depsides and depsidones, might be good candidates for the individualization of novel templates for the development of new antimicrobial agents or combinations of drugs for chemotherapy.

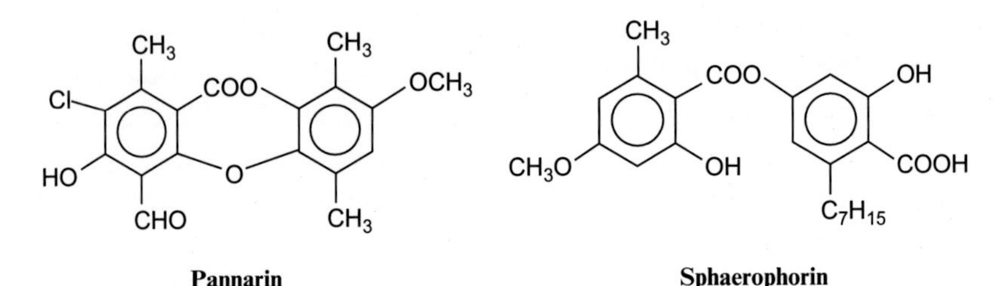

Fig. 2.3 Structure of two depsidones from lichens.

References

Bellio, P., et al., 2014a. Curcumin inhibits the SOS response induced by levofloxacin in *Escherichia coli*. Phytomedicine 21 (4), 430–434.
Bellio, P., et al., 2014b. Inhibition of the transcriptional repressor LexA: withstanding drug resistance by inhibiting the bacterial mechanisms of adaptation to antimicrobials. Life Sci. 241, 117116.
Bellio, P., et al., 2015. Interaction between lichen secondary metabolites and antibiotics against clinical isolates methicillin-resistant *Staphylococcus aureus* strains. Phytomedicine 22 (2), 223–230.
Celenza, C., et al., 2012. In vitro antimicrobial activity of pannarin alone and in combination with antibiotics against methicillin-resistant *Staphylococcus aureus* clinical isolates. Phytomedicine 19 (7), 596–602.
Celenza, G., et al., 2013. Antibacterial activity of selected metabolites from Chilean lichen species against methicillin-resistant staphylococci. Nat. Prod. Res. 27 (17), 1528–1531.
Celina, J., 2008. Inducible SOS response system of DNA repair and mutagenesis in *Escherichia coli*. Int. J. Biol. Sci. 4 (6), 338–344.
Davies, J., Davies, D., 2010. Origins and evolution of antibiotic resistance. Microbiol. Mol. Biol. Rev. 74 (3), 417–433.
Hayes, S.D., Wolf, C.R., 1990. Molecular mechanisms of drug resistance. Biochem. J. 272, 281–295.
Little, J.M., Mount, D.W., 1982. The SOS regulatory system of *Escherichia coli*. Cell 29 (1), 11–22.
Michel, B., 2005. After 30 years of study, the bacterial SOS response still surprises us. PLoS Biol. 3 (7), e255.
Natham, C., Cars, O., 2014. Antibiotics resistance—problems, progress, and prospects. N. Engl. J. Med. 371, 1761–1763.
Radman, M., 1974. Phenomenology of an inducible mutagenic DNA repair pathway in *Escherichia coli*: SOS repair hypothesis. In: Sherman, S., Miller, M., Lawrence, C., Tabor, W.H. (Eds.), Molecular and Environmental Aspects of Mutagenesis. CC Thomas Publisher, Springfield, IL, pp. 128–142.
Radman, M., 1975. Phenomenology of an inducible mutagenic DNA repair pathway in *Escherichia coli*: SOS repair hypothesis. Basic Life Sci. 5A, 355–367.
Sadashiv, S.O., Kaliwal, B.B., 2016. Resistance in bacteria. In: Trdan, S. (Ed.), Insecticide Resistance. InTech, Rijeka, Croatia (Chapter 15).
Segatore, B., et al., 2012. In vitro interaction of usnic acid in combination with antimicrobial agents against methicillin-resistant *Staphylococcus aureus* clinical isolates determined by FICI and ΔE model methods. Phytomedicine 19 (3–4), 341–347.
Vu, T.H., et al., 2015. Depsides: lichen metabolites active against hepatitis C virus. PLoS One. 10(3), e0120405.

CHAPTER THREE

Novel challenges require new solutions

Image courtesy of Shutterstock

Adaptive mutability

Antibiotics nowadays in use, just as insecticides do, all act with three or four mechanisms; without real alternative solutions, multiresistance will shortly be a reality. Multiresistance means that all the substances in use will lose any effect, independently of the structure and the action mechanism. Multiresistance is an important phenomenon first studied in medicinal drugs, such as those used against cancer. It has generated several novel concepts in treated complex pathologies. Five main approaches have emerged:

(a) Development of new antibiotics. This approach has been frustrating so far, since it needs a great deal of financial investment, which in a short time can be totally lost after the insurgence of resistance. Insisting on the same structural models and mechanisms of action can be frustrating and

Insect-Borne Diseases in the 21st Century
https://doi.org/10.1016/B978-0-12-818706-7.00003-6

can even reinforce resistance. It is necessary a good quantity of fantasy and persistence, beside the initial investment of money to support the project.

(b) Use of mixtures of substances with different effects. Recent advances in biology provide a rationale for intrinsic resistance and support the evaluation of combination therapy to reverse resistance. This approach is already in action for the treatment of several important pathologies, such as cancer and several infections, as well as being greatly explored in the case of malaria treatment, for example, as further reported. To overcome antibiotic-mediated resistance, a valuable alternative would be the use of a combination of drugs. Thus, substances that can increase susceptibility to currently licensed agents would be very attractive and valuable options. Several studies have demonstrated that a number of natural products, which failed as antimicrobials, are able to increase the effectiveness of chemotherapeutic agents against Gram-negative bacteria.

(c) Use of a limited quantity of the effective drug, not sufficient to display the necessary effect, but also unable to develop the resistance. This will be reported in detail, in the case of the use in African populations that treat malaria with limited use of chloroquine.

(d) Increase the life of currently available antibiotics by the use of substances able to interfere with the bacterial stress response system. This is the first step for the development of resistant organisms.

(e) Reverse the resistance process by agents able to counteract the mechanism of defense by the microorganism. This approach requires deep knowledge of the metabolism of the target, and important recent researches have been focused on this area.

Examples of the real consequences of these approaches will be reported and discussed, including the experiments performed during recent years by my group of researchers.

First, let us propose some considerations about insect-borne diseases. Nowadays, the above definitions of resistance risk being obsolete or insufficient to describe the complex system that generates the phenomenon that arose in insecticides, or the absence of any real efficacy in control of parasites and insect pests. The solution can no longer be considered simply as "find the best drug to kill the vector and solve the problem," as has been mainly considered so far. As in general medical treatment, where the "chemical magic bullet" was considered to be the central solution for any disease, this approach is nowadays in crisis, because the physiologic aspects are more complex and complicated by interactions at several levels. Resistance is a consequence of a series of events (Long et al., 2016; Fitzgerald and Rosenberg, 2019).

A small number of combinations are, to date, commercially available for the treatment of resistant infections. Waiting for new active drugs, and reserving the efficacy of those currently available, therefore remains a major goal. Furthermore, investigation of the potential antimicrobial activity of natural molecules from the secondary metabolism of living wild organisms could provide us, as in the past, with new templates for the synthesis of more efficient and potent antimicrobial agents. Finally, we could not solve the problem and try to control the phenomenon using a multiple approach: lower quantities of drugs, unable to generate resistance, interfere in the protection mechanism of the resister, and try to find new molecular solutions in their natural enemies, the natural products produced by millions of creatures and selected over many years. This time, no final solutions, no abnormal stratagems, no shortcuts can be intentionally constructed to solve the problem as soon as possible without any consideration of the consequences, in an attempt to bypass nature's rules easily.

We are focusing on the parasite, but many considerations can easily expanded on the vector and even the host. In particular, resistance to treatment is a unifying phenomenon. It is probably one of the gifts that we received from our first ancestor LUCA as essential behavior to survive. In the article "Adaptive mutability of colorectal cancers in response to targeted therapies" published in *Science* in November 2019, Alberto Bardelli of the University of Torino, Italy, and several co-authors evidenced the clear similarity of resistance emergence in bacteria and cancer cells (Russo et al., 2018, 2019). The paper states in the abstract that "the prevalent view is that resistance is a fait accompli: when treatment is initiated, cancers already contain drug-resistant mutant cells. Bacteria exposed to antibiotics transiently increase their mutation rates (adaptive mutability), thus improving the likelihood of survival. We investigated whether human colorectal cancer (CRC) cells likewise exploit adaptive mutability to evade therapeutic pressure." The research started from consideration of the development of secondary resistance. Primary resistance occurs when the initial treatment fails in the first place. Even after the right targeted therapy, treatment specific to that type of cancer often can work only for a limited amount of time. After a few months, or a year or two, the disease comes back, and it can be even more aggressive than it was the first time. Under stress, such as that imposed by clinical treatment, resistant cells of the parasite can temporarily increase the ability to mutagenize their DNA, developing a mutation necessary to allow them to survive, escaping the effects of targeted therapy, as reported by several independent researchers (Fitzgerald et al., 2017; Bjedov, 2003; Ghirighelli and Fumet, 2019). The teams of researchers report carefully the timetable of the resistance emergence after the treatment with cetuximab, a drug approved to treat patients with metastatic forms of the cancer: "most of the cells died after 96h of drug treatment, but a fraction of resistant cells survived through two weeks of cetuximab. When drug treatment was lifted after two weeks, the formerly resistant cells grew rapidly and became sensitive to the drug again. If the researchers continued drug treatment beyond two weeks, though, cells became permanently resistant." The paper is very interesting, since although the resistance phenomenon to anticancer drugs is already well-known, there is useful information clearly evidencing several aspects of the similar behavior between cells from very different organisms. The first author of the paper, Mariangela Russo, evidences the importance of this kind of investigation: "Finding the mechanism would allow [us] eventually to find new druggable hits," she says, opening up "the possibility to stop or delay this adaptive mutability and therefore delay the onset of secondary resistance." We shall return again to the uniformity of behavior in different organisms, as a key to interpret biological events. The relation between resistant strains of *Plasmodium* to chloroquine and cancer cells has been reported by Maud et al. (2006), evidencing the similarity between the resistance to quinolines or artemisinin derivatives and codon changes in a gene that encodes an ortholog of one of the P-glycoproteins expressed in multidrug resistant human cancer cells (ABC transporter) (Boumendjel et al., 2009).

The effects of environmental conditions will be continuously considered. Insect-borne diseases are and always will be present. We are not able to eliminate the etiological agents of a disease, but we can control and limit their insurgence. Bad living conditions and malnutrition are the ideal medium for their occurrence. Other responses are able only to limit the damages.

This book is dedicated to the possibility of radical changes in the current management of habitats—different habitats, from the micro to the macro. Reality may explained as the consequence of interactions between levels of existence, each characterized by different dimension. Therefore, multiple studies are necessary and a careful consideration of the possible interferences among the levels.

Consciousness of the current problems is increasing, but several events are misleading. The only hope is to inform everyone of what is really going on and work on possible solutions, considered as seeds for a new consciousness. Ordinary people can be the key. All these great meetings so far did not change the reality or influence the main factors of the planetary changes. Sometimes, they look more as a masquerade, to send a comforting episode. After the agreements of Paris in 2015, in 2016 the presence of CO_2 in the atmosphere registered a record increase of 3.3 ppm, and this trend has been confirmed. Periodically, experts and politicians sign all kind of documents and resolutions without any real effect. Meanwhile, something is moving in the right direction—single but significant actions that can be multiplied.

It is not possible to solve these problems with the same instruments and the same subjects that caused those problems.

References

Bjedov, O., 2003. Stress-induced mutagenesis in bacteria. Science 300, 1404–1409.

Boumendjel, A., Botomat, J., Robert, J. (Eds.), 2009. ABC Transporters and Multidrug Resistance. Wiley & Sons, Hoboken, NJ.

Fitzgerald, D.M., Rosenberg, S.M., 2019. What is mutation? A chapter in the series: how microbes "jeopardize" the modern synthesis. PLoS Genet. 15, e1007995.

Fitzgerald, D.M., Hastings, P.J., Rosenberg, S.M., 2017. Stress-induced mutagenesis: implications in cancer and drug resistance. Annu. Rev. Cancer Biol. 1, 119–140.

Ghirighelli, F., Fumet, J.-D., 2019. Is there a place for immunotherapy for metastatic microsatellite stable colorectal cancer? Front. Immunol. https://doi.org/10.3389/fimmu.2019.01816.

Long, H., et al., 2016. Antibiotic treatment enhances the genome-wide mutation rate of target cells. Proc. Natl. Acad. Sci. U. S. A. 113, E2498–E2505.

Maud, H., Alibert, S., Orlandi-Pradines, E., et al., 2006. Chloroquine resistance reversal agents as promising antimalarial drugs. Curr. Drug Targets 7 (8), 935–948.

Russo, M., et al., 2018. Reliance upon ancestral mutation is maintained in colorectal cancers that heterogeneously evolve during targeted therapies. Nat. Commun. 9, 2287.

Russo, M., et al., 2019. Adaptive mutability of colorectal cancers in response to targeted therapies. Science 366 (6472), 1473–1480.

CHAPTER FOUR

Bionetworks, system biology, and superorganisms

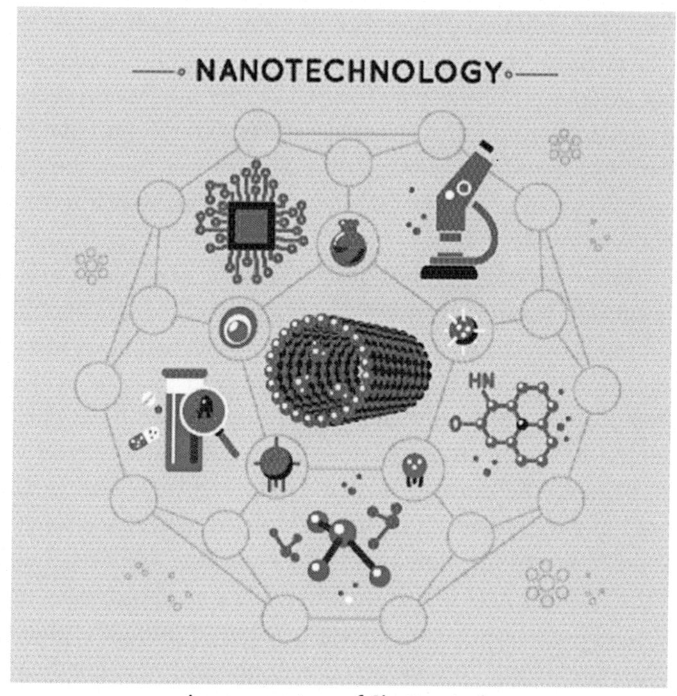

Image courtesy of Shutterstock

Looking for new solutions

What we call a "superorganism" is an integrated system involving several types of organisms, able to obtain a final result. The term superorganism was introduced in 2008 by the Human Microbiome Project to consider the human body as made up of both microbial and human cells, being considered as a unique ecosystem (Turnbaugh et al., 2007; Bik et al., 2006). This

Insect-Borne Diseases in the 21st Century
https://doi.org/10.1016/B978-0-12-818706-7.00004-8

approach evidences the relevance of microbiome in human homeostasis, accounting for the trillions of microorganisms from thousands of different species populating every part of the human body. A brave integration between the most primitive and the most advanced cells. Among them, some germs exert a positive effect on health and others are potential pathogens, even in the same population as effect of the environment pressures. This concept is here extended to report the interactions between organisms not necessarily living inside the same body at the same time, and the approach can result related to network of finalized activities, most of them unconscious and endured. Therefore, a common body is not necessary, because the same approach can be applied at an environmental level. In fact, in this case attention must also be focused on the interactions between the involved subjects, using the network model as recently developed in biology. In the case of insect-borne diseases, we can apply the concept of the superorganism in a sequel of bodies and habitats, which must generate a coherent final result.

Insect-borne diseases must be considered as the result of a series of coordinated events, wherein several organisms are involved in a complex situation. This implies some sort of coevolution between pathogen and vector, as part of the superorganism concept. Only the final effect is evident to us, whereas other steps are smartly crepitated or difficult to find. This explains why mankind has spent so much time on trying to understand what was going on, blaming the disease on divine interference or human fault. The concept of "superorganism" is useful not only to understand what is going on in an epidemic episode, but also to understand the role of each organism, and find the Achilles heel that is useful to select an appropriate strategy to control the phenomenon. So far, the approach of "limiting the phenomenon by any kind of wall" still dominates and is absolutely present in the protocols to face so-called emergencies (see the Xylella case in Chapter 5). It is a defense strategy, usually overcome in due course, like any immobilization strategy, and useful only as an early and immediate measure. Insect-borne diseases are not only emergencies but possibly represent daily life, knocking at the door of our future. They are not a novelty, but the recent additional episode of a long story. Someone could say (citing Billie Holyday), "It's the same old story," the continuous evolutionary ping-pong between competitive organisms and mankind. However, once again, we should find a way out with the help of science and technology—but we must find this way, and the target is even now moving and very active. Furthermore, it is necessary to consider the presence of several influences affecting seriously the natural tendencies by mankind's impact.

Mao and the sparrows

So far, the history of governmental measures to avoid the consequences of borne-insect diseases in order to defend and increase population health include plenty of failures and incredible errors. The main idea was to find an easy and rapid solution, usually based on killing the target enemy. In practice it is the usual shortcut to reach a rapid solution, apparently as simple as usually unsuccessful in the reality. In general, the negative effects were much more than the positive ones, when present. This is a point in common between communism, socialism, capitalism, and neocapitalism: the capacity to enact enormous disasters on the environment, presenting them as necessary for the community. There are several clear examples, but one in particular is considered here as evidence. The following case is interesting, at least for those new to it, and is also useful to learn how governments, which are usually accused of lack of action when it comes to environmental problems, can on the contrary be responsible for great perturbations.

Between 1958 and 1962, Mao Zedong, also known as Chairman Mao Tse Tung, was the leader of the Communist Party in China. The external enemies of national communism were defeated and it was time to shift attention to the internal enemies, any kind of enemies. Chairman Mao's first problem was the necessity of producing enough food for the immense population of his country; the second one was how to reinforce his leadership through the recall of the population to massive mobilizations. We have no reports about the advisors and reports that convinced Mao, but there was evidently an error in the information and a misrepresentation of the reality, a typical horrible mixture between stupidity and ignorance, the mother of pointless disasters. Scientists are in such cases stone guests, asked to quantify the entity of the announced tragedy. Thanks to his political superpower and undisputed charisma, Chairman Mao announced and led a giant economic and social movement throughout the country, known as the Great Leap Forward, calling all population to active participation. One of the movement's first initiatives was the Four Pests Campaign, also later passed through the history as the Great Sparrow Campaign or simply the Kill a Sparrow Campaign (Krupar, 2002; Butt and Sajid, 2018). The central idea of the Four Pests Campaign was a move toward public hygiene, identifying four targets considered to be major pests: rats, flies, mosquitoes, and sparrows.

Although very generic from a taxonomic point of view, three of the four pest targets may seem to us reasonable and in clear sympathy with this book, because of their role in spreading plague, typhoid, and the malaria—but what about sparrows? It is true that birds can be fundamental in the diffusion of insect-borne diseases in the case of viruses such as West Nile fever and St. Louis encephalitis, but these diseases are common in other regions of the world, and the focus on these birds was totally different. Sparrows were included in this list because they consume rice and other cereals from agricultural fields, stealing food essential to Chinese citizens. The Communist Party called people to act together against the pests, focusing on their extermination. The aims were publicized by the efficient capillary propaganda of the Communist Party, through posters illustrating the need for fly swatters, drums, gongs, and even guns as tools in the fight, with the publicized target focused on the need to improve public health. People reacted as expected and took all possible and available measures in order to kill these four types of creature. In the case of sparrows, any sort of means was encouraged: nests were torn apart, eggs broken, hunting trap bird nets were everywhere, and fledglings were killed. Farmers were convinced that killing sparrows would be a good way to protect the local harvest. Furthermore, citizens, scholars, work groups, and government agencies were encouraged to make any kind of noise, with pots, pans, and drums, to scare the sparrows continuously until they fell from the sky from exhaustion. In terms of accomplishing its objectives, the Four Pests Campaign was initially considered a success, with the deaths of 1.5 billion rats, 1 billion sparrows, more than 220 million pounds of flies, and more than 24 million pounds of mosquitoes. However, we must always remember that the real objective was the improvement of public health.

Today, we are used to calling events "fake news." There was a total error in assigning the role of sparrows in the environment, in particular as an integral part of crop protection. Not only do sparrows eat just grains when possible, but their main food consists of insects, as is the case for most birds. This was demonstrated by scientific reports evidencing the real role of sparrows. In 1959, researchers at China's Academy of Sciences performed autopsies on several dead sparrows. They discovered that the majority of their stomach contents was made up of insects and not grains, as previously claimed. The result of the campaign was that the insect population in China actually grew exponentially. We must note also that sparrows are the only natural predator of locusts. Locusts eat every plant and are very dangerous, as demonstrated by the biblical plague. The imbalance between predator and prey in the form of sparrow and locust was forcibly broken, obtaining an opposite result. A massive excess of locusts was able to proliferate and swarm, eating the majority of the agricultural production intended for human consumption. In addition, beside the hundreds of thousands of pounds of grain eaten by the locusts, they were able to profit from conditions of drought, flooding, and other changes due to agricultural policies. Besides rendering the sparrow nearly extinct in China and the reduction of crop production by 15%–70%, the Four Pests Campaign had as a collateral effect the starvation and deaths of humans. When the famine ended, and the equilibrium was partially restored, between 15 million and 36 million individuals had died.

Therefore, when sparrows were saved from the government's insanity, and when the government realized the important role of sparrows in pest control and successful agricultural harvests, another target was implemented. In 1962, when the Great Leap Forward campaign finally ended, this was not the end of the Four Pests Campaign. In 1998, the Chinese government revived a new version of the movement against pests. The campaign against sparrows was over, but the Chinese government replaced the sparrow target with bed bugs. This caused no agricultural damage but did nothing to prevent the Great Famine. Again, posters were seen urging citizens to kill the four pests, this time rats, flies, mosquitoes, and cockroaches. The campaign was a waste of time, considering that people had already been killing these pests before the posters were hung, off course without obtain their extermination.

In this case, as in others presented in many other parts of this book, on one side the butterfly effect it is evident. This effect is well-known, but less known is a corollary of this effect, considering an evaluative differentiation between the species. In the quantum theory interpretation of composition of matter, we must consider the need for an intimate diversity and the consequent difference in the impact. The concept of the central importance of each organism in the habitat (again derived by Darwin) must be in part revised. Some species, named keystone species, may have a more significant impact than others in the habitat equilibria, and attention should be focused on this aspect, since the conservation of all organisms is the ultimate aim. Thus, we have seen the superior impact of the Asian tiger mosquito, and the presence in a population of exceptional individual entities whose increase can change the development of all species and their impact.

This is only one of a plethora of cases evidencing governmental incapacity to identify adequate measures to solve environmental problems without political influences. On the basis of this consideration, what can be done? Wait until the boomerang effect restores the natural order of everything? Meanwhile, stand inert in the face of all the consequent disasters? Is it possible to develop reliable independent scientific models?

Computational reality

The main problem is that so far, everything has been perceived as a comparison between opinions. The real identity of an environmental emergency is usually not an object of sufficient consideration. It is necessary to distinguish the solutions of the environmental problems from political and profitable interferences, which are inevitably typical of human nature. In this confusion, the evidence and indications supplied by science must be absolutely convincing and solid. Nowadays, there is a new fundamental tool that has come out from the environmental cul-de-sac. The aid of an objective point of view is absolutely necessary and can be trusted, with awareness of its limits.

> *"Can you do Addition?" the White Queen asked. "What's one and one and one and one and one and one and one and one and one and one?"*
> *"I don't know," said Alice. "I lost count."*
> *"She can't do Addition," the Red Queen interrupted. "Can you do Subtraction? Take nine from eight."*
> *"Nine from eight I can't, you know," Alice replied very readily: "but –.*
> *"She can't do Subtraction," said the White Queen. "Can you do Division?"*
> ***Lewis Carroll,* Through the Looking Glass and What Alice Found There *(1871)***

In the extract above, Alice is in trouble with typical problems of arithmetic—a simple one concerning a calculation that is apparently elementary but beyond the limits of her memory and demonstrating the difficulty of adapting ordinary logic to any kind of calculation. For many years, people were convinced that computers were mainly useful to save them from the difficulties of calculus and would be able only to perform repetitive simple operations. This second point ignited an ongoing debate. Meanwhile, the scenario changed completely and rapidly overcame the limits of a presumed full human monopoly of any form of intelligence. The correct acknowledgment of the existence of another form of intelligence, and therefore the adequate utilization of its potentiality, is the real challenge of the 21st century. However, the reality claims its rights for an independent future, despite the human tentatives to realize an ideal result. Computerized Artificial Intelligence (AI) is fully integrated in scientific activity, as well in ordinary everyday life. However, since the beginning of the computer era, a fundamental debate about the possibility of trusting AI's capacity to help human judgment has been underway. As a confirmation, the story of the evolution of AI is full of uninspected upheavals, as consequences of responses to fundamental questions with spectacular contributions by some of the most talented minds of the last century. The pathway to the AI's achievement was full of twists and unexpected collateral effects. However, it is possible to coherent in a red thread leading to AI as we know now, such as androids, cyborgs, avatars, and cyclones. The debate on AI is nowadays more present and intriguing than ever. Waiting that the alter ego game furnishes the perfect symbiosis overcoming the Imitation Game, we can travel through the main steps of the AI spectacular outset.

First act: The stone in the pond was launched by David Hilbert at a conference in 1900. The famous German mathematician proposed to the scientific community 23 unsolved problems with the intent to define mathematics logically using the method of formal systems. In this way, the main goal was the developing of a finalistic proof of the consistency of the axioms of arithmetic by demonstration of a series of solutions to the problems. The formalist approach of Hilbert's school intended to develop in a more rigorous way the axioms of arithmetic proposed by Peano, showing the absence of any contradiction inside the system.

Second act: The challenge was accepted by a young and unknown Austro-Hungarian-born, later American, mathematician. A logic arguments enthusiastic, Kurt Friederick Gödel (1906–1978) occupied a central role in the debate. In 1930, Godel was present in a conference in Konisgerg and at the final discussion, in complete but firm quietness he made a complete revolution of the mathematic bases and exposed the demonstration that such a finalistic proof is provably impossible, being auto-referent. One year later, Gödel published his two theorems on incompleteness, when he was 25 years old, 1 year after finishing his doctorate at the University of Vienna (Tieszen, 1992). He stated in the First Incompleteness Theorem that "any effectively generated theory capable of expressing arithmetic cannot be both consistent and particular"—in simple terms, he accused mathematics and logics of working on circular hypotheses, meaning that the whole system suffers from auto-referentiality. The novelty was that Gödel's proposals had an enormous impact, even in mathematics, since probably the most important effect was that the operations of logic deduction can also be assimilated, in their intimate nature, to mathematical operations. In other words, the language of logic in several ways reassembles a particular form of calculation, changing symbols and operations. Gödel has been compared to a barbarian attacking the citadel of mathematics, and in fact his proposals were able to disrupt some precedent axioms and open the way to a new approach. He was quite conscious that the consequences were out of the mathematic themes, involving directly the semantic front, as evidenced in the correspondence between himself and Rudolf Carnap.

The castle of formalism had been seriously attacked, but was still not demolished. Therefore, the initial attempt supporting definitively the axioms of mathematics resulted in a complete crisis of the former system and the generation of a new logical approach. Although it was evident that the axioms of arithmetic were not consistent, however this as not sufficient to open the way to the new paradigm. It was necessary to solve the mother of any discussion in logic: the possibility to have a method able to decide in any situation if a proposition inside a formal system can be considered true or false. The shadow of the Epimenides paradox was still confounding the pathway of human logic.

Third act: The semantic front was at the same time the weakness of the old system and the starting point of the new one. The possibility to exit from the impasse, establishing a defined method, even consisting in a mechanic procedure, was proposed by a young scholar of Cambridge, Alan Mathison Turing (Gandy and Yates, 2001). In the summer of 1936, he published the definitive assault on Hilbert's mechanistic apparatus in a single determinant article consisting of the description of a hypothetic apparatus (Tseuscher, 2005; Soare, 1996). In his "On computable numbers, with an application to the *Entscheidungsproblem*," Turing not only completed the mission of demolishing formalism, but was able to go much further, and this is the central point for our discussion. First, we must consider that Turing was not exactly a mathematician, since he was arguably far more intrigued by chemistry and biology. Therefore, the end of a paradigm inevitably opens the door to other possibilities and a new reality not previously imagined. The paper is based on the description of a hypothetical mechanic apparatus, later named as the Turing machine, in principle able to perform analytically any calculation possible for a human being. The idea was not exactly original, having been already proposed by Euclid, Leibniz, Babbage, and even Hilbert. However, the tendency was negative, and the general idea, as stated by Andrew Hodges, considered that only an incompetent tremendously naive person could think that mathematics could make a discovery by just rotating a knob on some miraculous machine. Turing essentially was fighting against a prejudice deeply rooted, but he was also attempting to establish the existence of an algorithm able to solve any mathematic problem, which could be the solution of any human controversial by the adequate calculus. In other words, if the solution cannot be found inside the system, one should look outside of the system.

Using his machine, Turing's aim was to demonstrate that beside the classes of numbers and functions computable, there is another class with numbers and functions not computable, and these cannot be calculated by human beings or by the machine. In other words, the *Entscheidungsproblem* (Problem of Decision) (Turing, 1936) does not have a positive solution, but the Turing machine opened the route to new perspectives. Therefore, the Turing machine was imagined as a method to complete Gödel's deductions, but it was immediately clear that the implications were far beyond the initial scope. This was, beside the practical applications, the reason for the enormous success of the Turing machine. The soul of the Turing Machine is in the possibility of obtaining a total similarity between the logic approach of a computer and of a human mind, through the same semantic approach. The functionality of the machine generated the fundamental difference between hardware and software, causing the birth of the current computers and similar dispositive, shifting the debate from mathematics to the role and future of AI.

The moral of this story is that the effects of an invention can reverberate and be successfully applied in fields far away from the starting research point. The typical effect of application of technology to pure science, often with results totally unexpected and very distant from the starting point of interest. In my laboratory, we call it the Fosbury effect. Dick Fosbury could never have used his revolutionary back-first technique (the "Fosbury flop") in the high jump and won Olympic gold, without breaking his neck, if he had been forced to land on the sand instead of a soft mattress in plastic, recently invented at that time.

In addition, the introduction of vocals in our alphabet is due to the necessity of stonecutters in the ancient Greece to nominate with major precision the names of rich sponsors of their works. In this way, it was not the elaboration of a grammar study or the erudite elaboration of a minister, but the will to exhibit the gratitude by ex voto statues, dedicated to the gods' benevolence, that was at the origin of the complex alphabet. In such a way, the simplicity and utility of our alphabet symbols were considered more adequate for the working process of Turing's machine against other forms of writing, such as Chinese ideograms. An ideogram contains stylized direct information about the form and the nature of the object referred to, but by mixing consonants and vocals you can obtain infinite combinations, as well as with numbers, albeit etymology is lost. It is important to reflect on this point. Language started from the necessity to describe reality and transfer information about it, in a written or phonetic way. Therefore, everything starts from a sensorial perception. Ideograms are the product of visual perception and therefore are designs. Syllabic writing is the result of an acoustic experience, the conversion of phonemes, sounds, or noises in symbols. That the word A originally was the head of a bull, but it is out of any iconographic value when we use the letter, as well B was a house. The iconographic value of these letters is completely lost as regards their original importance. The alphabet is a series of atomic phonemes, which when combined can give the entirety of the phonetic possibilities, corresponding to parts of reality. In the alphanumeric system, the symbol itself is nothing and therefore can be utilized universally. The negative ontological essence of the alphanumeric symbolism, frozen inside a letter devoid of intrinsic meaning, found the epistasis of its performance in the possible representation of what is considered real and therefore identifiable in physic and metaphysic identification. A computable symbol structurally cannot represent more than nothing, but it is asked to find a precise identity when converted by an algorithm, i.e., consisting of a word inside a phrase or a number useful to express an exact quantity.

This logic was ready to be easily converted into the computer ratio, consisting of nothing more than a further symbolic metamorphosis. During the sequence of metamorphoses of the Occidental language, there was a progressive reduction of the number of symbols and a solution to the phonetic pronunciation problem. The ideographic system is based on thousands of graphemes; the syllabatic one needs 40–200, the consonantic-vocalic 20–30, and the binary one only two, but the latter is still able to represent and interpret reality. Deprived of any visual or phonetic linkage, the symbolism finally acquires its perfect universality necessary to integration. Thus, in AI is a correct evolution of the language system able to be useful to us and to the machine, in the time.

Semantic nihilism and the Turing machine

Among the aims of the Turing paper of 1936, as present in the original version, there was an analogy between its model of machine, named unorganized, and that of a neural network, as elaborated in the same period by McCulloch and Pitts (1943). Once we have accepted the analogy between computer and mankind, as evidenced by the ability to convert reality in symbols, like that of an object, in numbers able to reproduce the exact image of the object, more or less as we do looking at the object (as speculated by Steven Pinker since 2011), we must admit that this is only the simplest part of the problem (Pinker, 2011, 2016, 2009; Schlinger, 2004).

What about the interpretation of the object behind its form and appearance? There is a limit of complexity in this work of acquiring the reality around? Can a computer with AI give a solution to any problem using the appropriate algorithm or it is something more sophisticated necessary? Pinker's approach is related to the quality of the input, but what about the elaboration, or perhaps the manipulation, of the acquired data as the necessary output? Can everything be converted into numbers or other symbols without loss of information, and the conversion utilized adequately? Essentially, can computerized AI be applied successfully to biological problems? It is possible that the analogy between man and computer is pure fantasy, as stated by Gandy (1988): "There is not suggestion that our brains work like Turing machines." The same postulations of the conceptual Turing machine can be considered as the consequences of pure speculation, like the load-bearing columns of a building whose basement is totally unknown, as evident in the preference about the occidental nature of symbols, and computer and computor are only the appealing facades of the construction, but in any case adequate to obtain an useful result for the topic. In this incompleteness there is also the success of computer machinery (Cooper, 2012).

Although the limits of AI are still unknown and undetermined, being in continuous expansion, the general idea is that a biological process is too complex and variable, but this is a typical irresistible challenge for a scientist, and the Enigma machine is a wonderful example of the enormous potentiality of AI, solving problems and leaving aside any theoretical considerations. However, what you expect from a machine is essentially that it works to satisfy your requirements, regardless of the logical and philosophic problems.

Fourth act: Apparently, the Turing machine is very similar to a typing machine (Fig. 4.1). The device and mode of operation of the Turing machine are easy to imagine and understand, so simple yet so innovative

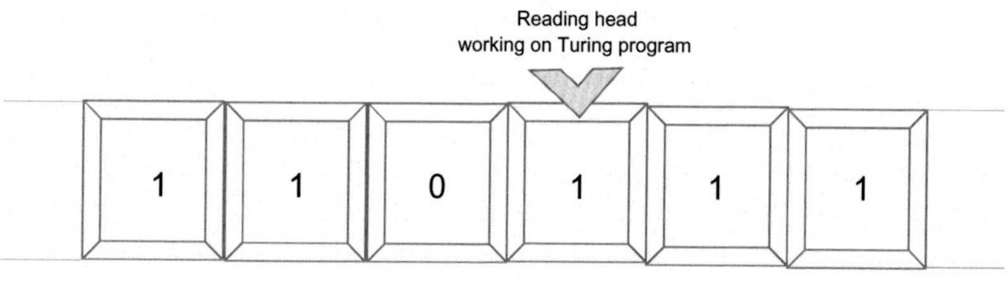

Fig. 4.1 The Turing machine as tape and reading head.

for that time. The heart of the machinery contains a heat and a tape. In an idealized computing device, a read/write head (the "scanner") operates with a paper tape passing through it. The tape is divided into squares, and each square bears a single symbol—in a binary system, 0 or 1, for example. The symbols are rewritten and converted into the machine language. In this way, the tape is also the machine's general-purpose storage medium and it is able to serve as the vehicle for both input and output and to store the results. Turing's aim was to show that the tape, and therefore the input inscribed on the tape before the computation, must consist of a finite number of symbols. However, the tape is of unbounded length, given the possibility of unlimited working memory and data storage.

The working memory can therefore store the results of intermediate steps of the computation as well as operate further manipulation in accordance with a programme. The operations consist of altering the head's internal wiring by means of a plugboard arrangement. The computing consists of inscribing the input on an infinite linear tape (e.g., in binary or decimal code), placing the head over the square containing the leftmost input symbol, and setting the machine in motion. Once the computation is completed, the machine will come to a halt with the head positioned over the square containing the leftmost symbol of the output (or elsewhere if so programmed). The Turing machine can be interpreted like a mechanical transposition of writing and reading done by a human being. The line of the computer screen in Microsoft Word is divided into single spaces, like a child's squared notebook. The eyes of the writer and its fingers, moving along the line, are the inputs whereas the outputs are the acquisition of data and the delivery of the machine. The writer and the machine are simultaneously flowing back and forth along the track of a well-defined pathway. When the thoughts and the way of thinking of the writer are transferred inside the computer, they are converted into an intangible matter able to be stored and immediately transferred by the computer network, after a deep change of the symbolism. The transfer generates a radical metamorphosis, although the writer is no longer aware of the transformation. Thanks to this manipulation, the ideas can be aggregated with others and converted by algorithms. The simple act of writing, for instance, to enquire about the planetary web system, constitutes the enquire part of the system, modifying in part the system itself and the writer. The litmus test is the simulated experiences in virtual reality, able to convert in part or completely the real world by a computer program. The parallelism computer/computor is the principle of operation in the Turing machine. In the itinerary dictated by the communication, the idea, generated by the mind, is crystallized in the sound of the phoneme and later in to calligraphy. In this way, the idea can go back finally to the mind (background effect), where it is sedimented as a modified representation of the reality. The representation is crystallized, transliterated, manipulated, and shared through an atomic system of alphabetic or numeric elements, but it seems that something is missing. The debate about the inner significance of the word has a long story connected with its de-somatization. Plato in Cratylus, reflecting about the theory of forms and elements, which is the theme of the dialogue, counterpoises the arguments of the sophist Ergomenes, which considered names as superficial and without value, like elements of a society game, and Cratylus, evidencing the ancestral nature of names, because "everything has a right name of its own, which comes by nature, and that a name is not whatever people call a thing by agreement, just a piece of their own voice applied to the thing, but that there is a kind of inherent correctness in names, which is the same for all men." The debate continued for centuries about the enigmatic *lekton* (whether expressible or significant, or simply referring) and has recently been revitalized.

The ticking of the Turing machine in its movement on the tape is the manifestation of the mechanic transposition of the passage of phoneme/writing into a digital version. In the writing and reading acts, there is a fundamental step. In the early days of writing, these acts were prerogatives of an élite: a communication necessary to exchange information and ideas among the élite, or to transfer orders to the population, giving again physical form. During these acts, pronunciation of the words is possible because the written words are transferred as images inside our mind, interfacing with what is already there, deeper and deeper, in an attempt to give physical form and manifest the immaterial, the *pneuma*, the *psyche*, the breath of soul, perhaps in a poetic form. If the writing form is the transaction of voice sounds in visible signs, *semeia anthropines phonées*, silent reading can make an immediate link to the eyes, establishing a path to the beginning. The art of thinking can be fully expressed and achieved by intelligent machines. Symbols, signs, and even images and sensations can be easily converted, stored, and shared independently by the chosen language. Again, parallelism emerges: electrons moving in a neural network and electrons running in software. The optimization of this process has a cost, like any evolutionary jump. Trusting a powerful tool like AI generates dependence. The clear general difference that has emerged during the last decade is the ability and confidence in the utilization of PCs and cellular phones. In a world running toward the dominance of apps, this is a line of demarcation. When your money is only a number inside a file and you must use an app to move it, if you are not able to, you cannot ask for help from the bank because it does not physically exist; there are no agencies, no bankers, no one to talk to and explain your case to—only an app. Right or wrong, this is the present and the future we must consider.

There is an ancestral resistance for novelties, in particular if they are difficult to understand and appropriate. Many people are dedicated to using AI, because it eases their life, but they are diffident about what they do not understand. When mathematics are involved, caution becomes suspicion and mistrust. This attitude involves also educated people and most politicians, and it can result in violent reactions. I am deeply convinced that the similarity between the tragic ends of Turing and Palamedes is not coincidence. Both crystalline heroes and models of pure intelligence, creators of novelties, inventors of numbers and symbols and of their significance and utilization. Both envied and persecuted until their dramatic ends. Both scapegoats sacrificed in favor of the dominance of models produced by perverse *callidus* Ulixes.

Once accepted the compatibility between the computer, the intelligent machine, and the computor, the model and creator (or vice versa), based on the same computational approach, the next step is the integration between the enormous computational capacity of the computer and the infinite intuition of man. The idea is that the sequence reality/symbol/computation needs further steps to avoid the limits of simple representation. This is possible thanks to the dual nature of the computor, the behavioristic one based on the direct relation between external stimulus and compartmental replay in a contest essentially deterministic and operationistic, similar to the computer operability, and the second mainly cognitive and representative of the inputs of the previous stage. The two states are in continuous interdependence in order to obtain the final result, consisting of the consciousness of reality and the decision about the adequate comportment. To maximize the final result, correct acquisition of as much environmental data as possible is preferred. However, among the condition for a useful interaction between computer and computor, the communality of symbols and arrangement of data in the best comprehensible manner are key to integration—in other words, whether the evolved form of the Turing machine can become bio-logic or remain bio-graphic. In this operation, the data are connected in a logic representation, but usually after a simplification, based on the determination of hierarchy of interest: a screening of what is of primary or secondary importance. The time is right for computational and modeling bioscience.

This long dissertation about AI is a necessary preamble to the next part of the chapter. About the digitalization we live a controversial experience. Most people are convinced of their absolute control over the computer, since it is possible to switch off the machine at any moment, but dependence is another matter. However, when we are not online, we are still likely to be connected and part of the net system somewhere.

The similarity of the computer and computor is not in the symbolism, but in the segmentation of the tape. In other words, the key is in the empty space separating the symbols. Reality is based on continuity, wherein the present does not exist, being a hypothetical passage to the future from the past. The Turing machine, as any alphanumerical system does, works in a discrete sequence. This is the original sin still marking the difficulty of any human in his tentative to represent and interpret the present as a continuous flow of the reality.

The alternative possibility of AI to nab the complexity of biological systems is if the biological reality is like a giant chess game, wherein each organism plays its game maximizing its capacities but according to rigid rules, or the flexibility and variability are so high that any prediction is impossible, like in the croquet game of Alice in Wonderland where the balls were live hedgehogs, the mallets live flamingos, and "the players all played at once without waiting for turns, quarrelling all the while, and fighting for the hedgehogs." The scientist may feel very uneasy, like Alice managing her flamingo. Therefore, AI is necessary—first, because general physical–chemical rules exist and living organisms cannot escape; second, because in the habitat all players play at once but rationally influence each other; and third, because systems can be simplified but variants need AI to be calculated as well human sensitivity and intelligence to be understood and verified.

AI is a powerful tool, but it also possesses independent capacities, which is both a problem and an opportunity for any technological advancement. So far the predictivity of AI in outbreaks was practically null, despite all the progress and tentatives. As evidenced clearly by the COVID-19 episode and the debacle of algorithms produced to interpretate epidemic trends, as evidenced by the GFT (Google Flu Trends) program about the trend of stagional flu in the world. In terms of the themes of this book, so far the help of AI in solving insect-born disease events was also very limited with regard of migration and boom of certain vectors. In Zika outbreak time, there was a titanic effort, to consider the simulation of a billion of people and six billions of mosquitoes utilizing 7500 computer only to evidence the pathway to the Zika virus from 2014 to 2016. However, so far these projects were not able to afford more results to understand the past than to predict the future. As for climate change, most of politicians, beyond the official crocodile tears, are doing nothing of practical effect, and several of the them still wave the flag of the typical propaganda that "the sky is always blue"—until the next catastrophe. However, scientists must continue with their research and their computerized models, and, very importantly, they have to work hard in transferring their knowledge to the ordinary people, who can assume responsibility for their own destiny. The solution is not to "ask the computer" and abdicate all personal mental judgment, but to be aware of the need for independent information and evaluation. Meanwhile, reassuring signs are coming from the scientific side.

The first attempt has been made on the cell, known as system biology (or biology of systems) (Breitling, 2010). Further advances were obtained by an exponential development in parallel models and highly advanced scientific devices. Over the past 30 years, medical research analytical technologies have arrived at the point where most (if not all) key molecular determinants deemed to affect human conditions and diseases can be scrutinized in detail. The final aim is that the utilization of these technologies, and others such as quantitative polymerase chain reaction (qPCR) and next generation sequencing, enables extracting information from complex datasets to obtain disease models to be developed and verified by wet-lab testing.

The study of biological systems, like those present in living organisms, are based on the following axioms: (a) a biological behavior cannot be reduced to the linear sum of their parts' functions, following a holistic approach instead of the more traditional reductionistic one; (b) the complexity of the system under study is not a limit but a positive aspect; (c) technology and computation are integrated in the approach to the elaboration of the solution; (d) data are obtained mainly by transcriptomics, metabolomics, and proteomics through high-throughput techniques, and dedicated to the construction and validation of models, the model being the real result of the process; and (e) contribute to stratified the so far dominating medicine en route to an approach of personalized health care.

In practice, system biology is simply the application of dynamical complex system concepts to the molecular metabolism using a sophisticated computerized approach, named the computational model (Baitaluk, 2009; Tavassoly et al., 2018; Zou and Laubichler, 2018). The theory of self-organized systems means the possibility to understand the internal logic of life, through the manifestations of cellular metabolism (Schrodinger, 1944). According to Sauer et al. (2007): "the pluralism of causes and effects in biological networks is better addressed by observing, through quantitative measures, multiple components simultaneously and by rigorous data integration with mathematical models." Among the aims of system biology, predictivity is the natural central result of all the process, generated by the integration of the available data, designing principles of biological circuits in the organic network. Systems medicine (Ayers and Day, 2015) is probably the most important application of system biology, serving to identify clinically important molecular targets for diagnostic and therapeutic measures, including the discovery of new pharmaceutical drugs for fighting, or at least controlling, the resistance phenomenon in insect-borne diseases.

Today, the cost of experiments, trials, claims, and marketing is a serious limit for the development of new medical drugs (Moore et al., 2018; Sinha et al., 2018). In 2018, a team including researchers from Johns Hopkins Bloomberg School of Public Health reported in *JAMA Internal Medicine* that clinical trials that are considered necessary to support the US FDA approvals of new drugs have a median cost of $19 million. However, the study suggests that these costs of the "pivotal" clinical trials contribute only modestly to the overall costs of developing new drugs, because it is necessary to consider all the previous investigation and study. Therefore, the $19 million median cost represents less than 1% of the average total cost of developing a new drug. This cost of pivotal drug trials is variable in the different classes of drugs, but in recent years it has been estimated at between $2 billion and $3 billion, including $2 million for a four-patient trial of a treatment for a rare metabolic disorder, up to a mean cost of $157 million in cardiovascular drugs, and just $21 million in endocrine and metabolic disease patients. The major cost is $347 million for a large study of a heart-failure drug. Although the market of an efficient drug in these diseases is enormous, it is evident that such costs, although totally justified by the need of health security, are killing off the pharmaceutical industry, in particular for antimicrobials. Although several studies are underway to lower costs and develop better trials (Moore et al., 2018), the main unsolved problem, as already mentioned, is multidrug resistance. In a few years, microbes can find a way to develop resistance, before the necessary return on the investment in terms of money and human resources. Again, computer aid can furnish at least partial solutions.

The final aim of system biology is a reliable model of the cellular system, wherein it would be possible to test the biological activity, avoiding the usual current tests based on living materials, i.e., cells, tissues, organisms, and consequent ethical and economic problems. In other words, a simulation approach, similar to that used for instance in the training of airplane pilots, involving a careful and methodical examination of all simultaneous molecular interactions at the occurring cellular levels. In this way, the costs of clinical tests could be avoided or decreased, as could the sacrifice of animals. However, so far, to obtain a reliable simulation in perfect synthesis with the Turing machine, a process of elimination of the details considered nonessential seems necessary, based on suppression of redundant and enucleation of invariant factors in search of the "essential character" of the living system under examination. This is probably the main limit and the most delicate aspect of any computerized apparatus when applied to a dynamic system. Once the coherence in functionality and compartmental features between computer and computor for input, storage, and utilization of the database is accepted, who takes responsibility for choices? Apparently, the computor possesses the necessary ratio and sensibility, but his or her capacity to manage the amount of data is limited and the objective capacity of analysis could be influenced by emotional factors. In contrast, the computer could be objective and respect the necessary scientific approach. In reality, all these considerations are wasting time: driving forces have been acting since the birth of the Computer Age for an integration between computer and computor, a hybrid naturally born and already affirmed in human society. The reality is that we cannot leave out the AI influence to impose the pathway of the rational use of AI versus the chaotic dangers of "big data" and G^{nth} in every type of our activity, in particular for the predictive ones.

The continued goal of AI is to trust the capacity of enormously complex machines to examine situations and produce reliable models, but they are ultimately machines, and to do this is possible because human behavior is considered intrinsically mechanic and computational just as that of intelligent machines is. Beside the large debate on the righteousness of this assertion, it is necessary to repeat that this assumption is becoming (or is already) ordinary and scientific reality. Computer and computor (in the original version is female) are still not married, but they are a couple in facts. On the other side, it is conceivable that at least in future the computational model will be able to perform a reliable and generally accepted simulation of a living system.

A cell, like an organism, is the result of an organized functional network of individual components. The aim of this organization is the correct and useful connection with the external conditions. This definition can be simply utilized to describe the environmental equilibria. The model simulation in this case changes the actors and the place, but in principle the approach can be similar. Therefore, the utilization of a computerized approach to produce reliable models is conceivable also in the case of the determination of factors influencing environmental changes and their effects. The main obstacle to realizing such models is not the complexity of the system, but the utility of such effort. Nowadays, there is a very low probability that such results could be the objective basis of a government policy on management of a habitat. However, it is the duty of science to continue in this direction. In this way, there is a chance of understanding what is going on in our changing world. Thus, in insect-borne diseases we fully understand the environmental abiotic factors affecting the migration of vectors or the pathway of emerging strains with genetic trait changes, and these data can be interconnected with the behavior of the vector and the biotic effects to obtain a model evidencing the future, such as the time and the possible areas of diffusion of the disease. Any kind of measure and intervention by governments and related agencies should therefore be undertaken considering carefully the results present in the models instead of the pressures of private or political interests.

While waiting for advancements in AI, there are other fronts offered by technology; in particular, general attention is focused on the innovations offered by nanotechnology, whose applications can be very important also for future control of insect-borne diseases. In particular, we are interested in the biological applications of nanotechnology, considering the potential environmental effects.

Nanobiotechnology

In this chapter, we are attempting to explore the last frontier of technology and its fundamental role in future control of insect-borne diseases. The argument is recent and still taking its first steps, but its potentiality cannot be underestimated. The relation between nanotechnology and insect-borne diseases control is already known, although so far it is limited to the experimental phase. Perspectives are concentrated on the control of the vectors, focusing on natural insecticide and the increase of activity obtained by the stabilization and protection of the active constituents (Mehlhorn, 2016).

In particular, I was involved in several investigations on the production of green nanoparticles, participating in research by a group led by Prof. K. Murugan (Thiruvalluvar University). The aim of the project was to check the activity of green nanoparticles on insect-borne disease vectors, but also the possibility to utilize different materials and determine the impact on the environment, in particular for predators that are the natural controllers of preparasites' diffusion. This approach was totally in accordance with my belief in research born from integration of different scientific experts and the need for useful integration between laboratory work and field experiments, as well as the persuasion of the utility of the investigation. The control of insect vectors is the cornerstone of exploration on environment equilibria and the interactions inside the habitat of several protagonists. The focus on the utilization of novel technological tools is clearly important for the future.

Technology consists of the application of scientific knowledge to solving challenges and practical aims of human life, usually related to changes of conditions in favor of human necessity. As slaves of Prometheus's prophecy and emulators of Palamedes, we trust technology to solve the problems of survival and maintain supremacy by advancing the utilization of available materials. Science and technology are interdependent but distinct, and both are consequences of innovation. In fact, although most people consider technology to be a practical and efficient application of scientific discoveries, the converse impact of technology on science opens up novel scientific questions and leads to new and more difficult arguments, enriching and reorienting the agenda of science. Thanks to technology, the full range of science discovery can find allocation and be addressed in an efficient and timely manner.

Let us consider another book. Probably the first convincing explanation of the potentiality of nanotechnology is contained in a short book written in 1944 by Erwin Schrödinger, one of the fathers of quantum theory, titled *What is Life?* (*Schrodinger, 1944*). The book is a compendium of a series of lectures given at King's College, Cambridge (UK) a few years previously. As evidenced by the title, the book is focused on questions typical of biology, but answers were explained by quantum theory. For instance, it contains an exact description of the molecular nature of DNA, 10 years before its discovery, as "a gene—or perhaps the whole chromosome fibre—to be an aperiodic solid." An aperiodic solid means a chemical compound made by minor molecules assembled in an unrepeated way.

The book starts with the question: "Why are atoms so small?" The answer, in a typical quantum style, is that in this way each atom is free to exert different contents of energy, without interfering with the total macroscopic result. Therefore: "If we were organisms so sensitive that a single atom, or even a few atoms, could make a perceptible impression on our senses—Heavens, what would life be like!" A similar consideration can be applied to cells, or to snow crystals. In this way, variability and constancy can coexist, without consequences about the confidence in general states of the matter. In other words, the physical laws are the same everywhere, but the mechanics of each phenomenon are not the same and are dependent on the scale.

The meter as a measure is familiar to us because it is closely related to the size of our bodies and to the reality expressed by our senses, but there are other realities, expressed by other units: the limit of our sensibility is 0.001 (10^{-3} m), the one concerning the cells is based on 0.000001 meter (10^{-6} m), and that of atoms and molecules is called the nanoscale and is based on 0.00000001–0.0000000001 (10^{-9} m) meters. In a single grain of sand, there are thousands of millions of atoms, all of them individually different and arranged in different ways. An atom is too small to be seen with any kind of instrument, and its presence can be deduced only indirectly with a special experiment and extreme difficulty. In atomic physics, it is necessary to use as a unit the so-called Angstrom (Å), which is the 10^{10th} part of a meter, avoiding the decimal notation of 0.0000000001 m, since atomic diameters range between 1 and 2 Å. The atomic world is impenetrable to our senses, but it influences our activity. Beyond this scale, we find the enigmatic level of quarks, where only mathematics and imagination are able to enter, and where something still not clearly understood, like strings, is acting. On the basis of these considerations, all of reality can be depicted as a series of concentric interacting worlds, wherein we can enter and interfere positively only by knowing and respecting the rules of each world. This is important also for the messages contained in this book. The world experimented on by a pathogen organism, like a bacterium (less than 1 μm = 0.000001), and even more by a virus (20–500 nm), is nearer to the nanoscale than our one; however, the parasite is obliged to interact with the world where we use to live and the solution is inserted into the intermediate cell world.

An insect-borne disease is the result of smart integration of concentric inclusive levels of reality, including also the giant environmental one, with its biotic and abiotic interferences. The knowledge of the homeostatic strategy of a parasite is crucial to understand and select the method to control it efficiently. The approach starts imaging to live in the world of the parasite or the vector, never forgetting that they make all their efforts to be an integrated part of our concentric world: the environment where we exist as a body made by cells wherein molecules act to make life. In nanotechnology, we work starting from the atomic world to influence all other states of matter. The last one can now be explored and influenced, although only partially, because of its intrinsic difficulties to interfere, but with awareness of its fundamental role. Previously, the molecular nanoscaled world has been out of our direct influence, but nowadays this limit can be overcome.

Let us consider now the problem of biological barriers. Most living organisms, such as bacteria, fungi, and plants, possess a cell wall; even virus have a robust envelope. However, it is absent in animals and the Protista, their precursors. This is the result of an evolutionary step. The cell wall ensures protection and individuality, but it is a source of difficulty for collaboration in the cellular network of the same tissue, and confers separation between the cytoplasm and the environment. Thanks to the absence of the cell wall, communication between tissue cells is allowed and preferred, and it is possible to run 100 m in less than 10′ or fly with the fantasy, but our cells are also more penetrable to parasites and other intruders. Arthropods are an exception, as they have protection in the form of an exoskeleton, which covers the whole body. This is necessarily a compromise between efficiency and protection. More protection means more separation and difficulties in the exchange of food and energy. An excess of reinforcement of the cell wall implicates the death of the cell, such as in the lignified vessels of shrubs and trees. There are several mechanisms for input and output, facilitated like osmosis, or active or passive transport, but these need energy. To avoid this waste of precious energy, in cell walls there are micro-holes that can be used as open doors for little molecules, such as water, that are able to seep through and enter the cell. Therefore, they are the simplest way to penetrate a bacterial cell. If we have a chemical weapon to kill the parasite, enabling the parasite to work, it is necessary to overcome the cell wall protection, and the micro-holes can be utilized for this purpose. Furthermore, the active substance must be protected during its trip until it reaches the cytoplasm. Until the organic substance is inside a cell, its integrity is usually assured, but outside, the instability is hiding. In particular, in the case of natural products, they are quite stable inside the cell, but they are rapidly oxidized and degraded outside.

Therefore, the key argument in nanotechnology is dimension. To have an effect, the active molecule must enter the cell, but there are protections. If the target is a microbe or a virus, there is a solid barrier constituted by an envelope to overcome; if the target is an insect, the exoskeleton preserves it from invaders. The nanoparticle must be under 50 nm diameter to utilize the nanoholes of the cell wall. In its journey, the active substance is protected inside the nanoparticle. Once that the nanoparticle is inside, its content should be able to realize the toxic effect against the pathogen, like Trojan horses containing the weapon against the responsible of the disease. However, it should be noted that empty nanoparticles also generate some perturbations in the target cell. The cell metabolism perceives the presence of the alien intruders and reacts in consequence. Therefore, it is necessary to check the mechanism of action to maximize the toxicity.

Recently, the ability to change the point of view reaching the nanoscale gave impressive results and applications, which is likely to be fundamental in the future of science and technology. Nanoscience is a multidisciplinary field based on the design and engineering of functional systems at the nanoscale. Nanotechnologies have the potential to revolutionize a wide array of biological applications, including drug delivery, diagnostics, imaging, sensing, gene delivery, artificial implants, tissue engineering, and pest management. However, probably the most important impacts must be expected in the habitat, due to its unique ability to interact at any level of the scale, from micro to macro, in magnificent continuity and efficiency.

The first step is focused on synthesis, characterization, and use of materials and devices that could have new properties and functions based mainly on their small size. The creation of nanomaterials is based on selection and assemblage of nanoparticles. The self-assembly is an automatic process, resulting in a geometric architecture obtained thanks to the reduced dimension and the intrinsic properties of the nanoparticle, including geometry and surface properties. Nanoparticles are designed to auto-arrange spontaneously in clusters or greater ordinated structures. Nanoparticles are collections of atoms arranged and bonded together with a structural radius of <100 nm, used as key fundamental components in the fabrication of a nanostructure, whose physical and chemical properties are dependent on the types and amounts of atoms in the material. Nanoparticles are selected on the basis of size, shape, large surface-to-volume ratio, absorption and surface functionalization, and utilized material. Nanotechnology acts selectively at an atomic/molecular level. Aims of nanoscience are related to the fabrication of cheap and more stable eco-friendly, easy-to-use marketed products, which are able to interact in atomic systems. Nanoparticles can be used directly to obtain new materials or to insert inside active principles, like in microencapsulation of drugs in pharmaceutical chemistry or the formation of films in intelligent packaging. The choice of appropriate materials is a prerequisite of any nanotechnological application.

In the last decades, noble metal nanoparticles have been preferred for synthesis of nanomaterials. Metallic nanoparticles are mostly obtained using noble metals, such as gold, silver, platinum, and lead. These metals are necessary to join together the ingredients of the nanoparticles and obtain their arrangement in self-assembly. Among metal nanomaterials, silver nanoparticles gained interest in bio-research due to their application and properties. Vast applications of silver nanoparticles include size-dependent interaction with HIV-1, bacteria, microorganisms, etc. The chemical reduction of aqueous solution of silver nitrate is one of the most widely used methods for the synthesis of silver nanoparticles.

However, redox synthesis methods generally use hazardous chemicals as reducing agents or require significant energy input. The consequent risk of pollution must be considered. There is thus a growing interest in the use of environmentally safer "green" reducing agents. Nanobiotechnology mimics natural processes. The self-assembly is typical of production of final 3D conformers of the macromolecules and the basis of supramolecular chemistry. Nanomaterial is obtained by self-assembly into ordered layers, whose structure is based on electrostatic forces, like hydrogen bonding, polar and dipolar interaction, hydrophilic or hydrophobic interactions, surface tension, and others, by analogy to the self-assembly of biopolymers, like those present in membranes, vesicles, nucleic acids, polysaccharides, and proteins.

Nanobiotechnology, also named green nanotechnology, is focused on interaction in biologic systems using eco-friendly and safe materials, which means novel and cheap bioreducing agents for eco-friendly nanosynthetical routes. Nanobiotechnology is a new and growing science, focused on eco-friendly green-mediated synthesis of nanoparticles. It is a fast-growing research in the field of nanotechnology with relevant applications. This technology has already been used in several areas of science, such as chemistry, biology, physics, materials science, and engineering, and applied for innovative solutions in drug delivery, diagnostics, gene delivery, and tissue engineering. Nanobiotechnology research in principle can be applied in any field and opens newer avenues for unraveling a wide array of applications in production of biomedical sensors, antimicrobials, catalysts, electronics, optical fibers, agricultural products, bio-labeling, and other items.

Nanoparticles can be classified on the basis of their chemical composition, divided mainly into metal and/or organic constituents. In several cases, the materials can be used at the same time, since nanocrystals of gold, silver, and their alloys have been synthesized with the assistance of organic materials of different types. In insect-borne diseases, nanobiotechnology has the potential to revolutionize the agricultural and food industry with novel tools that enhance the ability of humans and plants to confront the new challenges successfully. The plant-mediated fabrication of nanoparticles is advantageous over chemical and physical methods, since it is cheap, single-step, biodegradable, and environmentally friendly. Furthermore, living organisms can also be used, since microbes are used in nanotechnology for producing nanoparticles and current green synthesis has shown that environmentally friendly and renewable sources of fungi can be used as effective reducing agents for the synthesis of silver nanoparticles. Green nanotechnology synthesis also comprehends plant-mediated biosynthesis of metal nanoparticles. This nanotechnology is advantageous over other chemical and physical methods, as it is cheap, single-step, does not require high pressure, is biodegradable, saves energy and temperature, and avoids the use of highly toxic chemicals, also in field conditions.

The aim of green nanotechnology is focused on the production of new products with high selected activity and ecofriendly properties. Plant-mediated fabrication of nanoparticles is advantageous over chemical and physical methods, since it is cheap, single-step, and environmental care. As a matter of fact, the impact of long-term exposure to low, highly dispersed doses of nanoparticles on ecosystems cannot be determined easily. Among the aims of green nanotechnology is the utilization of organic materials, avoiding or limiting the utilization of metals or potentially toxic materials in the manufacture of nanoparticles (Govindarajan et al., 2016). Several materials have been successfully used to obtain green nanoparticles, developing efficient and rapid extracellular syntheses of silver and gold nanoparticles, which showed excellent mosquitocidal properties (Murugan et al., 2015d).

On the other hand, nanobiotechnology can be used to improve our understanding of the biology of vectors and parasites, as well as developing improved systems for monitoring environmental conditions and enhancing the ability of plants and other hosts to resist microorganism attacks. Therefore, the goal is not the production of further lethal weapons against the target enemy, but considering several aspects of the action. Nanocrystals of gold and silver, and their alloys, have been synthesized with the assistance of various bacteria as well as the environmentally friendly and renewable sources of fungi used as effective reducing agents for the synthesis of silver nanoparticles. Plant-mediated biosynthesis of nanoparticles may be advantageous as successful alternatives to classic chemical and physical methods and in pest management and parasitology.

Among the experimental results on the basis of appropriate utilization of nanoparticles obtained through the green synthesis route, the necessary evidences of the insert and presence of the active ingredients by biophysical characterization. Nanoparticles can be obtained independently from the addition of the selected active ingredient. The best solution to this key issue is to adopt a mix of experimental approaches, each confirming the effective internal presence of the organic material. The morphology can be determined by transmission electron microscopy, showing spherical shapes, with an average size of 3–18 nm. X-ray powder diffraction can highlight the crystalline nature of the nanoparticles, ultraviolet-visible absorption spectroscopy is used to monitor their synthesis, and energy dispersive spectroscopy is used to confirm the presence of elemental elements, such as silver or gold. Finally, Fourier transform infrared spectroscopy shows the main reducing groups from the organic content. The last analysis is the most important, showing absorption bands typical of secondary natural products, such as OH or CH or CO stretching. However, my opinion is that IR evidence is quite general and not sufficiently specific to each plant ingredient. Waiting for help from the advancements of technology in analytical chemistry, a solution could be to obtain an IR fingerprint characterizing the plant extract to be compared with that obtained from the nanoparticles, as already obtained by high-performance thin-layer chromatography (HPTLC), as well as referring to information obtained by other analyses.

In insect-borne diseases, nanoparticles possess peculiar toxicity mechanisms, due to surface modification, and they are synthetized to perform selected biological activities. In particular, several reports are present in the literature concerning the potential antibacterial, antifungal, and antiplasmodial and mosquitocidal properties of silver and gold nanoparticles containing natural materials (Suresh et al., 2015; Sujitha et al., 2015). Silver and gold nanoparticles synthesized using *Chrysosporium tropicum* became active against the Aedes aegypti larvae; the larvicidal activity of silver nanoparticles synthesized using Pergularia daemia latex was screened against the *Ae. aegypti*; and silver nanoparticles synthesized using *Feronia elephantum* plant leaf extract showed adulticide activity against *Anopheles stephensi*, *Ae. aegypti,* and Culex quinquefasciatus (Murugan et al., 2016a). Larvicidal and antimicrobial activity of silver nanoparticles synthesized using marine fluorescent pseudomonads was positively tested against various human pathogenic bacteria, such as *Aspergillus foetidus*, and fungal pathogens of plants.

The choice of material is key, considering the large range of useful species and the relative kind of physic-chemical properties (Subramaniam et al., 2015, 2017). In particular, fungi can produce significant amounts of nanoparticles because they can secrete larger amounts of proteins, which directly translate to higher productivity of nanoparticles. The mechanism of silver nanoparticle production by fungi is based on the following steps: trapping of Ag^+ ions at the surface of the fungal cells and the subsequent reduction of the silver ions by the enzymes present in the fungal system.

Biosynthesis of nanoparticles using biomass was used as a profitable tool, but again the properties of the materials must be considered carefully. In this case, the material needs time to act, and this goes against their natural degradation. In contrast, for the same reasons, natural substances are eco-friendly and must be preferred to synthetic products. It is necessary to obtain an equilibrium between these two opposite strains—a typical situation that needs a technological answer. In recent years, a growing number of plant-borne compounds have been proposed for efficient and rapid extracellular synthesis of metal nanoparticles effective against mosquitoes at very low doses (1–30 ppm).

Making green nanoparticles

Nanobiotechnology is the last frontier for utilizing natural products. Nanotechnology has gained popularity in recent years due to its application in various fields. Nanoemulsion consists in kinetically stable droplets prepared using stable particles of oil and water, and requires lower surfactant concentrations against other products based on microemulsion. These characters are suitable for drug delivery in consideration of the smaller droplet size, larger surface area, and lower surface than the most common other products, based on micro and not nanoscale. It contains antimicrobial activity against bacteria, fungi, and virus.

Other considerations in research on new green nanotechnology approaches lead to choices of material to use in the production of nanoparticles owing to possible environmental effects. Nanotechnology can be important for future treatment of the human habitat. The quality of environment in which we live is a real challenge for everyone. People in towns may worry about air pollution due to chemical agents, while people in rural areas may be more concerned about water and its contamination by microorganisms. Among the most relevant aspects of human evolution, we must consider the progressive tendency to utilize very pure water for alimentation. We need large quantities of pure water, since every 20–22 days, the water in our body should be replenished. However, if urban populations can rely on efficient methods of purification, in rural places pathogens and parasites can pollute waters and constitute a severe plague, especially in developing countries. The consequences of lack of access to safe drinking water are significant. Intestinal infection causes diarrheal diseases and affects about 1.6 million people globally (about 90% of whom are children), including cholera. A total of 133 million are suffering intestinal helminthes infection, including 1.5 registered cases of hepatitis. The bacterium *Chlamydia trachomatis* causes active eye infections in more than 80 million people and can induce trauchoma in about 500 million. The treatment for hygienizing water is usually achieved by addition of chemical coagulants and chlorine, as effective bactericide agents. An alternative green tool to conventional chemical treatments relies on the utilization of plants and plant extracts to clarify and purify water. However, two important points must be considered: the choice of the plant material and the utilization method. *Datura metel*-synthesized silver nanoparticles were utilized to magnify predation of dragonfly nymphs against the malaria vector *Anopheles stephensi* (Murugan et al., 2015d).

In addition to plants of several botanical families, natural raw materials of different origins were used to synthetize nanoparticles utilizing marine organisms, such as seagrass (Mahyoub et al., 2017), and spongeweed (Murugan et al., 2016a,b,c,e,f,g,h), and also fern, chitosan, earthworm (Jaganathan et al., 2016), and others, and their activity against vectors was tested successfully.

In a 2015 paper (Dinesh et al., 2015), the plant *Aloe vera* was studied with green-synthetized silver nanoparticles to obtain mosquitocidal and antibacterial activity in field conditions. The genus *Aloe* comprises about 500 species, most of them present in South Africa, but some species are distributed almost globally, either for their beauty as ornamental plants or for their medical and cosmetic benefits. Among the *Aloe* species, the most important is *A. vera*, for its more than 200 active constituents and multiple utilizations as an extract or gel. The medicinal extract, containing anthraquinones, has a long medicinal tradition as a laxative, whereas the gel is a cicatrizing and a skin protective agent that is currently widely utilized. The reported activities of the gel include strong antimicrobial properties against bacteria, fungi, viruses, and yeasts. It is rich in hydrocolloids, since *Aloe* species are succulent plants, adapted to live in arid and hot climates. As other succulent plants do, they survive thanks to the water stored in the mesophyll inside a special tissue, named aquifer parenchyma. To perform this job, this tissue contains mainly hydrocolloids, which are complex polysaccharides, made by a polymerization of an uneven pool of sugars. The main activity of hydrocolloids, as the name suggests, is to hold on to the water molecules. These molecules are polyols and thanks to the large presence of hydroxyls, they are able to produce a great quantity of H-bonding with the water molecules. Hydrocolloids change the lipophobic nature of water, interacting with the intimate network of water clusters and giving rise to a sort of "quarter state of water," although this is unstable. They act as a bridge between the water and fats, giving rise to a coexistence between polar and nonpolar compounds. Thanks to hydrocolloids, we have "miracle" products such as soap bubbles, cosmetic creams, mayonnaise, ice-creams, and many more that are consumed in everyday life. Hydrocolloids have very important biological properties, including also toxic effects on aquatic organisms, directly on the target or changing the environmental parameters.

The green-synthetized silver nanoparticles, containing a cell-free aqueous leaf extract of *A. vera*, became toxic to larvae (I–IV instar) and pupae of the malaria vector *An. stephensi* in vitro and in the field, and showed antibacterial properties against *Bacillus subtilis*, *Klebsiella pneumoniae*, and *Salmonella typhi*. In addition, the efficiency of the micro-emulsion, prepared using *A. vera* extract and silver treatment in water treatment, was reported. The water quality was tested by analysis of parameters, such as color, turbidity, and pH, and analyzed over the different water reservoirs, acting as breeding sites of mosquitoes in pre- and post-treatment phases. The results obtained indicate the possible utilization by uniform dispersion to clean water from parasites and bacterial pathogens.

Among the plant materials we utilized in nanotechnology, fern-synthesized nanoparticles were utilized in the fight against malaria. The obtained silver nanoparticles of *Pteridium aquilinum* leaf extract were analyzed by LC/MS analysis and they showed high mosquitocidal activity (Panneerselvam et al., 2016). In another study, the potentiality of fern-synthesized silver nanocrystals was studied as a possible new class of mosquito oviposition deterrents (Rajaganesh et al., 2015).

Another research front was the experimental search for very unusual materials. Again attention had to be focused on our ancestors and their incredible properties, lost during the evolution of complex organisms. The story of the discovery of magnetotactic bacteria is an example of what was and still is going on in academic scenarios. These wonderful creatures were first discovered in 1963 by Salvatore Bellini at the University of Pavia, Italy, and rediscovered in 1975 by Richard Blakemore. It is relevant to report the story of this rediscovery. Bellini, when observing under his microscope a group of bacteria, noticed their evident orientation in a unique direction. Thus, these bacteria, instead of being statically disposed in a circle, appeared well-ordered in a single queue, like soldiers or people waiting for the bus. Nowadays, we believe that his alignment aided these organisms in reaching regions of optimal oxygen concentration. His explanation was that these microorganisms moved according to the direction of the North Pole, just like the needle of a compass, and, considering the phenomenon to be a consequence of magnetism, called them "magnetosensitive bacteria." As was usual in European universities at the time, he communicated the results of his experience in Italian in a local academic scientific journal, the *Istituto di Microbiologia*. I well remember that at the beginning of my career, we could access only short summaries in English of research in Russian or in Chinese, published in the official journal of a well-known institution. Thus Bellini's discoveries remained practically unconsidered until they were brought to the attention of Richard Frankel in 2007. After Frankel's translation and its publication in the *Chinese Journal of Oceanography and Limnology*, Bellini's work was visible to the international community. Therefore, a microbiology graduate student at the University of Massachusetts at Amherst, Richard Blakemore, published in *Science* the same observations about magnetosomes, changing the name in magnetotactics, but using an electron microscope. Blakemore was working in the Woods Hole Oceanographic Institution, in whose collections the pertinent publications of the Institute of Microbiology of the University of Pavia were extant, but he did not mention Bellini's research in his own report. Unfortunately, these kind of accidents, like Rosalind Franklin's one, are common in the scientific community, also in times of peer reviews, especially when the discovery is important. Nowadays, magnetosomes are the object of several applications, such as in chemotherapy, immunotherapy, and gene therapy, at pre-clinic stage also in patients. Their structures have been differently functionalized in accordance with the target, utilizing the presence of various chemical groups at their surface.

We were interested in the production of magnetic nanoparticles, consisting of suspensions containing chains of magnetosomes. A magnetosome is defined as an intracellular, membrane-bounded magnetic iron-bearing inorganic crystal. In particular, we considered chains of magnetic nanoparticles that were extracted from magnetotactic bacteria. These magnetotactic microorganisms are interesting and unique, since they possess flagella and other typical characters, but also contain structured particles, rich in iron, within intra-cytoplasmic membrane vesicles. These contents impart a magnetic response to bacterial cells. Furthermore, though bacterial magnetosomes are true prokaryotic organelles, they display a comparable degree of complexity to their eukaryotic counterpart. In fact, magnetosome formation requires a complex process, involving vesicle formation, extracellular iron uptake by the cell, iron transport into the magnetosome vesicle, and biologically controlled magnetite or greigite mineralization within the magnetosome vesicle. Magnetic nanoparticles may be preferred for their unique characters, like high field irreversibility, high saturation field, super paramagnetism, extra anisotropy contributions, or shifted loops after field cooling. These phenomena arise from narrow and finite-size effects and surface effects that dominate the magnetic behavior of individual nanoparticles. When tested, magnetotactic bacteria evidenced larvicidal, pupicidal, and dengue viral effects of extracellular synthesized magneto-nanoparticles against *Ae. aegypti* mosquitoes. Magnetic nanoparticles proved to be highly toxic to chloroquine-resistant *Plasmodium falciparum*, dengue virus (DEN-2), and their mosquito vectors (Murugan et al., 2016d).

Following the research line the selection of the material necessary to obtain the nano-mosquitocites. Another important task in green nanotechnology is. In this way, several materials were tested. In particular, mangrove was elected to obtain green-mediated synthesis of silver nanoparticles with high HIV-1 reverse transcriptase inhibitory potential (Kumar et al., 2016a, b). In another paper, the mangrove *Sonneratia*'s alba-synthesized silver nanoparticles reported the capacity to magnify guppy fish predation against Aedes aegypti young instars (Murugan et al., 2017b).

Among them, we proposed (Roni et al., 2016; Murugan et al., 2016a,b, c,e,f,g,h) a novel method of biofabrication of silver nanoparticles with insecticide activity against malaria vectors using eco extracted from crab shells. Again, it was decided to explore a multiple approach regarding the activity of the nanoparticles, including the impact on non-target organisms in an aquatic environment. First, the insecticide effects were experimented on against larvae and pupae of *Anopheles stephensi*, obtaining LC_{50} ranging from 3.18 ppm (I) to 6.54 ppm. Second, antibacterial properties of Ch-AgNP were proved against *Bacillus subtilis*, Klebsiella pneumoniae, and *Salmonella typhi*, while no growth inhibition was reported in assays conducted on *Proteus vulgaris*. In terms of non-target effects, in standard laboratory conditions the predation efficiency of Danio rerio zebrafishes was tested, obtaining 68.8% and 61.6% against I and II instar larvae of A. stephensi, respectively. However, when chitosan-nanoparticles were added to the environment (Murugan et al., 2016b, Murugan et al., 2017a,b,c), fish predation was boosted to 89.5% and 77.3%, respectively. However, deleterious effects of aquatic environment on the non-target crab *Paratelphusa hydrodromous* were observed when quantitative analysis of antioxidant enzymes SOD,

CAT, and LPO from hepatopancreas of fresh water crabs exposed for 16 days to a nanoparticle-contaminated aquatic environment was conducted, highlighting some risks concerning the use of nanoparticles in aquatic environments at least for some organisms present in that environment (Murugan et al., 2016b). The biosynthesis, mosquitocidal, and antibacterial properties of *Toddalia asiatica*-synthesized silver nanoparticles were reported, and these did not impact predation of guppy *Poecilia reticulata* against the filariasis mosquito *Culex quinquefasciatus* (Murugan et al., 2015f).

A delicate argument concerns the effects of the release of nanoparticles on the habitat. Most concern is nowadays focused on the toxicity on predators of mosquito vectors, but other consequences cannot be excluded. In another paper, we reported utilization of seaweed in the synthesis of silver nanoparticles. The nanoparticles were thus obtained and their impact against the filariasis vector *Culex quinquefasciatus* was observed. Lymphatic filariasis is a parasitic infection that leads to a disease commonly known as elephantiasis, affecting nearly 1.4 billion people in 73 countries worldwide. Also in this case, negative effects on human health and the environment must be considered, since control against mosquito larvae is obtained by treatment with organophosphates and insect growth regulators. The frond extract of the seaweed *Caulerpa scalpelliformis* was utilized in the formation of nanoparticles (Murugan et al., 2015c). Once the toxicity of the seaweed extract and silver nanoparticles had been assessed against the filarial vector *C. quinquefasciatus*, the experiments were focused on the predatory efficiency of the cyclopoid crustacean *Mesocyclops longisetus* (Murugan et al., 2015e). This a copepod predator of mosquito larvae, as in the case of already reported. In a nanoparticle-contaminated environment, predation efficiency was 84% and 63%, respectively, on I and II instar larvae of C. quinquefasciatus, compared to 78% and 59% in the control treatment. The study showed that seaweed-synthesized silver nanoparticles use lead to little detrimental effects against aquatic predators, such as copepods (Govindarajan et al., 2015; Murugan et al., 2015a), including other *Mesocyclops* species (Chandramohan et al., 2016; Murugan et al., 2015c, 2015g).

An attempt to examine different effects on the habitat was made in a study of control of dengue by nanoparticles (Murugan et al., 2017a,b,c), in order to obtain the potential of mangrove extract (*Sonneratia alba*)-synthesized silver nanoparticles (AgNP), as eco-friendly nanoformulations effective against dengue virus and its mosquito vectors, as well as the effects on usual predators of the same habitat. AgNPs were obtained by a cheap method relying on *S. alba* extract as a reducing and stabilizing agent. First, the activity of the *S. alba* extract alone against *Aedes aegypti* was obtained, with LC_{50} values that ranged from 192.03 ppm (larva I) to 353.36 ppm (pupa). However, the AgNP toxicity turned out to be much higher, since the LC_{50} ranged from 3.15 (I) to 13.61 ppm (pupa). Later, effects of predators of the vectors were tested with sublethal doses of AgNP, obtaining magnified predation rates of guppy fishes, *Poecilia reticulata,* against *Ae. aegypti* and *Chironomus kiiensis* larvae. Mangrove-fabricated AgNPs were also evaluated for their antimicrobial potential against *Bacillus subtilis, Klebsiella pneumoniae*, and *Salmonella typhi. S. alba*-synthesized AgNPs tested at doses ranging from 5 to 15 μg/mL downregulated the expression of the envelope (E) gene and protein in dengue virus (serotype DEN-2), while only small cytotoxicity rates (<15%) were detected on Vero cells when AgNPs were tested at 10 μg/mL (Meenakshi and Jayaprakash, 2014).

After the tiger mosquito, it is time to meet the tiger frog. This study is based on a synergistic approach using biocontrol agents and botanical nano-insecticides for mosquito control. With single-step green synthesis, silver nanoparticles were obtained using the extract of *Artemisia vulgaris* (the same genus of the antimalarial *A. annua*) leaves as a reducing and stabilizing agent (Govindarajan and Benelli, 2016; Benelli et al., 2016). The mosquitocidal properties of *A. vulgaris* leaf extract inserted in green-synthesized nanoparticles against larvae and pupae of *Ae. aegypti* were investigated. The nanoparticles proved to be highly toxic to *Ae. aegypti* larval instars (I–IV) and pupae, with LC_{50} ranging from 4.4 (I) to 13.1 ppm (pupae). To evaluate the habitat disequilibrating effects, it was also necessary to investigate the ecologic effects of the green nanoparticles on the predatory efficiency of Asian bullfrog tadpoles, *Hoplobatrachus tigerinus*, against larvae of *Ae. aegypti*, under laboratory conditions and in an aquatic environment treated with ultra-low doses of the nanoparticles (Murugan et al., 2015b). *Ae. aegypti* and the bull frog actually share the same aquatic habitat, and the frog is fundamental in the biological control of this vector of dengue and chikungunya diseases. In the laboratory, the mean number of prey consumed per tadpole per day was 29.0 (I), 26.0 (II), 21.4 (III), and 16.7 (IV). After treatment with nanoparticles, the mean number of mosquito prey per tadpole per day increased to 34.2 (I), 32.4 (II), 27.4 (III), and 22.6 (IV). Despite the positive data, there are several aspects to be considered before any further steps. We do not know the reason for the increase in predatory activity and possible collateral effects. Any information on the behavior of any participant in the project must be considered on the basis of previous experiences. The Asian bullfrog (*Hoplobatrachus tigerinus*) is native to the Indian sub-continent and it is rapidly invading the Andaman archipelago, the Bay of Bengal, after its recent introduction. Rapid predation by larval *H. tigerinus* resulted in no survival of endemic frog tadpoles previously present in that habitat. The tadpoles of the tiger frog are voraciously carnivorous, attacking any prey, including two species of endemic anuran tadpoles, *Microhyla chakrapanii* and *Kaloula ghoshi*, in a sort of cannibalism. Survival of *H. tigerinus* was density-dependent. Therefore, invasive populations, or potential induced incursions of alien species for biological control, can generate important ecological variations. Already Darwin had observed many years previously the potential impact caused by the loss of even a single species.

In the case of mosquito control, these activities must also consider recent important challenges, such as the recent outbreaks of novel arbovirus, including the development of resistance in several Culicidae species and the spreading of alien invasive mosquitoes in several regions. In particular, microbiopesticides must be joined with other measures, such as personal defenses by insecticide-treated bed nets and repellents made from natural products. Further strategies may consider bio-agents active against mosquito young instars (i.e., selected species among fishes, amphibians, and copepods), sterile insect techniques (SIT) including boosted SIT, symbiontic-based methods, and transgenic-produced mosquitos. These measures must be integrated, since no single method has produced fully successful or satisfying results.

The neem nano-emulsion can be easily obtained thanks to the oily nature of seed extract and used as an antimicrobial agent. However, the structure of the active components, such as the limonoids of neem (see Chapter 7), are subjected to rapid attack by oxygen and humidity when exposed to air, and their permanence is important to obtain a good insecticide effect. Green-fabricated nanoparticles have been studied as promising toxic agents against mosquito young instars, and as adult oviposition deterrents in aquatic environment merits. The idea was to treat the water with ultra-low doses (e.g., 1–3 ppm) of green-synthesized nanomosquitocides, which reduced the motility of mosquito larvae. Nanoparticles are important to avoid degradation of limonoids in the open air, since azadirachtines are reported to have a half-life of a few days, giving rise to the need for costly several treatments. Biodegradability is mainly due to exposure to the sun and to humidity. This is an advantage, avoiding the problems like DDT due to the accumulation in the habitat, but it is one of the more important limits in neem utilization. Therefore, nanotechnology can be used to protect active constituents and also to facilitate interaction with the target organism.

At the end of this short review of selected examples of utilization of nanoparticles in control of insect-borne disease vectors, it is necessary to underline some points of interest. Nanotechnology is still in its infancy, and its real perspectives are still in progress. In particular, the quantity of positive results obtained using organic matter of different origins can be suspicious. However, parasitism is relevant to all kinds of organisms, in particular plants that must depend on their chemical arsenal. Therefore, the real task is the exact chemical determination of the active constituents, often largely lacking, and the mechanisms of action necessary for the certification of their utilization. Only through these advancements will the reality of nanotechnology in insect-borne diseases achieve completely its fundamental role and importance, moving on from the initial step of the pure experimental phase. Meanwhile, it is necessary to join the market and obtain applications of green nanotechnology products on a large scale, once their safe and eco-friendly effects have been confirmed and certified.

Integrated protection programs

As previously reported, in-depth knowledge of the habitat network is one possible tool to develop an efficient and reliable pest control system. As evident in the Xylella episode (see Chapter 5), all the components must be considered interconnected and treated in the logic of the superorganism: (a) the soil, subjected to radical changes in its composition by the climate impact or human abandon or intervention; (b) the host, whose natural defenses are the first and best bulwark against infection and disease development; (c) the etiological agent, considered not only as the responsible of the disease but as a voluble environmental agent whose actions are moved by homeostatic reasons, which can be also considered as part of an useful control strategy; and (d) the vector, the ideal target being visible and accessible, but also acting as a delicate part of the network. In this scenario, the reason for the adopted measures cannot be reduced to the wellness of the host, and information about the tendency of the phenomenon must be obtained, including the main factors governing the changes. It is complicated, but it is the only way to avoid disasters. The programme must consider the key importance of the vectors. Monitor the harmful organisms, selecting different methods and appropriate times, as well as the appropriate chemical groups, taking into account the growth period and the presence of natural enemies. Selective insecticides should be developed and areas where harmful insects susceptible to insecticides can reproduce should be provided. The expected result is that the harmful organisms should mate those resistant and in this way decrease the portion of the resilient genes inside the population.

Regarding integrated control programs in agriculture, many different conventional and sustainable strategies can be incorporated, including application of insecticides correctly and at the right time, pest monitoring, resistance management, protection of beneficial insects, cultural practice, crop rotation, pest-resistant varieties (even transgenic when allowed), and chemical attractants or deterrents derived where possible by natural products to avoid an insurgence of resistance.

Great attention about novel methods of control avoiding synthetic insecticides is focused on targeting the causative vectors and pathogens by biological control. The aim is avoid the consequences of massive utilization, as environmental damage, destruction of useful insects, and resistance decided the decline of such substances to combat mosquitoes that were once considered the most effective way of controlling mosquito-borne diseases. The current, simplest, and generally accepted definition for biological control is an environmentally sound and effective means of reducing or mitigating vectors and vector effects through the use of natural enemies. The different biological control agents being studied in different parts of the world for the control of vector mosquitoes include many naturally occurring predators, parasites, and pathogens of vector insects, including fungi and bacteria. However, although biological control should provide an effective and eco-friendly approach, which can be used as an alternative to minimize the mosquito population and can provide an effective and eco-friendly approach to mosquito control, the results in many cases were not satisfactory. The selection of natural enemies can include also alien species, the autochthonous ones not being sufficient. However, the effects of such introductions must be carefully considered.

Phylum	Arthropoda
Class	Insecta
Order	Hemiptera
Family	Pentatomidae
Genus	*Halyomorpha*
Species	*H. halys*

If you have occasion to travel in Italy in the central region, Emilia-Romagna, you may well notice the presence of large cultivations of fruit trees, mainly apple and pear trees. It is a flat land, traversed by the river Po, the largest in Italy, where important and historic towns such as Bologna, Ferrara, and Padova alternate with cultivated fields. In these territories, the apple trees, which are nowadays more profitable for the market, have substituted hemp, tobacco, and mays, generating a monoculture, or better a monovariety cultivation, with very few utilized cultivars among the cultured apple trees. Simplified habitats are more vulnerable. Furthermore, this situation is nonsense, considering the hundreds of varieties characterizing Pyrus malus, as the result of careful and patient work of farmers and the plastic genome of this species. In the Arboretum of Alma Ata in Kazakhstan, considered the homeland of the apple, the precious genomes of hundreds of cultivars are still served, but the world market is focused on a few types, such as Golden Delicious and Fuji.

This part of Italy was always rich and productive, in terms of ideas and art as well as delicious food, but in the last 8 years the situation has changed under the real threat of an alien uninvited guest. The brown marmorated stink bug (*Halyomorpha halys*) is menacing 60%–70% of fruit production, gorging on unripe fruit. The bugs have sunk their needle-sharp stylets into the pear, apple, and peach fruits, creating wounds evident on the surface, producing a clear, sugary goo. They form corky brown blemishes and leave the trees more vulnerable to infection. This insect is native to Orient Asia, but since 1998 has been accidentally introduced first in many parts of North America, and has recently become established in Africa, Europe, and South America. One hypothesis is a transportation of eggs from China or Japan occurred as a stowaway in packing crates or on various types of machinery, again from China or Japan. The nymphs and adults, feeding on more than 100 species of plants, have seriously affected many agricultural crops in the Mid-Atlantic United States, where in 2010, $37 million dollars in apple crops were lost, and some stone fruit growers lost more than 90% of their crops, including, in addition to fruit trees, corn, soybeans, berries, and tomatoes in 43 states of the USA. The brown marmorated stink bug is a sucking insect, belonging to Hemiptera, the "true bugs." Through its proboscis, it pierces the host plant to feed, causing leaf stippling, seed loss, and possible transmission of plant pathogens causing several devastating diseases. The formation of dimpled necrotic areas on the outer surface of fruits causes the main damage in terms of market appeal, with consumers looking for perfectly shaped fruits in their shops.

The diffusion of the brown marmorated stink bug, like many invasive species, is benefited by the absence of major enemies in its new home to keep its population in check. The description "marmorated" is derived from the gray color of the insect, although there are common species colored in green. These insects are known for their tendency to escape the cold season by finding shelter inside farms and houses, where they are sound when they are flying and for their pungent odor when disturbed or crushed or simply when one attempts to move them. This shield-shaped stink bug alien invasive pest is different from the indigenous species due to its indiscriminate appetite and its coriander-like odor, due to two simple aldehydes: trans-2-decenal and trans-2-octenal. The stink bug emits its smell through holes in its abdomen, as a defense to prevent it from being eaten by birds and lizards. However, simply handling the bug, injuring it, can trigger it to release the odor. After several conventional attempts to stop this pest, all hopes now focus on biological control, consisting in the release of the stink bug's natural enemy. The choice is the samurai wasp (*Trissolcus japonicus*), which despite its very small shape is very effective, being a pitiless and systematic killer. The stink bug's nemesis, the wasp injects its own eggs into the stink bug's, leaving larvae that eat the developing bugs before chewing their way out. Currently, the samurai strains of these wasps are imported from Asia, but in 2005 the taxonomist Elijah Talamas, at the Florida Department of Agriculture and Consumer Services in Gainesville, identified native wasps parasitizing stink bug eggs in Maryland, USA. The most convincing hypothesis is that bug and wasp both migrated on their own, and in future they will participate their struggle for life. However, this is not automatic, since new habitat conditions are different from the starting ones. Therefore, many nations have begun to regulate strictly the release of biocontrol agents, which can include insects, fungi, and bacteria, and require studies to predict potential "nontarget effects." Meanwhile, the only real measure is severe and strict controls on imports of goods and organisms. Thus, a whole cargo of Ferrari cars (the house of Ferrari is at Modena in Emilia-Romagna) was prevented from entering in Australia because of three exemplars of marmorated stink bug found in the cargo.

References

Ayers, D., Day, P.J., 2015. Systems medicine: the application of systems biology approaches for modern medical research and drug development. Mol. Biol. Int. 2015, 698169.

Baitaluk, M., 2009. System biology of gene regulation. Biomedical Informatics, Methods Mol. Biol. 569, 55–87.

Benelli, G., Pavela, R., Maggi, F., Petrelli, R., Nicoletti, M., 2016. Commentary: making green pesticides greener? The potential of plant products for nanosynthesis and pest control. J. Clust. Sci. 28 (1), 3–10.

Bik, E.M., et al., 2006. Molecular analysis of the bacterial microbiota in the human stomach. Proc. Natl. Acad. Sci. U. S. A. 103, 732–737.

Breitling, R., 2010. What is systems biology? Front. Physiol. 1, 9.

Butt, K.M., Sajid, S., 2018. Chinese Economy under Mao Zedong and Deng Xiaoping. J. Pol. Stud. 25 (1), 169–178.

Chandramohan, B., et al., 2016. Do nanomosquitocides impact predation of *Mesocyclops edax* copepods against Anopheles stephensi larvae? In: Nanoparticles in the Fight Against Parasites, pp. 173–190.

Cooper, S.B., 2012. The Incomputable Alan Turing—arXiv. https://arxiv.org.

Dinesh, D., et al., 2015. Mosquitocidal and antibacterial activity of green-synthesized silver nanoparticles from *Aloe vera* extracts: towards an effective tool against the malaria vector Anopheles stephensi? Parasitol. Res. 114 (4), 1519–1529.

Gandy, R.O., 1988. The confluence of ideas in 1936. In: Herken, R. (Ed.), The Universal Turing Machine: A Half-Century Survey. Oxford University Press, New York, pp. 51–102.

Gandy, R.O., Yates, C.E.M. (Eds.), 2001. Collected Works of A.M. Turing. vols. I–IV. North Holland, Amsterdam-London, UK.

Govindarajan, M., Benelli, G., 2016. *Artemisia absinthium*-borne compounds as novel larvicides: effectiveness against six mosquito vectors and acute toxicity on non-target aquatic organisms. Parasitol. Res. 115 (12), 4649–4661.

Govindarajan, M., et al., 2015. Facile synthesis of mosquitocidal silver nanoparticles using *Mussaenda glabra* leaf extract: characterisation and impact on non-target aquatic organisms. Nat. Prod. Res. 30 (21), 2491–2494.

Govindarajan, M., Nicoletti, M., Benelli, G., 2016. Bio-physical characterization of poly-dispersed silver nanocrystals fabricated using *Carissa spinarum*: a potent tool against mosquito vectors. J. Clust. Sci. 27 (2), 745–761.

Jaganathan, A., et al., 2016. Earthworm-mediated synthesis of silver nanoparticles: a potent tool against hepatocellular carcinoma, *P. falciparum* parasites and malaria mosquitoes. Parasitol. Int. 65, 276–284.

Krupar, J.N., 2002. Mao's war against nature: politics and the environment in Revolutionary China (review). J. Cold War Stud. 7 (4), 174–176.

Kumar, P.M., Murugan, K., Madhiyazhagan, P., Kovendan, K., Amerasan, D., Nicoletti, M., et al., 2016a. Biosynthesis, characterization, and acute toxicity of Berberis tinctoria-fabricated silver nanoparticles against the Asian tiger mosquito, *Aedes albopictus*. Parasitol. Res. 115 (2), 751–759.

Kumar, S.D., Singaravelu, G., Ajithkumar, S., Murugan, K., Nicoletti, M., Benelli, G., 2016b. Mangrove-mediated green synthesis of silver nanoparticles with high HIV-1 reverse transcriptase inhibitory potential. J. Clust. Sci. 28 (1), 359–367.

Mahyoub, J.A., et al., 2017. Seagrasses as sources of mosquito nano-larvicides? Toxicity and uptake of Halodule uninervis-biofabricated silver nanoparticles in dengue and Zika virus vector *Aedes aegypti*. J. Clust. Sci. 28 (1), 565–580.

McCulloch, W., Pitts, W., 1943. A logical calculus of the ideas immanent in nervous activity. Bull. Math. Biophys. 5, 115–133.

Meenakshi, S.V., Jayaprakash, K., 2014. Mosquito larvicidal efficacy of leaf extract from mangrove plant *Rhizophora mucronata* (Family Rhizophoraceae) against Anopheles and Aedes species. J. Pharmacogn. Phytochem. 3 (1), 78–83.

Mehlhorn, H., 2016. Nanoparticles in the fight against parasites. In: Parasitology Research Monograph. vol. 8. Springer, NY.

Moore, T.J., Zhang, H., Anderson, G., Alexander, G.C., 2018. Estimated costs of pivotal trials for novel therapeutic agents approved by the US Food and Drug Administration. 2015–2016. JAMA Intern. Med. 178 (11), 1451–1457.

Murugan, K., et al., 2015a. Toxicity of seaweed-synthesized silver nanoparticles against the filariasis vector *Culex quinquefasciatus* and its impact on predation efficiency of the cyclopoid crustacean. *Mesocyclops longisetus*. Parasitol. Res. 114 (6), 2243–2253.

Murugan, K., et al., 2015b. Predation by Asian bullfrog tadpoles, *Hoplobatrachus tigerinus*, against the dengue vector, *Aedes aegypti*, in an aquatic environment treated with mosquitocidal nanoparticles. Parasitol. Res. 114 (10), 3601–3610.

Murugan, K., et al., 2015c. Seaweed-synthesized silver nanoparticles: an eco-friendly tool in the fight against *Plasmodium falciparum* and its vector Anopheles stephensi? Parasitol. Res. 114 (11), 4087–4097.
Murugan, K., et al., 2015d. Datura metel-synthesized silver nanoparticles magnify predation of dragonfly nymphs against the malaria vector Anopheles stephensi. Parasitol. Res. 114 (12), 4645–4654.
Murugan, K., et al., 2015e. Toxicity of seaweed-synthesized silver nanoparticles against the filariasis vector *Culex quinquefasciatus* and its impact on predation efficiency of the cyclopoid crustacean *Mesocyclops longisetus*. Parasitol. Res. 114 (6), 2243–2253.
Murugan, K., et al., 2015f. Biosynthesis, mosquitocidal and antibacterial properties of Toddalia asiatica-synthesized silver nanoparticles: do they impact predation of guppy *Poecilia reticulata* against. Environ. Sci. Pollut. Res. 22 (21), 17053–17064.
Murugan, K., Benelli, G., Panneerselvam, C., Subramaniam, J., Nicoletti, M., et al., 2015g. *Cymbopogon citratus*-synthesized gold nanoparticles boost the predation efficiency of copepod *Mesocyclops aspericornis* against malaria and dengue mosquitoes. Exp. Parasitol. 153, 129–138.
Murugan, K., et al., 2016a. Eco-friendly drugs from the marine environment: spongeweed-synthesized silver nanoparticles are highly effective on *Plasmodium falciparum* and its vector Anopheles stephensi. Environ. Sci. Pollut. Res. 23 (16), 16671–16685.
Murugan, K., et al., 2016b. Magnetic nanoparticles are highly toxic to chloroquine-resistant *Plasmodium falciparum*, dengue virus (DEN-2), and their mosquito vectors. Parasitol. Res. 116 (2), 495–502.
Murugan, K., et al., 2016c. Hydrothermal synthesis of titanium dioxide nanoparticles: mosquitocidal potential and anticancer activity on human breast cancer cells (MCF-7). Parasitol. Res. 115 (3), 1085–1096.
Murugan, K., et al., 2016e. Rapid biosynthesis of silver nanoparticles using *Crotalaria verrucosa* leaves against the dengue vector *Aedes aegypti*: what happens around? An analysis of dragonfly predatory. Nat. Prod. Res. 30 (7), 826–833.
Murugan, K., et al., 2016f. Carbon and silver nanoparticles in the fight against the filariasis vector *Culex quinquefasciatus*: genotoxicity and impact on behavioural traits of non-target aquatic organisms. Parasitol. Res. 115 (3), 1071–1083.
Murugan, K., et al., 2016g. Chitosan-fabricated Ag nanoparticles and larvivorous fishes: a novel route to control the coastal malaria vector Anopheles sundaicus? Hydrobiologia 797 (1), 335–350.
Murugan, K., Aruna, P., Panneerselvam, C., Nicoletti, M., et al., 2016h. Fighting arboviral diseases: low toxicity on mammalian cells, dengue growth inhibition (in vitro), and mosquitocidal activity of Centroceras clavulatum-synthesized silver nanoparticles. Parasitol. Res. 115 (2), 651–662.
Murugan, K., et al., 2017a. Nanofabrication of graphene quantum dots with high toxicity against malaria mosquitoes, *Plasmodium falciparum* and MCF-7 cancer cells: impact on predation. J. Clust. Sci. 28 (1), 393–411.
Murugan, K., et al., 2017b. Mangrove helps: *Sonneratia alba*-synthesized silver nanoparticles magnify guppy fish predation against *Aedes aegypti* young instars and down-regulate the expression of envelope (E) gene in dengue virus (Serotype DEN-2). J. Clust. Sci. 28 (1), 437–461.
Murugan, K., et al., 2017c. Fabrication of nano-mosquitocides using chitosan from crab shells: impact on non-target organisms in the aquatic environment. Ecotoxicol. Environ. Saf. 132, 318–328.
Panneerselvam, C., et al., 2016. Fern-synthesized nanoparticles in the fight against malaria: LC/MS analysis of *Pteridium aquilinum* leaf extract and biosynthesis of silver nanoparticles with high mosquitocidal activity. Parasitol. Res. 115 (3), 997–1013.

Pinker, S., 2009. How the Mind Works (1997/2009), 2009th ed. W. W. Norton & Company, New York.
Pinker, S., 2011. Steven Pinker—Books—The Blank Slate. Pinker.wjh.harvard.edu. Archived from the original on 2011-05-10. Retrieved 2011-01-19.
Pinker, S., 2016. The Blank Slate (2002/2016). Viking, New York.
Rajaganesh, P., et al., 2015. Fern-synthesized silver nanocrystals: towards a new class of mosquito oviposition deterrents? Res. Vet. Sci. 109, 40–51.
Roni, M., et al., 2016. Characterization and biotoxicity of Hypnea musciformis-synthesized silver nanoparticles as potential eco-friendly control tool against Aedes aegypt4i and Plutella xylostella. Ecotoxicol. Environ. Saf. 121, 31–38.
Sauer, U., Heinemann, M., Zamboni, N., 2007. Genetics. Getting closer to the whole picture. Science 316 (5824), 550–551.
Schlinger, H.D., 2004. The almost blank slate. Skept. Mag. 11(2).
Schrodinger, E., 1944. What Is Life? The Physical Aspect of the Living Cell. Cambridge University Press, Cambridge.
Sinha, M.S., Najafzadeh, M., Rajasingh, E.K., Love, J., Kesselheim, A.S., 2018. Labeling changes and costs for clinical trials performed under the US Food and Drug Administration. Pediatric exclusivity extension, 2007 to 2012. JAMA Intern. Med. 178 (11), 1458–1466.
Soare, R.I., 1996. Computability and recursuin. Bull. Symb. Log. 2 (3), 284–321.
Subramaniam, J., et al., 2015. Eco-friendly control of malaria and arbovirus vectors using the mosquitofish *Gambusia affinis* and ultra-low dosages of *Mimusops elengi*-synthesized silver nanoparticles. Environ. Sci. Pollut. Res. 22 (24), 20,067–20,083.
Subramaniam, J., et al., 2017. Do chenopodium ambrosioides-synthesized silver nanoparticles impact *Oryzias melastigma* predation against *Aedes albopictus* larvae? J. Clust. Sci. 28 (1), 413–436.
Sujitha, V., et al., 2015. Green-synthesized silver nanoparticles as a novel control tool against dengue virus (DEN-2) and its primary vector *Aedes aegypti*. Parasitol. Res. 114 (9), 3315–3325.
Suresh, U., et al., 2015. Tackling the growing threat of dengue: Phyllanthus niruri-mediated synthesis of silver nanoparticles and their mosquitocidal properties against the dengue vector *Aedes aegypti*. Parasitol. Res. 114 (4), 1551–1562.
Tavassoly, I., Goldfarb, J., Iyengar, R., 2018. Systems biology primer: the basic methods and approaches. Essays Biochem. 62 (4), 487–500.
Tieszen, R., 1992. Kurt Godel and phenomenology. Philos. Sci. 59, 176–194.
Tseuscher, C., 2005. Turing's Connectionism. An Investigation of Neural Network Architectures. Springer, London.
Turing, A.M., 1936. On computable numbers, with an application to the Entscheidungsproblem. Proc. London Math. Soc. 42 (2), 230–265 Reprinted in A. M. Turing, Collected Works: Mathematical Logic, pp. 18–53.
Turnbaugh, P.J., et al., 2007. The human microbiome project. Nature 449, 804–810.
Zou, Y., Laubichler, M.D., 2018. From systems to biology: a computational analysis of the research articles on systems biology from 1992 to 2013. PLoS One. 13(7), e0200929.

CHAPTER FIVE

Three scenarios in insect-borne diseases

Image courtesy of Shutterstock

https://doi.org/10.1016/B978-0-12-818706-7.00005-X

Insect-borne diseases affecting mankind

Reading in the books the history of each country, at any time we can find sequences of episodes about outbreaks due to the plague and other epidemics. Traces of parasitic diseases on humanity can be easily found in ancient Egyptian civilization, where pests and plague were considered as inherent and ineluctable parts of Creation:

> *"Who creates that on which the mosquito lives, worms and fleas likewise, who looks after the mice in their holes and keeps alive the beetles in every timber."*
>
> ***(From the Hymn to Amen-Re, c.1600 BC, after Jan Assmann Ägypten*—Theologie und Frömmigkeit einer frühen Hochkultur, *p. 73)***

Also in the Bible, the devastating effects of pests and plagues are clearly present, but the concept regarding the origin and the acceptance are very different from the Egyptian ones.

> *"I will send swarms of flies on you and your officials, on your people and into your houses. The houses of Egyptians will be full of flies, even the ground will be covered with them."*
>
> ***(Exodus 8: 21)***

Among Egypt's plagues in the Bible, at least four can be directly related to parasites and their vectors. The description of these episodes is full of a sentiment of collapsing certainty in the face of unchained nature's forces. There is a feeling of mankind's impotence in particular against insect-borne diseases until an endless darkness.

> *"Stretch out your hand toward the sky so that darkness will spread over Egypt—darkness that can be felt."*

A very impressive description of the plague's (or a similar disease) incumbency is contained in the description of the Four Horsemen of the Apocalypse in the last book of the New Testament of the Bible, the Book of Revelation by John of Patmos, at 6:1–8. "I looked and there before me was a pale horse! Its rider was named Death, and Hades was following close behind him." The identities of the four horsemen are then summed up as follows: "They were given power over a fourth of the earth to kill by the sword (war), famine, and plague and by the wild beasts of the earth."

Other references to the deep relationship between mankind's life and insect-borne diseases can be found everywhere, especially in poetry. In these texts, we can find desperation and horror about terrible pain, but also descriptions of how mankind can transfigure even terrible pests in beauty, conferring on the disease the respect due to a shining divinity. Homer in the *Iliad* refers to "the star they call Orion's Dog—brightest of all but a fatal sign emblazoned on the heavens, it brings such killing fever down on wretched men." We find again fever and Troy in Dante's *Inferno*: "Master Adam yawns and introduces them to Potiphar's wife [who falsely accused Joseph of raping her] and Sinon of the Greeks, who tricked the Trojans into taking the Trojan horse inside their city walls. Both are afflicted by a fever so fierce that it makes their skin smoke." But he also gives a precise description of the symptoms: "Like those who shake, feeling the quartana fever coming on—their nails already blue, so that shiver at the mere sight of shade—such was I then."

Nowadays, the impact of insect-borne diseases can be monitored and determined. In the period 2010–2013, 832,900 annual deaths were estimated for parasitic infection, including: malaria, 584,000; cryptosporidiosis, 100,000; amebiasis, 55,000; leishmaniasis, 51,600; schistosomiasis, 11,700; Chagas disease, 10,300; cysticercosis, 1200; and food-borne trematodiases, 7000. However, these numbers are only indicative, since most infections do not show symptoms. Also in this aspect, there are immense variations in many parts of the world. In any case, it is evident the impact of insect-borne diseases in indelibly passages in the story of the Anthropocene Age.

Lessons from the past

There is a sort of ancient memory transmitted through generations about dangers around us. Some people exhibit ancestral well-founded fears against some animals. Africans do not like snakes, whereas Europeans are horrified by rats, which are considered primarily responsible for diffusion of plague (McCormick, 2003). In human memory, the most devastating arthropod-borne disease is not malaria or dengue or Ebola, but the plague named the Black Death and in Europe the Black Death is still considered the plague par excellence (Carter and Mendis, 2002; Benedictow, 2004).

The plague is caused by the Gram-negative bacillus *Yersinia pestis*, only 2 μm long, which is a coccobacillus. The infection to humans takes a disease in three main forms: pneumonic, septicemic, and bubonic plague. The last one is the most evident and famous. Its transmission is in accordance with classic sequence of arthropod-borne disease: pathogen, vector, target. In this case, the vector is *Xenopsylla cheopis*, the Oriental flea, which feeds on an infected animal and ingurgitates the blood containing the parasite cells. Inside the intestine, *Y. pestis* starts to multiply and create a mass of cells, which appears as a dark mass in the gut. Several proteins then contribute to the maintenance of the bacteria in the flea's digestive tract, among them the hemin storage system. This is a mechanism to increase the infection: when the foregut of the flea is blocked by the *Y. pestis* biofilm complicating the digestion of the nutrient, the flea needs to eat again, and makes multiple attempts to feed. In this way, *Y. pestis* is regurgitated into the wound of another host, causing infection.

Y. pestis is deprived of any form of locomotion and unable to propagate by spores, but its method of transportation is very efficient. Initial acquisition of *Y. pestis* by the vector occurs even during feeding on an infected animal and *Yersinia* murine toxin (Ymt) plays its role. Ymt is highly toxic to rodents and was considered important to ensure reinfection of new hosts and for the survival of *Y. pestis* in fleas. Rodents are actually not highly affected by *Y. pestis* and other similar parasites, whereas rats remain the ideal hosts. Focusing on human targets, it is necessary to note the central role of the flea: our skin is usually a good barrier against infection, but the bite overcomes this barrier.

As with other epidemics, this plague is characterized by a turnover of expansion and contraction, long periods of silent inactivity followed by furious activity. Two in particular were described by direct reliable reporters, which depicted the facts and their astonishment at the entity of the phenomenon. The first one is located at the decline of Roman Empire in the 6th century CE, and the second in the 14th century. Histories of the Great Plagues have been recently rewritten thank to the analyses of buried victims' DNA. Michael McCormick, an archeologist and medieval historian at Harvard University, reported in a published paper that the year 536 was the worst one in history to be alive (McCormick, 2003). His consideration is based on the documents of that time and confirmed by the analysis of ice carrots collected from the top of Mont Rosa, in the Colle Gnifetti permanent glacier, evidencing the effects of a giant eruption. Similarly, in Antarctica and Greenland glaciers showed volcanic debris from 540, indicating a second eruption. The ice layers tell us of a blanket of fog and ash darkening the sun for years in Europe, the Middle East, and large parts of Asia, causing temperatures to plummet in the summer to between 1.6 and 2.5 degrees across these continents. Day and night in a perennial eclipse, the abnormally cold summer spurred the coldest decade in the past 2300 years. Production of food, with no bread, was affected for at least 5 years. This is the time necessary for the environmental *sudarium* to leave its definite mark on an afflicted and desperate humanity.

Once again, we entrust our possibilities to understand the disease's start and spreading, through the correlation with the sequence of the environmental events occurred in the considered period. In 535, in the volcanic archipelagos of Indonesia, including the ill-famed Krakatoa, a giant explosion occurred, followed by enormous eruptions, and generated a gigantic cloud of ash and lapillus. The cloud moved to the Occident, reaching China and later Europe. We have no direct historic reports of that time from Java, but this is considered a sign of the scale of the disaster. We can only imagine what kind of tsunami and devastation were generated. First, the Chinese historians wrote of a terrible thunder coming from the south and shaking the very ground. This was only the beginning of the catastrophe. Later, an immense cloud raised and covered the sun and the darkness reached Europe, where several historians of the time directly reported the experience. About 10–80 cubic km of incandescent material was projected into the sky, producing a cloud 20–150 m thick, which slowly covered most of the planet. Michael the Syrian reports that the sky became dark and the darkness lasted for 11 months, like an eclipse. The dendrochronology of fossil wood confirms this event. Krakatoa is far from Europe, but another contemporaneous eruption took place in Island. The environment was highly affected twice. According to the current opinions of scientists, the first consequence was a global decrease in temperature: 5–10 degrees in about 10–20 years. Suddenly, the next year (536) everything changed. A warming, caused by greenhouse gases and loss of ozone protection, burned the same places. Later, until 565 the temperatures were very low, but later again there was a hot period, with temperatures much higher than those we are experiencing now. To have an idea of how rapid the change was, between 1929, which had a freezing winter, and the heatwave of 2009, we had an average annual difference of half a degree. In Italy, the weather conditions between 540 and 590 looked completely unstable. In 590 abundant and continuous rains caused inundations, even in Rome as never before ("rivers devastated Flaminia route as ploughing, the banks of greatest bridges joined together, pressed by an inundation extremely virulent from woods," as observed by Cassiodorus). The account of Paulus Diaconus is impressive:

> *In hac diluvii effusione in tantum apud urbem Romam fluvius Tiberis excrevit, ut aquae eius super muros urbis influerent et maximas in ea regiones occuparent. Tunc per alveum eiusdem fluminis cum multa serpentium multitudine draco etiam mirae magnitudinis per urbem transiens usque ad mare discendit. Subsecuta statim est hanc inundationem gravissima pestilentia, quam inguinariam appellant. Quae tanta strage populum devastavit, ut de inaestimabili multitudine vix pauci remanerent. Primumque Pelagium papam, virum venerabilem, perculit et sine mora extinxit.*

In such greatness of water produced by the deluge, the river Tiber inundated Rome, so that the water overcame the town's walls and was able to occupy all the areas. Thus, from its riverbed a great number of snakes, together with an enormous dragon, crossed the city until they descended to the sea. Then, a very serious epidemic immediately followed the flooding, which was called inguinariam *[bubonic]. Such great carnage devastated the people, so that those from the incalculable multitude, barely a few people survived. The Pope Pelagius, the venerable, faltered and without delay extinguished.*

In contrast, the following year saw a terrible drought, with no rain between January to September and a devastating invasion of locusts in the north-east regions of the peninsula. Temperature and moisture are the most important factors in a habitat, modeling presence of living organisms, and causing migration, usually in combination with biotic influences.

A new pope was immediately elected and he took the name of Gregorius I, later named as the Great. The importance of Gregorius was immediately clear to the population of Rome. Considering the catastrophic situation, the pope called the entire population of the town to participate in a historic procession. From seven parts of Rome, all people converged in Santa Maria Maggiore Cathedral, asking for God's help and compassion. From a medical point of view, an overcrowded square was not the best idea against the plague, but, according to the legend, the archangel Michael appeared to Pope Gregory I and the Roman population. Whether one believes this or not, the pestilence in Rome stopped, and the castle's name was changed in honor of the episode. Now, if you go to Rome, you may well visit St Angel Castle, which is topped by a bronze statue of an angel sheathing his sword, as a sign of the end of the plague. This is also where Tosca completed her tragic destiny in Puccini's opera.

However, Rome's pestilence was not the worst episode of that year. After the fall of the Roman Empire, the Pax Romana was followed by a period of conflicts and confusion. Wars are a tremendous well of inspiration for writers and a source of precious information for us about what was going on in ancient times. Thanks to *De Bello Gallico*, we know something about Celtic civilization, before its leveling by Romans and its integration into the Empire. Homer's *Iliad*, Thucydides' *History of the Peloponnesian War*, other reports from Hippocrates (Pappas et al., 2008), and even Tolstoy's *War and Peace* and several of Shakespeare's plays are fascinating accounts of wars full of information and descriptions of life at that time. During the military campaigns, besides the combating troops there is a plethora of writers, experts, and counselors of different types, which reported the facts and events with their own impressions and points of view, such as in the Second World War, when the conflict in Europe was fully documented by photographers and film makers.

Procopius of Caesarea was a historian living in the period of the great plague in Europe in the 6th century. He wrote in seven books, named the *History of the Wars* (or simply *Wars*), the events of that period, describing the conflicts which characterized that convulse years. In that period, wars were the result of the confrontation between the waves of Barbarians, coming from the North and the Orient, and the remaining part of the Roman Empire under the Emperor Justinian, whose explicit intent was the reunification of the Empire. During wars there are people moving to conquer and people moving to defend. The result is that incoming people can bring some diseases, but usually they are easy prey for pathogens. Contrary to a popular opinion, there are evidences that the mass migration of people moving from Africa to Europe is not bearing diseases, on the contrary these people are subjected to local strains of pathogens. At that time, ordinary people had to find refuge in the fortified towns, which thus became overcrowded and faced long sieges. It was necessary to store a great quantity of food, mainly as cereals; more grains meant more rats, and more rats meant more fleas, everywhere. Rats are very intelligent and well-adapted to live in our habitat. In our towns, there are more rats than humans, but these millions of rats are not visible and they are able to conceal themselves and come out under cover of darkness. We observe their presence through their excrement and the consequences of pathogenic activity related to them. According to reports, a virulent strain entered Rome from Egypt, probably derived from China and dormant for a long time, incubating in the central region of the Great Lakes. In Roman times, Egypt was called the granary of Rome and it supported for a long time the needs of imperial Rome. A boat full of grain was probably the transportation through which infected rats moved from the Nile to Gaza and later the European continent. In *The Persian War* (2–22) Procopius described carefully the beginning and the lent progression of the pestilence touching any village or island, until it arrived in Byzantium, the capital of Justinian's Empire. In a few weeks, the plague found its ideal habitat, where density of human settlement and lack of hygiene played vital roles for the spread of the disease. At that time, the town contained 500,000 citizens, with enormous naval traffic.

Once the parasites overcame the initial difficulties, the consequences were a series of devastating attacks. The first period accounted for 5000 deaths every day, rising to 12,000, until after 12 days the town had lost 10% of its population. The Justinian plague raged for 3 months causing an emergency in terms of managing the enormous number of corpses (Rosen, 2007). Cadavers were thrown out from the walls and amassed on the rocks below until their removal by the sea. Byzantium was a putrescent grave in the open air. Later, all of central Europe was affected, including Rome, as we have seen.

This is the story, but the interpretation of facts is different. In his book *The Secret History*, Procopius describes Justinian as like a devil and the source of any damage, accusing him of being responsible for the fall of the Roman Empire, but the plague was probably a fundamental cofactor. It continued for centuries until 700 CE and the victims are estimated at 40%–50% of the population. Finally, the plague receded, but lurked for centuries until in 1347 it hit Europe again.

In the mid-14th century, Europe again experienced a great plague, carried in the guts of fleas that rode on the backs of black rats from China. The port city of Caffa in Crimea on the Black Sea was thought to be the location of the infestation's origin (Fig. 5.1). Again, a merchant boat from Caffa returned to Italy, infecting Genoa, and the disease spread throughout Italy and later the Mediterranean Basin and all of Europe.

After eight centuries and many generations, the pestilence reappeared with the same virulence. The new outbreak, named the Black Death, is estimated to have killed 30%–60% of Europe's total population, affecting this

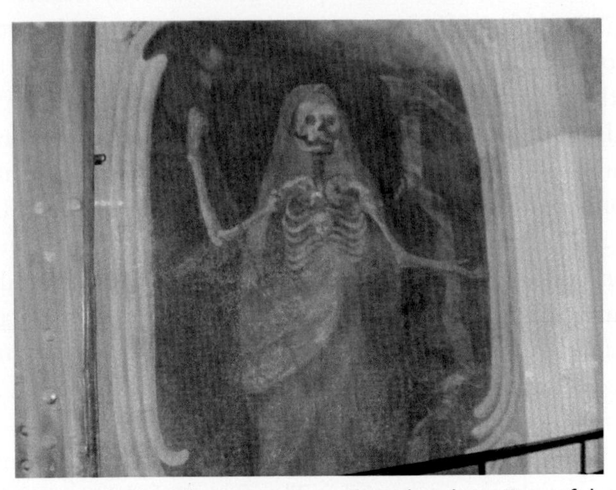

Fig. 5.1 Pestilence and other diseases were attributed to the actions of devils and considered to be a consequence of some sin or deviance.

time both towns and rural sites. It is considered the most devastating pandemic in human history, causing 75–200 million of deaths in Eurasia from 1347 to 1351. The horror of that period is still evident outside the cathedrals of central Europe, with the gargoyles threatening a dispersed and frightened humanity, and inside dominant pictures of triumphant Death armed with his terrible scythe (Fig. 5.2). Again, effects were cultural and social, changing the course of history, and again, most people considered the plague to be God's punishment. Few other than those in religious orders dared to nurse the sick. After just 4 years, two houses out of every three were totally empty and the astonished survivors scattered around without any hope in the future. Finally, the epidemic curve changed and the future became brighter. However, this tragedy launched an evolutionary transformation in Europe that deviated the course of history, causing social and cultural revolutions such as Lutheranism. People expected protection from the Church, and paid for expensive indulgences. The Church shockingly failed to influence human affairs and its authority was challenged by many, including Chaucer in his mocking *Canterbury Tales*. In a sense, the Black Death was the incubator of enclosure and of the Reformation. On October 31, 1517, Martin Luther nailed his 95 Theses to Wittenberg's cathedral doors, asserting that indulgences were not valid. The tripartite medieval division of society, the revision of that of our Arian precursors—those who fought (the nobility

Fig. 5.2 Death dominated without any possibility of interfering in its course. The personification of death is common in the imagery and art of that period.

Fig. 5.3 The masks used for centuries in the past as protection against the plague and now sold as souvenirs in the shops of Venice.

and knights), those who prayed (the churchmen), and those who labored (the peasants)—was revolutionized, never to be the same again (Fig. 5.3).

The scenario of the next episode is London, the new capital of Europe. The sequence of events is more or less the same, as well as the virulence. From the south, probably Netherlands, a boat carrying cotton arrived at the docks of London, which was the first area to be affected by the pestilence, now named Bubonic Plague. The date April 12, 1665 marks the first record of a victim, Rebecca Andrews, but the plague was already affecting thousands of very poor workers living in crowded and bad conditions, whereas those who could, including most doctors, lawyers, and merchants, fled the city; even the King and Parliament went to Oxford. Nobody knows how many of these people may have died from the plague. Between 1665 and 1666, 200,000 deaths are estimated, corresponding to one-quarter of the population, until the Great Fire of London destroyed much of the city, but also killed most of its rats and fleas. The Great Plague also hit other parts of Europe; in Amsterdam, the disease killed about 50,000 people. The plague killed 30%–60% of Europe's total population, reducing the world's population to 125–150 million (Benedictow, 2004). After such virulence, too many deaths, and immense devastation, what stopped such immense biological power? So far, no one has been able to furnish a convincing explanation. A.R. Appleby (1980) explained everything by the key role of the rat-flea-human pathway. When the rats acquired immunity, the plague suddenly disappeared. Others observed that when the plague spread over London during 1665, there was a higher prevalence of blood-sucking bugs in humans at the time of the Black Death than in recent times. On the other hand, the human-to-human transmission by way of these ectoparasites has been advanced as a route that could speed up the spread of plague: fewer humans, less infection.

We are living periods characterized by an alternance of violent outbreaks, under the attack of virulent parasites, and relative peace, when the pathogens are silent. In 1855, the third pandemic plague began in Yunnan province of China. As expected, Bubonic Plague spread from the Orient to other parts of inhabited continents. However, other acts must be expected. There are reports that more than 12 million people died, with 10 million people killed in India alone. According to the WHO, thousands of cases of the plague are still reported every year, although with proper treatment, the prognosis for victims is now much better, thanks to treatment with antibiotics. In 2000, the WHO's Report on Global Surveillance of Epidemic-prone Infectious Diseases stated:

> *"Plague has declined dramatically since the early part of the twentieth century, when outbreaks could cause tens of millions of deaths. This is due primarily to improvements in living standards and health services. However, a substantial number of countries continue to be affected by plague, case fatality rates remain high and antimicrobial resistance has begun. Therefore, continued vigilance is required, particularly in human populations living near natural plague foci. Plague foci are not fixed, and can change in response to shifts in factors such as climate, landscape, and rodent population migration. Natural foci of plague are situated in all continents except Australia, within a broad belt in tropical, subtropical and warmer temperate climates, between the parallels 55° N and 40° S."*

Finally, we are ready to use knowledge about the Great Plague as a classic model of insect-borne disease for some considerations.

A never-ending story

This is a never-ending story, every time different and still evolving. What is the current real impact of insect-borne diseases? What are the forces on the battlefield, and the alliances? Is it a global challenge or we are going to solve

Table 5.1 Most important diseases caused by infections due to microorganisms or viruses and their incidences in mortality.

Disease	Annual global cases (millions)	Annual global deaths (millions)	% Between death and cases
Malaria	600	3–8	0.005–0.013
Dengue	96	0.02	0.0002
Tuberculosis	8.6	1.300	0.151
Influenza	4	0.375	0.093
HIV	2.3	1.600	0.695

this eternal problem eventually? Although these are the key questions, replies are not easy to find.

Data reporting the current situation concerning insect-borne diseases are summarized in Table 5.1. They can be divided into two main groups on the basis of the etiological agent, which could be a virus or a bacterium. In any case, a specialized flying mosquito acts as an efficient vector, transferring the agent by bite into the host. Therefore, several aspects of the diseases are common, although important differences must be evidenced, as symptoms, diffusion, incidence, and mode of action, influencing also the strategy and method of control.

In 2006, the World Health Organization (WHO) declared the mosquito vectors for the transmission of diseases as "public enemy number one." The concern of the WHO was based on numbers indicating global impact and the need for new solutions. Insect-borne diseases are prevalent in more than 100 countries across the world, infecting more than 700 million people every year globally, with India alone accounting for 40 million and resulting in more than 1 million deaths each year. The main diseases transmitted by blood-feeding mosquitoes, according to the WHO, include dengue fever, dengue hemorrhagic fever, Japanese encephalitis, malaria, and filariasis. In particular, lymphatic filariasis has been reported to affect at least 120 million people in 73 countries, including Africa, India, Southeast Asia, and the Pacific Islands. In India, global filariasis constitutes around 40% of global filariasis. Japanese encephalitis accounts for an annual incidence of 30,000–50,000 with a mortality estimate of 10,000. These diseases are increasing in prevalence, particularly in tropical and subtropical zones, causing morbidity, mortality, economic loss, and social disruption.

However, the above data are not enough to evidence the impact of insect-borne diseases. As with any science, but in particular social sciences, data can be considered, or interpreted, in different ways, i.e., in total amount or in relative percentages, but usually the data are manipulated to be coherent with the idea to be supported. In addition, the interpretation can be affected by the intention of the writer or the reporter. In this way, if we consider the percentage of mortality of worldwide important infective diseases, as reported in Table 5.2, we find that there are more dangerous sicknesses than the diseases considered in this book. However, these data are also very partial considering that in many cases, most people at risk or infected do not become sick or are able to coexist with the disease.

What is the truth about the future of insect-borne diseases? The only possible consideration is that in this scenario, to control mosquitoes and mosquito-borne diseases, and decrease health and economic impacts worldwide, synthetic insecticide-based interventions are necessary, particularly in situations of epidemic outbreak and sudden increases of adult mosquitoes. However, the utilization of insecticides is going to change or disappear, in the next 20–30 years for reasons related mainly to the resistance phenomenons.

Table 5.2 Major vector-diffused diseases.

Disease	Vector	Pathogen	Impact and other data
Dengue, chikungunya fever, urban yellow fever	*Aedes aegypti*	Chikungunya virus	3.5 Million people at risk in 128 countries
Zika	*Aedes aegypti*	Chikungunya virus	About 80% people at risk or infected, do not become sick
Lymphatic filariasis and others	*Culex*		*quinquefasciatus*
Arboviruses	1 million people at risk in 71 countries (at least 36 million disfigured and incapacity		
Malaria	*Anopheles* sp.	*Plasmodium* sp.	600 million at risk (3–8 million deaths)
West Nile virus	*Culex pipiens* and others	*Flavivirus* sp.	Difficult to determine, most cases being asymptomatic

Eco-friendly control tools are urgently needed. In recent years, huge efforts have been made to investigate the efficacy of botanical products against mosquito vectors; many plant-borne compounds have been reported as effective against Culicidae, acting as adulticidal, larvicidal, ovicidal, oviposition deterrent, growth and/or reproduction inhibitors, and/or adult repellents.

The insect world

It is necessary to explain why most disease vectors are insects. The most important vectors are reported in Table 5.3. If microorganisms, like bacteria, are the most relevant component of biomass, insects are the predominant part of biodiversity. We are talking about more than a million described species, accounting for more than half of all known living organisms on the planet. However, the total number of extant insect species is estimated at between 6 million and 10 million. Potentially more than 90% of animal life forms on Earth are insects and these numbers increase if we consider the arthropod phylum, of which *Insecta* are the largest class. Therefore, they take myriad different forms, but to recognize an insect at a glance you must remember that they have six legs (hexapod), an external skeleton (exoskeleton), the body is divided in three parts, and the development involves a series of molts (metamorphoses). People imagine insects walking, flying, or sometimes swimming, but many insects spend at least part of their lives under water, in the larval state. Therefore, water is very important in the pathway to the adult and at larval and egg states, the insects are water-dependent and unable to move around easily. In other words, they are easy to find and vulnerable in the early stages. Insects are common in nearly all environments, although only a small number of species may be found in the oceans and in ice lands.

Table 5.3 The most important vectors in insect-borne diseases.

Vector	Disease
Anopheles sp.	• Malaria • Lymphatic filariasis
Aedes sp.	• Chikungunya • Dengue fever • Lymphatic filariasis • Rift Valley fever • Yellow fever • Zika
Culex sp.	• Japanese encephalitis

The science of studying insects is named entomology, from the Greek word ἔντομον (*éntomon*), meaning "cut into sections," an etymology present also in the Latin word *insectus*, literally meaning "cut into" and therefore "with a notched or divided body." Segmentation of the body is more or less widely present in many forms of pluricellular living organisms. Segmentation in animals typically falls into three types, characteristic of different species of arthropods, vertebrates, and annelids. Segmentation is important for allowing independence of the parts, consisting of free movements, special development of certain parts, and in some individuals the regeneration of certain parts. It is also conservation of energy by fractal repetition of the part, like in our segmented vertebral column. This is probably why the millipede, *Illacme plenipes*, presents 170 segments and 662 legs! Therefore, segmentation may result in the repetition of organs or in a body composed of self-similar units, or both. The taxonomic consequence is the classical splitting of segmented organisms into eusegmented and pseudosegmented. Eusegmented organisms are said to be comprised of "segments," as in the case of arthropods and annelids, which are said to be distinguished by their mode of formation from a posterior growth zone. When this characteristic is absent or does not affect all organ systems, there is a distinction in the generation, named "pseudosegmentation" or "metamerism." All insects have three main body regions: head, thorax, and abdomen. The head consists of the main visible parts on the head as the large compound eyes, the antenna (feelers), and the mouthparts. The thorax is the middle region of the body and bears the legs and wings, if the latter are present. The abdomen, usually deprived of outstanding features on most insects, looks like a series of similar segments. The mouth and the abdomen play key roles in insect-borne disease. In particular, the abdomen, often considered of secondary importance, is a necessary reservoir for the parasite. In insect-borne diseases, the segmentation of the vector is important for the role of each segment in the transmission and development of the disease, as well as the consequent adaptation of the parasite. Arthropods may be a good vector due to their passive role in parasitism, considering their protection by the exoskeleton and the segmentation, which creates an efficient container, but the best vectors are certainly the biting ones, because the key strategy is transmission.

Therefore, among the vectors responsible for diseases of major public health significance globally, a special place must be assigned to mosquitoes. Taxonomically, mosquitoes belong to phylum Arthropoda, class Insecta, order Diptera, family Culicidae. Culicidae, divided into three subfamilies—Anophelinae, Culicinae, and Toxorhynchitinae—accounts for 3450 recognized species of mosquitoes in 38 genera. In some cases, like malaria, the etiologic agent is localized in one genus, like malaria in *Anopheles*; in other cases, a single vector can be a vector of several diseases, like *Culex* sp. Arthropods, in addition to insects, include arachnids, myriapods, and crustaceans.

However, there is probably another similar answer regarding their prevalent role as vectors, relating to their incredible variability. Why are insects such a reservoir of biodiversity compared to any other form of organisms? Probably because they are extremely able to adapt and interact with different kinds of environments. A great deal of evidence of the special adaption capacities of insects can be seen around us. Insects are important pollinators of the most widespread and advanced plants, the Angiosperms, and contributed heavily to their success by a natural mechanism called coevolution. Angiosperms, also known as flowering plants, developed a close alliance with insects based on the characters of flowers, including coloration and flavor. In other words, the distinct and special coloration of a flower is the consequence of a relationship between that particular type of plant and that particular type of insect, a consequence of different environmental and evolutionary advancement. Modern insects appeared in the Devonian era (about 400 million years ago). In pollination, in addition to nutrition (nectars) and smell (attractant odors), colors are important as one of the distinctive signaling characters of the plant species. Flowers, as the reproductive part of the plant containing micro- and macrospores in sporangia, were already present in Gymnosperms and in the primitive Angiosperms, but not colored, and early insects (such as beetles) acting as pollinators had little vision, did not fly, and visited flowers with pale colors (white, cream, brown). The advent of Angiosperms dates to 130 million years ago, in the early Cretaceous period, and changed the scenario: flies appeared, able to fly and with modified eyes to see colored flowers, including purple and red. Later, in the Tertiary Age, the coevolution generated the sequence: wasps seeing up to yellow flowers, bees to purple, yellow, blue and even UV flowers, and finally butterflies from red to blue.

This is only one example, probably the clearest one to us, of the result of a strong selective environmental pressure and the ability of insects to respond efficiently. Insect-borne diseases are just one consequence of such efficiency. Insect vectors of diseases are numerous and varying in their characteristics as dependent on their habitats of origin. Generally, among the four stages, the first three (egg, larva, pup) need particular conditions to develop, being aquatic or needing stagnant water. Adults are small (varying in length from 5 to 13 mm), with slender and fragile bodies, a pair of narrow wings, and three pairs of long legs. They can fly for very short distances and survive for long journeys, but being light they can be transported by the wind, performing long journeys and therefore migration is not a problem. If the distance is too great, the resistant and invisible eggs can utilize any sort of transportation. Their key organ for the disease is an elongated proboscis with mouthparts adapted for piercing, being useful for feeding. This organ is also in charge of transmitting the infection.

The starting point

Only a century ago, Ronald Ross discovered that malaria is caused by the bite of an infectious mosquito, evidencing that mosquito-to-human malaria transmission occurs when sporozoites from the salivary gland of the mosquito are injected into the skin during blood-feeding.

The bite of a mosquito must be considered the starting step of an insect-borne disease. It's a minor event, involving minimal parts of our body and of the vector, but it can be the start of a sequence of relevant steps causing serious damage to our health. However, we must consider what happened before the bite and that not any bite is equally impacting. Therefore, it is important to focus on the aspects (usually underestimated) that can differentiate a mosquito's bite and its consequences.

In the case of an insect-borne disease, the situation is complex and the bite is an essential but not unique part of a sequence of episodes, which are elements of cycles, and these cycles can repeat indefinitely. Before the bite, several other actions are necessary to start the infection, including a complex cycle of infection inside the vector, that precedes the further cycle in the host. The bite is the exact point of contact between the two cycles. Female mosquitoes bite humans and other animals without any intention of propagating the infection, being also victims of the disease, but in this way they obtain nutrients and energy contained in human blood. This energy is necessary to make the eggs they need to reproduce. To obtain this result, a mosquito uses the sharp tip of its straw-like mouth (proboscis or stylet) to pierce a person's skin and therefore locate the blood vessel and draw blood up through its mouth. This is the exact moment of the contact between the three actors: vector, host, and parasite. As the mosquito does this, its saliva and the prey's blood come into contact, exchanging contents. It takes some time, because of our first defensive line. The invasion carried by the bite is perceived and the body reacts, trying to stop the intruder. Thanks to an anticoagulant, the saliva stops the person's blood from clotting and, in the case of an infected mosquito, the cells of the etiologic agent of the disease (perhaps received by us) also flow into the blood, identifying the jump from a host to another. The new host, such as a human body, is usually preferred by the parasite, if not already infected. If the blood were to clot around the mosquito's mouth, it might get stuck. Mosquito bites itch and swell because of the body's histamine response. The swelling around the bite is caused by histamine, which is produced in response by the immune system. If you do not itch after being bitten, there may be something wrong with your immune system. In other words, an itch is a natural and expected reaction, showing us that our body has detected a possible harmful intruder (Lindsay et al., 2012).

We shall now focus briefly on histamine and its numerous metabolic roles, because it is not only a matter of itching. Histamine (Fig. 5.4) acts as a signaling molecule in the stomach, skin, and immune and nervous systems. Histamine-containing neurons, whose sole source is the posterior hypothalamus, innervate the whole central nervous system. They are active exclusively during walking, since first of all, this biogenic amine is one of the most important our neurotransmitters, carrying chemical messages between nerve cells. Generally, in this way signals travel from the periphery to the brain and conversely. However, histamine is also stored in the blood cells called basophils, which harbor histamine-containing special granules. Once released from its granules, histamine produces many varied effects within the body, including the dilation of blood vessels, which increases permeability and lowers blood pressure, facilitating the work of the vector. On the other side, the effect of histamine on erythrocytes is crucial for the body's health because of its role in the immune response, our main defense against the disease. Histamine is derived from the amino acid histidine by loss of the

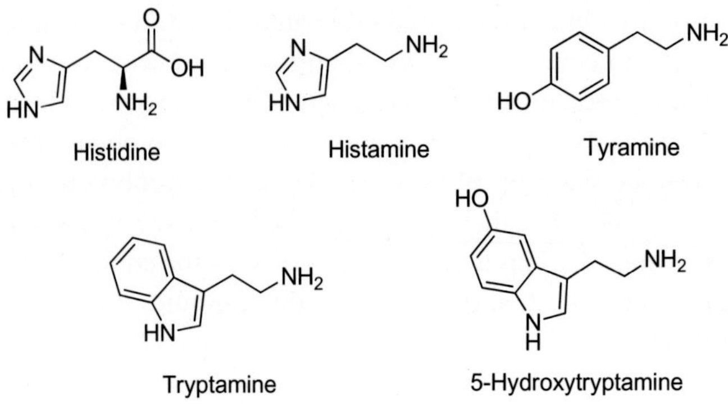

Fig. 5.4 Structures of histamine and related amines acting on the central nervous system (CNS).

carboxyl group (decarboxylation). In fact, chemically, histamine is 2-(4-imidazolyl)ethylamine. The ethylamine "backbone" is a common feature of many bioamines, acting as transmitters (e.g., dopamine, norepinephrine, and serotonin). The structure is based on the general rule for neurotransmitters, like serotonin, with a flat aromatic part and a nitrogen functional group at certain distance. Despite its very simple structure, it is a bioamine that has great importance in the regulation of metabolism and appropriate response to environmental pressure. Therefore, it is distributed widely (albeit unevenly) throughout animals; it can also be found in many plants and even in bacteria. Histamine is strictly related to the nervous system, being stored within and released from neurons. Once released, histamine activates both postsynaptic and presynaptic receptors, although histaminergic nerve terminals do not exhibit a high-affinity uptake system for histamine as other neurotransmitters do, suggesting a wider functionality for this molecule.

Histamine also induces antinociceptive (pain-relieving) responses in animals after microinjection into several brain regions. In this short dissertation about the consequence of a mosquito bite, usually considered a mundane occurrence, we must consider the local reaction of bodily tissues first to injury caused by physical damage, as clearly observed in the inflammation sequence, second as agent against the infection, and third as allergic reaction. Injured tissue mast cells release histamine, causing the reaction of fluid and cells of the immune system, such as leukocytes (white blood cells) and blood plasma proteins. Together with other immune factors, they leak from the bloodstream through the vessel walls and migrate to the site of tissue injury or infection, where they begin to fight the infection and nourish and heal the injured tissues. This process is normally able to defend us, but in the case of infirmity or recurrent and massive attacks, it is not sufficient to avoid the disease. Histamine, being related to the body's immune response, can cause many symptoms of allergies, such as a runny nose or sneezing, in a person's eyes, nose, throat, lungs, skin, and/or gastrointestinal tract.

In other words, when a mosquito bite breaks the skin, a person's body recognizes the mosquito's saliva as a foreign substance. The bite causes a response of the immune system, which aims to flush out the intruder by an increase of histamine in the blood flow, meanwhile white blood cells crowd around the affected area, and this causes inflammation or swelling. There is another reason why mosquito bites itch, since histamine also sends a signal to the nerves around the bite. Nerve endings in the skin send messages to the spinal cord, telling it that histamine has been released. Histamine causes a number of immediate effects, including (if we are fortunate0 widening of the blood vessels close to the bite. Otherwise, the bite gets warm and begins to swell. Fluids escape from the single-cell layer of the capillaries, bringing with them clotting agents and white blood cells. These cells start to wrap around and digest the mosquito proteins, but there are chemicals in mosquito saliva that slow down the response of white blood cells. This effect can last up to 7 days, which is why it takes days for a mosquito bite (and the associated itch) to go away. For the same reason, sicknesses like malaria, dengue, and West Nile virus are so widespread and difficult to control. However, the chemicals in mosquito saliva only work on our immune response, and not on the process of histamine release. From the spinal cord, these signals reach the brain where they are translated into an itchy sensation.

The alerted brain send orders of reaction. Therefore, scratching is the first immediate natural reflex to an itch, and very hard to control. The itch-scratch reflex has been extensively studied, mainly in cats and dogs, and we must differentiate between an acute itch and a chronic itch. A chronic itch has very little to do with histamine, instead of the acute itch, caused by things like mosquito bites, which is regulated through histamine release. When histamines start to act, they are also pruritogens. These cause the persistent itchy sensation which typically follows a mosquito bite. The more mosquito bites you have, the more you itch. As many female mosquitoes take advantage of night hours to feed, this often leads to increased itching at night, but this depends on the species of biting mosquito. During the night when we are sleeping, the bedroom is a favorite place of female mosquitoes, not only due to our prone and unconscious state. When we are lying unprotected in our beds at night, breathing out lots of carbon dioxide and at just the right temperature, we attract many female mosquitoes, which can perform their work without interferences or hazards. Our bodies tend to have higher amounts of carbon dioxide as the night goes on, released via our lungs during respiration. Sweat and perspiration are concentrated around us and work as attractants, since CO_2 is the most efficient mosquito attractant. In this way, female mosquitoes will easily make their way to the bedroom and to us when we are resting. This also explains the sensation of being followed and persecuted by mosquitoes. More bedroom rest means more bites, and mosquito bites are favored if you are the type of person who falls into a deep sleep. If one bite releases a few granules of histamine, 10 bites will release 10 times that many. In this way, the intoxication becomes persistent and the itchy sensation will increase, although in the case of insect-borne disease, this is just a collateral effect.

A further reason favoring biting activity during the night is because in the bedroom, we have fewer distractions. The itch-scratch reflex is controlled by the brain. Mosquitoes adapt to our daily rhythms to increase their chances of getting a good meal and they can live through winter in the right environment. During the day, our brain is usually busy and distracted, but active, and it can tell us to scratch or react against the mosquito bite.

In many places in the world, mosquito-borne diseases are endemic, and at the sunset this compact army of flying vectors moves to invade and conquer any place in search of human targets. The greater the number of parasites within the salivary gland of the mosquito following blood-feeding, the more likely it is to have transmitted the disease. In other words, highly infected mosquitoes are more likely to cause infection (and to do so quicker) than lightly infected mosquitoes. This suggests that mosquito-based methods for measuring transmission in the field need to be refined as they currently only consider whether a mosquito is infected or not (and not how heavily infected the mosquito is).

Our reaction is important. The type of mosquito bite does not make a difference to the itch, although the larger the mosquito is, the more saliva is injected, and the more histamine is released. Frequent mosquito bites mean more itching. Histamine can be released over days, with multiple bites over multiple days having a cumulative effect. Whether that pesky mosquito is an *Anopheles* or a *Culex*, its bite is going to itch. An itch-free mosquito bite simply does not exist—unless, of course, we do something about it. Stopping the itch does not just mean applying a cream or taking a pill. As in all good medicine, prevention is far better than cure. A vicious circle of bites, involving histamine release, itch-scratch reflex and local inflammation, can be counteracted by use of a repellent or antihistamines, hydrocortisone creams, and ammonia solutions. However, if the mosquito was infected, the consequences are more important.

Mosquitoes have been with us for millions of years and are probably one of the most studied insects on the planet, thanks to their carrying a huge number of very serious diseases, which are still no closer to being eradicated completely—quite the opposite, again with the decisive help of humanity. Many species are learning to live indoors, adapting their behavior to our living conditions and acquiring survival strategies to overcome the chilly winter temperatures.

However, the mechanism of transmission by mosquitoes is not automatic and mosquitoes are not always efficient vectors. Novel transmission reducing drugs and vaccines are tested by experimentally infecting people using infectious mosquitoes. The importance of parasite number has widespread implications across malariology, ranging from our basic understanding of the parasite, how vaccines are evaluated, and the way in which transmission should be measured in the field. The relation between the number of parasites and infection provides direct evidence to suggest that the world's first licensed malaria vaccine may be only partially effective because it fails to provide protection against highly infected mosquitoes. It also provides direct evidence for why the only registered malaria vaccine RTS,S was only partially effective in recent clinical trials.

Leaving histamine for now, it should be noted that all this is probably just a small part of the histamine story, that connects to the arguments of this book. Histamine's influences were restricted to the simple sequence of sensorial experience-perception-acquisition-response-consciousness-memory, but it is possible that histamine and other similar bioamines are involved in other levels, including emotions.

Malaria generalities

If plague is the most impressive insect-borne disease in terms of impact and devastation, malaria was, is, and probably will be the most important tropical disease affecting mankind, and is certainly nowadays the most diffused globally. The story of malaria is interesting due to its central relevance in the progress of medicine, as well as the manifestation of resistance and the consequent need of new strategies. The word malaria is derived from the Italian *mala aria*, literally meaning bad air, as the result of the association with humid and moist areas where life is difficult and air is of bad quality. The idea was that something in the air could infect people. Nobody knew about invisible dangerous creatures. In Rome, ancient hospitals were constructed with the doors angled toward the Vatican, to allow entry of the wind transporting the Holy Spirit.

The word testifies the old origins of the disease and its ancient presence mainly in central and south Italy, but this disease, in contrast to emerging ones like Zika or Ebola, should probably be considered a case of coevolution with the genus *Homo sapiens*, still in action. In other words, species of malaria agents were probably already present in the early Tertiary period some 60 million years ago; they were derived from a common ancestor, but rapidly found ideal hosts in humans, and this coexistence is destined to continue.

The wide diffusion of this disease is based on a complicated infection and propagation strategy. Again, we must refer to the sequence parasite-vector-host. Incidence of malaria differs by area. Although 40% of the world's population can be considered at risk, most transmission of *Falciparum malaria* occurs in Sub-Saharan Africa. In this area, mainly children under the age of 5 years are most affected. The incidence of disease declines in older children because of increasing immunity. In contrast, in South-East Asia, the disease occurs more commonly in adults and the clinical features are different. However, another character of distinction of malaria, different from other vectored diseases so far mainly limited to tropical and sub-tropical regions, is the historical occurrence in temperate latitudes—probably another effect of coevolution.

Regarding epidemiology, we must consider that if there are more than 500 million clinical cases every year, only 1% of symptomatic infections may develop into severe malaria. In such cases, malaria symptoms consist of anemia, hypoglycemia, and metabolic acidosis, leading to coma or multiple organ failure, which may cause more than 1 million deaths annually.

The etiological agent

The parasite is a unicellular protozoan of some species of the genus *Plasmodium*. Protozoa are considered the precursors of animals and pertain to the Protista (or Protoctista) kingdom, which is the most complex macrotaxon, wherein a plethora of different eukaryotic organisms are located. Protista were created by a need for simplification: a eukaryote, which is not an animal, a plant or a fungus, is a Protista. In the Protista biological kingdom, there are eukaryotes very different in life cycles, trophic mode, locomotion structures, areal distribution, and organization. Among them, we can find several important parasites, affecting humans, animals, and plants. Besides malaria, a number of human parasitic diseases are caused by protozoa, such as amoebiasis, giardiasis, toxoplasmosis, cryptosporidiosis, trichomoniasis, Chagas disease, leishmaniasis, African trypanosomiasis (sleeping sickness), amoebic dysentery, acanthamoeba keratitis, and primary amoebic meningoencephalitis (naegleriasis). In all these cases, the parasite is a microbe and the mechanisms of infection and disease are quite similar, and there is no definite treatment. However, malaria wins for diffusion and importance. We must consider carefully the reason for malaria's success, which involves the congruence of several key factors. In a superorganism, the mixing of viruses, prokaryotes, and eukaryotes is now substituted by all eukaryotes, although pertaining to very different organisms: the pathogen as the protozoan *Plasmodium*, the vector as the insect *Anopheles*, and the host as humans or other animals (Table 5.4).

Table 5.4 Scientific classification of *Plasmodium falciparum*.

Dominium	Eukaryota
Regnum	Protista
Subregnum	Chromalveolata
Clade	SAR
Infrakingdom	Alveolata
Phylum	Apicomplexa
Class	Aconoidasida
Order	Haemosporida
Family	Plasmodiidae
Genus	*Plasmodium*
Species	*Plasmodium falciparum* Welch 1879
Synonyms	*Oscillaria malariae* Laveran 1881, *Plasmodium malariae* Marchiafava and Celli 1885, *Laverania malariae* Feletti and Grassi 1880

The genus *Plasmodium* includes more than 170 different species that infect mammals, reptiles, birds, and amphibians. However, only four species have long been known to cause malaria in humans: *Plasmodium falciparum*, *P. vivax*, *P. ovale*, and *P. malariae*, with several subspecies and varieties. *P. falciparum* is considered responsible for the majority of the morbidity and mortality attributed to malaria, especially involving the central nervous system (CNS). It is also possible that something is changing since more recently, *P. knowlesi*, which normally infects long-tailed and pig-tailed macaque monkeys, has also been implicated as a cause of human malaria in South-East Asia: Borneo, Thailand, Singapore, and parts of the Philippines.

The vector

Anopheles, like any other flying insects, are efficient vectors and by their biting the infection can be easily transferred. There are approximately 3500 species of mosquitoes grouped into 41 genera, but of the approximately 530 *Anopheles* species, only 30–40 transmit malaria in nature. The rest, even biting humans frequently, cannot sustain development of malaria parasites. They can easily be present in rural and urban habitats, needing only small quantities of water to survive and being able to reproduce in large numbers. Malaria is transmitted by different *Anopheles* species in several geographic regions, since different environments support different species. Male mosquitoes do not bite, so cannot transmit malaria or other diseases. The biting adult females are generally short-lived, with only a small proportion living long enough (more than 10 days in tropical regions) to transmit malaria. These limitations are not sufficient to restrict malaria diffusion anywhere. In fact, *Anopheles* are present worldwide except Antarctica, and we have long known that malaria is present also in temperate regions, although the vector is active only in hot seasons.

In more detail, the taxonomic studies of entomologists and geneticists have led to the knowledge of 537 species of *Anopheles* currently known and formally named (87%). Genus *Anopheles* is disproportionately divided between seven subgenera with different geographic distributions and numbers of species: *Anopheles* (cosmopolitan, 182 species), *Baimaia* (Oriental, one species), *Cellia* (Old World, 220 species), *Kerteszia* (Neotropical, 12 species), *Lophopodomyia* (Neotropical, six species), *Nyssorhynchus* (Neotropical, 39 species), and *Stethomyia* (Neotropical, five species). Four of the subgenera, *Anopheles, Cellia, Kerteszia,* and *Nyssorhynchus*, include the species that transmit human malarial parasites. In several cases of *Anopheles* vectors, complexes of sibling species were reported, like in *A. gambiae*, the most important vector of malaria in Sub-Saharan Africa, which consists of at least seven morphologically indistinguishable species. With respect to plague, malaria presents some analogies and several differences. In particular, mosquitoes guarantee a better and wider distribution, although *Anopheles* is mainly restricted to hot countries, because its species are not able to survive easily in cool weather. Owing to their impact on human health, concerning malaria and filariasis, *Anopheles* is the most studied and best-known genus of mosquitoes. As vectors, *Anopheles* mosquitoes have affected the lives of more humans than any other insect and have had an impact on human genome evolution, such as emergence of sickle cell anemia as a mode of resistance to malarial protozoa and others.

The host

In addition to mosquitoes and humans, malaria can affect several animals, such as reptiles, birds, and various mammals. Although the data must be considered with caution, considering the difficulties to obtain reliable information and the wide diffusion of the disease, it is estimated that malaria could affect 550–600 million people worldwide and cause an estimated 1–3 million deaths each year. In addition, individuals with asymptomatic malaria (carriers of *Plasmodium* species) are significantly under-identified and thus represent a large unknown transmission factor for malaria, since the majority is often undeclared or unknown. In many countries, malaria is considered endemic and in some way populations and parasites have developed some kind of mutual cohabitation.

The disease is widely distributed in subtropical and tropical developing countries, particularly in Sub-Saharan Africa, Central America and the Caribbean, South America, Central and South Asia, temperate parts of East Asia and South-East Asia, and parts of Oceania. Although the vectors are present in all these areas, the types of malaria are differentiated, since the *Plasmodium* reported above are unequally distributed. *P. falciparum* infection is predominant in Sub-Saharan Africa, South-East Asia, and some parts of the Caribbean, especially Haiti and the Dominican Republic. *P. falciparum* and *P. vivax* are concomitantly encountered in South America and the Indian subcontinent. *P. malariae* is found most commonly in Sub-Saharan Africa, but may be encountered in most endemic areas.

Malaria in detail

Every step of the mechanism of malaria transmission and disease effects has been carefully elucidated, although several aspects concerning its control are still unclear. The parasite is a protozoan, which allows a large possibility of metamorphosis, which is the main reason for the success of malaria. The capacity of adaptation of *Plasmodium* to different microhabitats is the key. The term "metamorphosis," meaning transformation, modification, and change of form, may have different interpretations in biology, depending on the discipline. In botany, the term is used in the case of alteration or degeneration of an organ or a part of a plant in which tissues are changed, like a leaf changed into a thorn, causing a definitive change. Thus, the spinification of leaves in succulent plants, like Cactaceae and Euphorbiaceae, is a metamorphosis of an organ into something adapted to arid and hot conditions. In zoology, the term is strictly utilized for the process of transformation from an immature form to an adult one in two or more distinct stages, like in an insect or amphibian. The change can be radical in these organisms, whereas in humans we have mutation but not mutants. However, in the human fetus there are a series of transformations and in our genome there are evident traces of the sequences of other organisms.

During the occurrence of malaria, *Plasmodium* undergoes a programmed series of metamorphoses, selected to obtain the best adaptation to host cell environments. We must remember that each organism can be identified by and in its DNA. Our life starts with parental DNA and ends with our own

DNA if we have progeny. Life starts from another life, through nucleic acids transmission. Evolution is concentrated on adaptation to the environment and homeostasis of this unique molecule. DNA is an extraordinary macromolecule, being the most complex that we know. However, it has a handicap: this molecule is delicate and needs protection, consisting in an involucre, more or less complex, like a cell or a body, in order to work and survive. However, in the viruses only a protein involucre is present. Among the consequences, the metamorphosis of the vector and the parasite means that, despite the radical changes of form, the individual is the same, being identified in the genome. Emerging from temporary packaging, the DNA can revive, maintaining its identity. The second advantage consists of a smart alternation between different modes of reproduction, including a turnover between asexual and sexual ones. Furthermore, *Plasmodium* is able to remain dormant inside the host, waiting for an opportune moment to diffuse and maintaining a situation of misleading equilibrium of the disease. In comparison with the catastrophic explosions of plague, the situation in malaria can be maintained at a minimal level for long periods, allowing the possibility of endemic disease and recrudescence. Finally, the symptoms and health effects of various malaria diseases are the results of previous differences, and include also important targets, like the nervous system.

The main consideration is that the parasite is able to conform not only to the different hosts, but also to the hosts' different microhabitats. *Plasmodium* jumps from mosquitoes to mammals and vice versa, and adapts its morphology in harmony. The reason for these changes is the need to ensure and optimize alimentation for the Protista, as any other living organism. In particular, the female mosquito is looking for energy, and the most profitable and tempting sources of energy are our red blood cells and its precious content, hemoglobin. Generation of a plethora of new lives urgently requires great quantities of energy. Our blood is a valuable reservoir of energy; the female vector feeds on our blood to aliment its eggs and the microorganism feeds on blood cells to fuel its frantic meiosis.

The whole process of the infection and reproduction is quite complex, comprehending two cycles, one in the vector and another in the host. Let us now concentrate on the human cycle, and therefore on malarial disease. When the female *Anopheles* mosquito bites someone, infection is transmitted from the saliva of the vector to the patient for its blood, wherein *Plasmodium* cells, as sporozoites, follow the bloodstream. Only a few sporozoites are injected directly into the blood vessels via the mosquito bite. Therefore, the parasite first needs to reproduce. At that time, *Plasmodium* is a sporozoite, the best form for rapid and successful reproduction, but it is also protected by a solid cell wall during the dangerous journey from one organism to another. The sporozoites efficiently leave the skin, localize in the liver, and invade hepatocytes. Within 8 h, sporozoites are carried rapidly to the liver where they multiply asexually and in approximately 7–10 days they are able to transform, becoming liver schizonts.

The third phase of the *Plasmodium* path involves the blood after the liver step. When they mature, these schizonts release thousands of merozoites into the bloodstream, where they cause the infected hepatocytes to burst. Inside the sporozoite cell, the single nucleus generates many other similar nuclei. The cell then explodes, liberating its content when nucleu have become cells. One single *Plasmodium* sporozoite in one liver cell multiplies into tens of thousands of exoerythrocytic merozoites, each of which is able to invade a red blood cell, initiating the stage of the infection that causes the fever of malarial disease. However, if the conditions are not appropriate, alternatively dormant hypnozoites are produced by *P. vivax* and *P. ovale*. In this way, the whole potentiality of the parasite can be totally maintained and preserved, ready to act in better times.

The merozoites are the best form to benefit from the treasured energy contained in the erythrocytes. Merozoite's invasion of red blood cells involves multiple interactions and events to satisfy the conditions of the new microhabitat. Merozoites first recognize their target by contact, then reorient and attach to surface receptors on the erythrocytes, and finally penetrate and infect individual erythrocytes (Fig. 5.5). A substantial processing of merozoite surface proteins occurs before, during, and after invasion. Merozoites possess a particular sensitive fibrillar coat of surface proteins. This coat appears remarkably during the first phase of the erythrocytes invasion and allows the merozoite to recognize the target, relaying the more active merozoite (Fig. 5.6).

The blood stage of *P. falciparum* merits a little focus in consideration of the further arguments. Inside the red cell, the bacterium utilizes the host's hemoglobin as a food source, to obtain the amino acids derived from proteolytic digestion for the parasite's biosynthetic requirements. Hemoglobin degradation involves several proteases. Once the denatured globin is produced, a cysteine protease, falcipain, degrades this denatured globin, which by the action of plasmepsins is further degraded into small peptides by other proteases until the final production of amino acids. During this process, large amounts of free nontoxic heme are released from hemoglobin as a by-product. The released heme is autoxidized into hematin, hemin, or aquaferriprotoporphyrin IX, consisting in a ferric form (Fe(III)) that is a

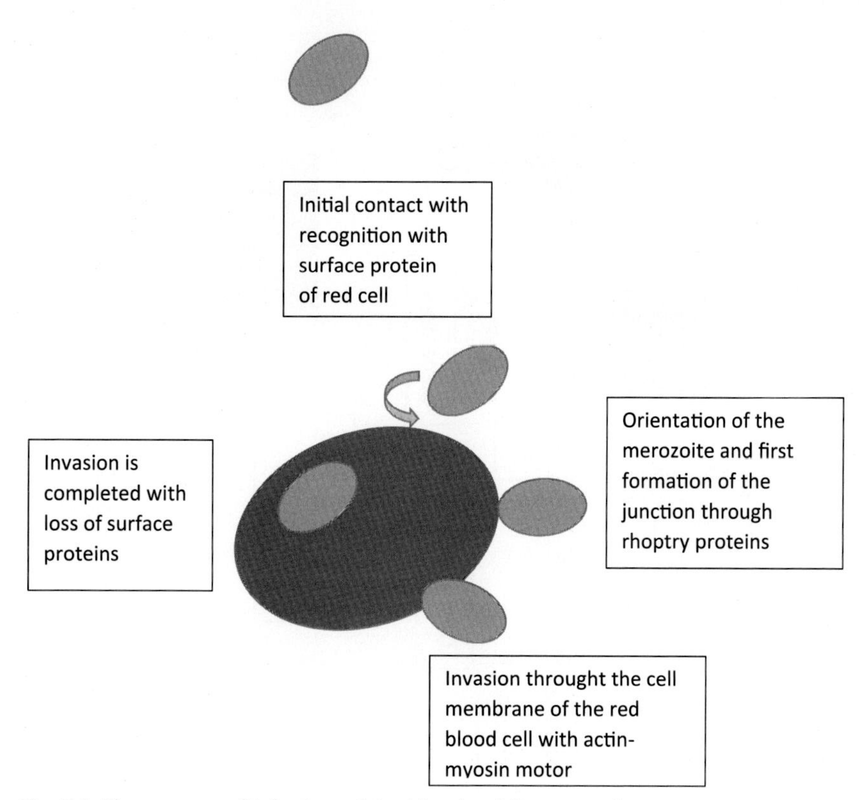

Fig. 5.5 The process of infection of the blood cell by merozoites.

highly toxic product for the parasite, inhibiting vacuolar proteases and damaging its membranes. In practice, there is a danger of autointoxication by the parasite during its alimentation and therefore a detoxification of heme is necessary for the survival and growth of malaria parasite. Enzymes convert heme into hemozoin, popularly known as malaria pigment, and consisting of a dimer of heme units linked through an iron-carboxylate bond. In the dimer, a bond is formed by the linking of central ferric iron of one heme unit with the propionate side chain of another heme. This pigment is inert in the parasite and released into the host blood supply after infected erythrocytes burst open at the end of the parasite's life cycle. Many researches have evidenced the capacity of quinoline antimalarial drugs to inhibit formation of hemozoin and therefore limit parasite growth and reproduction.

The cell surface receptors necessary for attachment are specific for the *Plasmodium* species. For example, *P. vivax* attaches to the red blood cell via a receptor related to the Duffy blood-group antigen. Thus, individuals

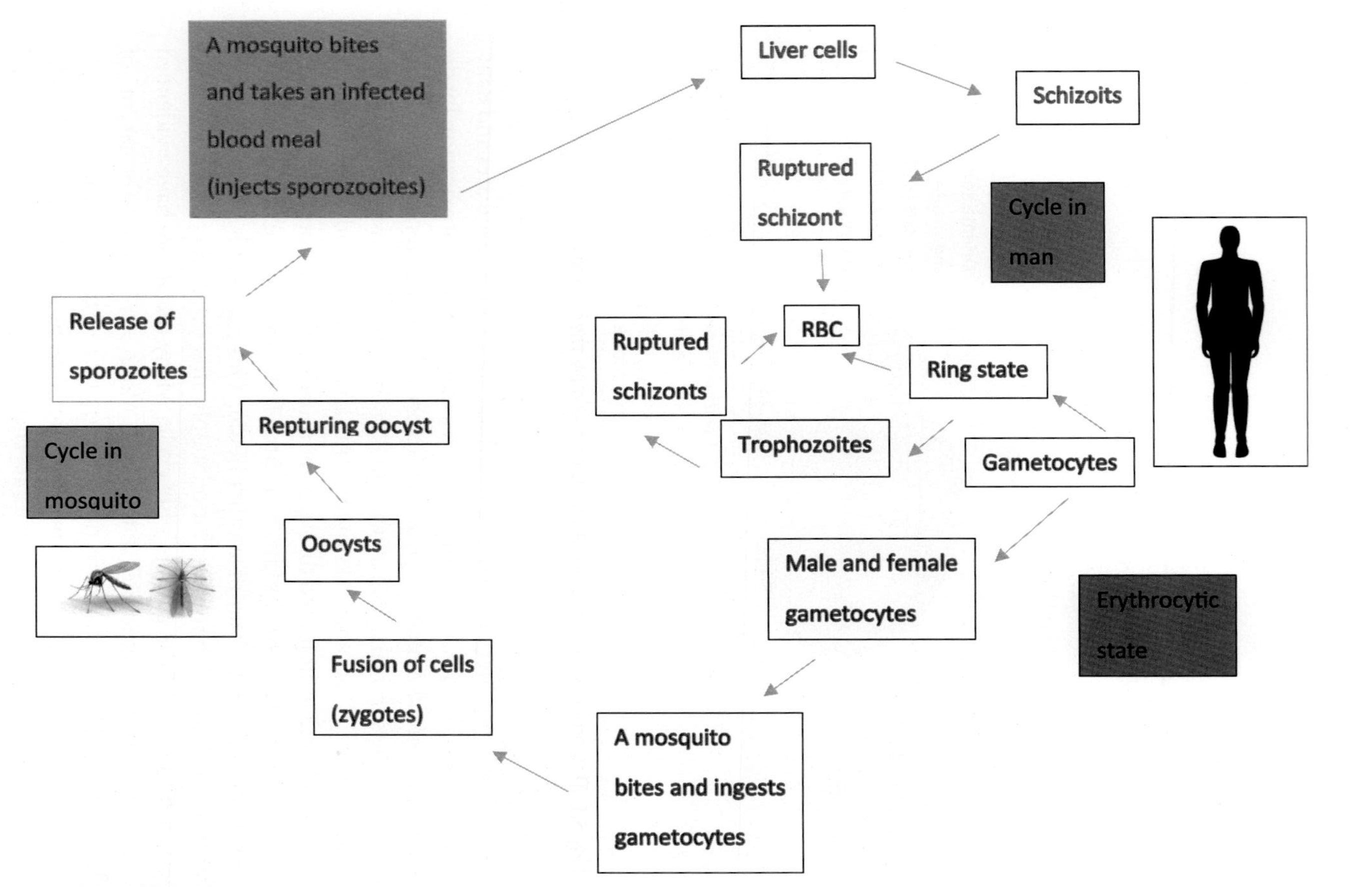

Fig. 5.6 The malaria cycles.

from West Africa, who generally have Duffy-negative blood, are resistant to invasion and infection by *P. vivax*. Even now, *P. vivax* infection remains uncommon among indigenous West Africans.

Once inside, merozoites reside in a vacuole of the host red blood cell, where they transform in the trophozoites. The trophozoites are able to feed efficiently on hemoglobin. Inside the vacuole, another transformation: the early trophozoite manifests as the classic ring form, then the trophozoite gets fat, and grows to occupy most of the erythrocyte's lumen. After about 24–36 h, it enters a second stage of asexual division, schizogony, to form an erythrocytic schizont, each containing 12–24 merozoites per infected erythrocyte. In comparison with the sporozoite, the merozoite is able to prepare the conditions for the passage to sexual reproduction. This reproduction is necessary to maintain the parasite for longer periods and to improve its genome.

The time interval between mosquito bite and entry of merozoites into the bloodstream is about 10–14 (range: 7–28) days. This is a latent period, known as the prepatent period. Continued asexual replication in the bloodstream through repeated cycles of maturation and rupture of red cells with release of merozoites eventually results in symptomatic infection. During this process, a fraction of the merozoites undergo sexual differentiation and develop into sexual forms called gametocytes, which produce no symptoms themselves, but may circulate for a prolonged period of time.

There are different types of malaria with different symptom of the disease. *P. vivax* and *P. ovale* generally have a predilection for young erythrocytes, while *P. malariae* infects old cells. For this reason, these three species seldom manifest a parasitemia greater than 2%. In contrast, *P. falciparum* infects erythrocytes of all ages and can therefore manifest with parasitemia levels often above 5%. Relapse, recurrence, and recrudescence are three important aspects of malaria, which are well known by the patients in case of untreated malaria. For malaria caused by the two relapsing malaria—*P. vivax* and *P. ovale*—treatment with a blood schizonticide invariably results in an immediate response. In relapse, a resumption of replication of previously dormant hypnozoites in the liver starts the cycle again, with development into pre-erythrocytic schizonts that produce merozoites, and these reinvade the bloodstream. A new inoculation of sporozoites from a mosquito bite may generate a recurrence. Following a single exposure/mosquito bite, the malaria-naïve patient may perish during the initial disease or develop both a humoral and cellular immune response, which after recurrent infections could lead to a phenomenon known as premunition. A recurrence is called a recrudescence if it is caused by the persistence of blood forms in small numbers between attacks. Recrudescence may occur over a period of many years, causing a replication of the disease's symptoms.

In 1970, the WHO officially declared Italy free of malaria. However, from 1985 hundreds of new cases were registered yearly, increased to 1000 by the year 2000, with seven deaths. These cases concerned import diseases with Italian citizens travelling for tourism or migrating, but also diseases contracted in Italy.

As we have seen, the situation differs between developed and developing countries, but there is a common feature: both situations are in evolution.

The state of the art of malaria

Malaria is the most important parasitosis and the second most infective disease in the world for morbidity and mortality, after tuberculosis. It is considered endemic for 40% of the population in subtropical and tropical areas, at altitudes below 1800 m. Officially, every year 10,000–30,000 European and American travelers, using air or sea transportation, are affected by malaria. However, these data must be considered underestimated since most of them solve the disease with autodiagnosis, assuming antimalarial drugs.

Malaria is popularly known for the typical cyclic intermittent fevers, corresponding to cycles of reproduction of the parasite. The infection by *P. falciparum* is known as malignant tertiana or tropical disease, whereas that from *P. vivax* and *P. ovale* is the benignant tertiana, and finally that from *P. malariae* is the quartana intermittent fever malaria. In malaria, the terms "tertiana" and "quartana" refer to the intermittent occurring of the symptoms, like the fever. However, these terms are misleading, considering that only a minimal number of cases present intermittent fever. Every 48 h in tertiana (meaning every 3 days) or every 72 h in quartana (meaning the 4th day), respectively. The fever is caused by the presence of various generations of the parasite in the host blood. However, the initial symptoms of malaria are similar to those of common flu, such as cephalic pain, shivers, alternant feeling of cold and hot, and general illness and discomfort. As already considered, a *P. falciparum* infection may cause kidney distress, lung edema, endocranic hypertension, until coma and final *exitus*, consisting in the death of the patient. Death is a consequence of damage by the package of the parasited cells inside vital organs, in particular in the case of the very dangerous cerebral malaria (Chen et al., 2000).

Cerebral malaria is the most severe neurological manifestation of cases of malaria infection by *Plasmodium falciparum*. The clinical syndrome of cerebral malaria is characterized by coma and asexual forms of the parasite on peripheral blood smears. The incidence is considered to be 1120/100,000 each year in endemic areas of Africa, with 575,000 children in danger of developing cerebral malaria annually and peak incidence in the pre-school period. Therefore, African children under the age of 5 years are most affected and the incidence of this disease declines in older children with increasing immunity. The coma develops suddenly, often with seizure onset, following 1–3 days of fever. A few children develop coma following progressive weakness and prostration and other important symptoms. In adults, cerebral malaria is part of a multiorgan disease. Patients develop fever, headache, body aches, and progressively delirium and coma, but compared to African children, papilledema seizures and retinal changes are less common and time to reach coma is slower. Mortality is high, but surviving patients can also present pathogenesis with relevant brain injury and long-term neuro-cognitive impairments. The clinical symptoms appear rapidly after the infection and are attributed to parasitized red blood cells sequestered in cerebral microcirculation.

However, recent reports suggest that the incidence of severe malaria is on the decline, as in general for malaria. Recent data about current incidence of malaria can be obtained from the WHO World Malaria report (WHO, 2019). Considering the geographic distribution and the reported severe cases, it was suggested that in many areas where malaria is endemic, as a consequence of repeated infection, people are able to develop a high level of antibodies, generating a form of resistance to the infection. Another interpretation of the phenomenon considers an evolution of the parasite toward a less virulent form, allowing an increased resistance to survive together with the host. This is an important aspect considering the future development of the disease.

As a matter of fact, malaria is widely spread, but neither its distribution nor its consequences are homogenous. Central Africa is most affected by malaria, so let us consider Nigeria. Regarding malaria in this country, Dr. Elvis Eze, from Nigeria but currently working in Bart's Hospital in London, witnessed first-hand the malaria presence in Nigeria, observing that the disease killed more than 100,000 people in 2016 alone. He stated that "In Nigeria you never escape malaria." However, Nigeria is arguably one of the leading countries in Africa and it is considered an emerging country. In fact, despite the challenges caused by the insect-borne diseases and a never-ending bloody war between the Muslim North and the Catholic South, and the expectation medium age of 18 years, in the last 15 years the population rose to 30 million, and in the next few years Nigeria is predicted to become one of the most populated countries in the word. On the light of the occidental parameters, this can be considered a paradox, in consideration also of the bad life conditions of most population in Nigeria and the incidence of diseases. On the contrary, in developed and advanced countries, where wellness is ensured, the population is decreasing everywhere due to the low birth rate. Looking at these data, malaria is not the most important problem for Nigerian people and similar considerations can be used for neighboring countries. In any case, malaria cannot be considered the sole reason for the current migration from Central Africa to Europe.

Signs from the past

Molecular biology gives us precious information, looking for ancient traces in fossils of our parasitized forefathers—in this case, through analysis of genomes of the protozoaires of the Laverania family, whose members could be considered the ancestors of *Plasmodium falciparum*, or in any case closely related to *Plasmodium* (Otto et al., 2014). Dr. Thomas Otto reports:

"Using the parasites' genomic data, their family tree was constructed and identified major genetic events that led to their emergence (Otto et al., 2018). The movement of a single cluster of genes was an early crucial event that enabled the malaria parasites to infect the red blood cells of a new host species. After reconfiguring and fine-tuning the repertoires of genes that interact with the host and the vector, the parasites were able to establish long-term transmissible infections in humans."

The first evidence of malaria parasites was found in mosquitoes preserved in amber and dated to the Palaeogene period, meaning they are approximately 30 million years old (Poinar, 2005). This founding does not mean the presence of malaria as we know, since human malaria, which likely originated in Africa, coevolved for a long time passing through a series of its hosts, including nonhuman primates. In fact, protozoa, responsible for malaria, are diversified into primate, rodent, bird, and reptile host lineages.

Probably, everything started 50,000 years ago, although the first real evidence of the presence of malaria parasites dated 4000–6000 years ago, and this is probably in connection with the first great human concentrations. The research was able to design the genealogical tree of the Laverania until the appearance of the cluster of genes responsible for the transmission of malaria to humans. *Plasmodium falciparum* has a common ancestor with the gorilla parasite *Plasmodium praefalciparum*; however, the latter is not capable of establishing repeated infection and transmission in humans. Otto's team was able to generate multiple genomes from all known Laverania species. The complete genome sequence of the closest ancestor of *P. falciparum* enabled researchers to estimate the timing of the beginning of speciation at 40,000–60,000 years ago.

According to the paper published in *Nature Microbiology* by the researchers of the eminent Wellcome Sanger Institute in Britain, *P. falciparum* was the only parasite of the Laverania family able to adapt its transfer successfully from gorillas to humans as reported by Dr. Matt Berriman (Otto et al., 2018): "We sequenced the genomes of all known species in a family of malaria parasites that gave rise to the deadliest form of human malaria. We estimated when *Plasmodium falciparum* and its relatives diverged and found evidence that the recent expansion of modern humans created the home in which the parasites irreversibly evolved into a human-specific form." This research is important and the obtained indications have current relevance. Many infective dangerous diseases can remain sequestrated into a preferred host for a long time, but environmental forces can generate some genetic changes and induce the jump that infects mankind.

Today I have two maps on my desk. One contains the diffusion of *Homo sapiens* from a point in Central Africa to everywhere. Geneticists were able to sign pathways and dates, evidencing the progressive spread of our species to any possible part of the planet. Our ancestors were brave explorers and many lands, such as America or Oceania, were discovered many times before the dates officially reported in history. These explorers could trust only a few natural instruments and their insatiable thirst for knowledge. The second map is the phylogenetic tree of our species. This tree presents many branches. At the end of each branch there is an ancestor more or less similar to us, but only one contains a surviving species, since all the others were extinguished, including, very recently, Neanderthal man. The unique remaining species is *Homo sapiens*. Perhaps there is a connection between the two maps. The disappearance of the other human forms and the persistence of only one is a peculiarity. It is very unusual in comparison with the genealogy of other species, otherwise Darwin never could have speculated on the origin of the species, working on the minor differences between birds of nearby islands in Galapagos archipelago. What caused their extermination? And more important, *who* was responsible for their disappearance? The reply is: we are the ones responsible. Our ancestors lived in different regions, diversified by habitats and separated by geographic barriers. In principle, all the conditions were present for them to continue their evolution pathway. The explanation is therefore probably in the other map. *Homo sapiens* was an insatiable explorer, never satisfied with his current conditions. An eternal migrant, once arrived in a place, he first used to kill off the autochthonous people living there, save part of the women, conquer the territory, and utilize its sources. Later, as soon as the situation was stabilized, with villages and organization, a new vanguard started to explore and conquer new places. The species *Homo sapiens* was obsessed by an unsatisfied desire for new experiences and knowledge, and a continuous need to travel. As Ulysses in Dante's *Inferno* states: "Ye were not form'd to live the life of brutes, but virtue to pursue and knowledge high." Nowadays, practically every part of the geography of our planet has been visited and reported on, and we are striving to go to Mars, even if we expect to find there only dust and stones. And we have yet more worlds to visit and explore. Many secrets are waiting, from the subatomic world to the magnitude of the universe. Many experiences are still available to extinguish and satisfy our thirst for new discoveries. Scientists are pushed by such a desire, and some, probably the most imaginative ones, like to explore through time, looking for traces of past events, to explain the present and predict the future. In genomic studies, the traces of genome changes are followed and connected with the spread of the disease (Sallares et al., 2004) and the migratory pathways (Rolshausen, 2019; Teri et al., 2009; Birney et al., 2007; Jackson, 2015).

Again by analysis of DNA bodies in graves, we know that malaria was present in Italy at least 2000 years ago. DNA is a delicate molecule, being made by the junction of monomers, in this case consisting of the nucleotides. Furthermore, the conformational configurational structure can be degenerated by several agents. Even the nucleotides are made up of three different components. In any junction, hydrolysis, the most common reaction in metabolism, can affect the stability of a nucleic acid. Therefore, it is very difficult for DNA to remain unchanged during centuries, unless it can benefit from a special protection.

In the pulp inside human teeth, genomic traces of *Plasmodium falciparum* were determined, confirming many reports that malaria was endemic in Imperial Rome. The reports of that time state that several districts of the town were considered highly dangerous because of the disease, facilitated by the presence near the town of a wide swampland, the Agro Pontino, an ideal reservoir for mosquitoes. The situation continued without changes until 1623. In that year, there was a conclave in Rome to elect the new pope.

A conclave in Rome was considered by the cardinals and their entourages as a calamity, since in such occasion many of them were likely to be victims of the malaria. However, at the conclave the Spanish cardinal Juan De Lugo distributed Cinchona bark, saving the lives of the cardinals. Afterwards, the bark was utilized in the Arciospedale di Santo Spirito in Rome with great success. The bark of Cinchona trees was an old and popular remedy in Peru, where the trees are common in the mountains of Andes (Baliraine et al., 2011). The name *Cinchona* for the genus was created by Linneus in 1742 in honor of Countess Cinchòn and her story linked to malaria. The story goes that when she was in Lima in the 12th century after her husband, the viceroy, had been nominated by the King of Spain to rule in those distant conquered territories, the countess fell ill with a terrible fever. No remedy could be found, until an old Peruvian servant used the bark of a local tree to save the countess. Therefore, the remedy became known as "countess's bark." In fact, the *cascarilleros*, populations of some tribes of the mountains of Andesm used to collected the bark, which was well-known in local popular medicine for its antipyretic properties. The *cascarilleros* (*cascara* means bark in Spanish) used to keep secret the place and type of plant used, so that the bark was for a long time veiled in mystery. According to some reports, the origins of the bark was kept secret by the *cascarilleros* until one of them was corrupted by Dutch merchants. The traitor paid with his life, but the seeds crossed the world, reaching Indonesia, where hybrid Cinchona trees, carefully obtained by agronomists, produced more active barks.

The story of Countess Cinchòn is a beautiful episode, a fascinating example of ethnopharmacology adequate for the birth of a potent and unique remedy, as described to Linneus by Le Condamine, but the problem is that it is not true. The viceroy had two wives; the first never left Europe and the second died in Panama during a journey to South America. However, another name for the bark, "Jesuit's bark," is more credible, since this religious order was fundamental to the diffusion of the remedy, as it was for many other vital herbal drugs. The Jesuits noticed that the Indians in Peru used to chew the bark to prevent shivering with cold and to reduce labor pains in women (a utilization recalling that of coca leaves). They also considered the bark useful to cure shivering caused by malaria and brought the bark back to Europe, where at that time the disease was a significant problem, with a large part of its population suffering from malaria. From that time, correct dosing of the bark or extracts of the bark was evidenced to cure several cases of malaria. It is noteworthy that the first effective drug against malaria came from South America, a continent in which malaria had not been a relevant problem before the arrival of the Europeans, and from trees growing in high mountains where both the parasite and vector are absent. This means that the resources of plants cannot be rationalized by our rational approaches, but the traditional use represents an important starting point for scientific evaluation of natural drugs.

In 1643, Cinchona bark was first officially mentioned in medicine as being used for the treatment of fever in Europe, and in 1677 it was included in the London *Pharmacopeia* as "Cortex Peruanus." In the 19th century, the bark of Cinchona trees was also known in all Europe as a raw material to obtain a powerful elixir, and considered as a panacea against fever and other diseases, including the famous mysterious vine used by the English James Talbor (or Tabor or Talbot in France). Talbor was a questionable personage who acquired great fame for his miraculous cure of malaria using a mysterious elixir. He became very famous when he was asked to intervene at the bedside of important personalities such as Charles II of England, the Dauphin of France, and the Queen of Spain. It is believed that Talbor's elixir contained also opium, as in another pharmaceutical panacea, laudanum, but the main ingredient, as revealed after Talbor's death, was Cinchona bark. Nowadays the elixir of quinine is also produced and considered a potent tonic. Cinchona bark extracts are used as an ingredient in many aperitifs, digestifs, and soft drinks, daily consumed everywhere. Quinine is a flavor component of tonic water, bitter lemon, vermouth, and cocktails. The USA Food and Drug Administration (FDA) limits quinine in tonic water to 83 ppm, due to its side effects. Quinine is used instead of strychnine as a standard substance for a bitter taste (threshold of sulfate salt: 0.000008) in gustatory physiology.

About 35 quinine type alkaloids have been reported from *Cinchona* species. The antimalarial activity of Cinchona bark must be assigned to four complex quinoline alkaloids, all used as a mixture together. The four alkaloids, i.e., quinine, cinchonine, cinchonidine, and quinidine, are present in different yields in various species and varieties (Fig. 5.7). The most utilized species are *C. calissaya* (known as yellow quine), *C. succirubra* (red quine), *C. ledgeriana* (gray quine), and their hybrids, also differing in alkaloid contents. The structures of these alkaloids are based on a quinoline moiety attached through a secondary alcohol linkage to a quinuclidine ring having a vinyl group. The first part is bicyclic and aromatic, whereas the second is an unusual heterocyclic system, in some way recalling the tropane unit of some Solanaceae species. The stereogenic center at C9, consisting of a secondary alcoholic group, affords two pairs of epimers: quinine and cinchonidine

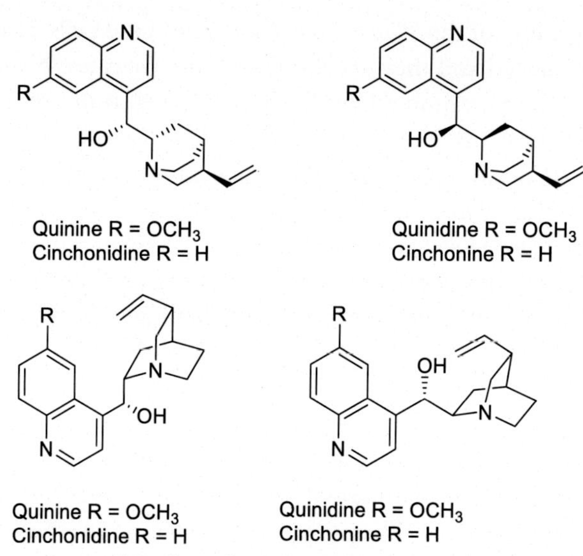

Fig. 5.7 The structures of Cinchona bark's main alkaloids. In the figure, the same structures of natural alkaloids are presented in two molecular writing forms. In such way, some 3D particulars can be evidenced.

versus quinidine and cinchonine. Therefore, quinine and cinchonidine are (−)-isomers, and quinidine and cinchonine are (+)-isomers. However, quinine and quinidine each contain a methoxy group, which is absent in cinchonidine and cinchonine.

Quinine has been the leading compound for the development of a large number of antimalarial compounds, where the functional groups, like vinyl and hydroxyl, have been used to modify the structure or add other units. Quinine is less potent as an antimalarial but less toxic than quinidine. It has schizontocidal and gametocidal activities against *P. vivax* and *P. malariae*, but not for *P. falciparum*. Therefore, quinine has been used as a suppressive and therapeutic agent, but not as a prophylactic agent. In any case, widespread use of quinine is limited due to toxicity and adverse effects. The continuous use of quinine causes side effects, which can be severe in some cases and include cardiotoxicity, visual disturbance/blindness, deafness, convulsions, and hypoglycemia. The side effect consequent of an overdose of quinine is named cinchonism, and this can be expressed through various symptoms including rashes, headache, confusion, and vomiting.

The most widely accepted hypothesis of quinine's action is based on the well-studied and closely related chloroquine. This model involves the inhibition of enzyme heme polymerase avoiding hemezoin biocrystallization.

During the feeding of the *Plasmodium* inside red blood cells, free cytotoxic hemes are produced and they accumulate in the parasites. Free hemes are toxic to the parasites, causing their deaths. However, using heme polymerization, *Plasmodium* is able to convert the toxicity to a form of protection by the heme detoxification pathway, which facilitates the aggregation of cytotoxic heme. Chloroquine's mechanism of action, also hypothesized for quinine, against *Plasmodium falciparum* is theorized by interfering with the parasite's ability to dissolve and metabolize hemoglobin. When the parasite enters red blood cells and starts to feed hemoglobin, the heme (the porphirinic portion) is realized.

Heme is toxic to *Plasmodium*. Therefore, more feeding by the parasite means also more production of toxins, but the *Plasmodium* is able to solve the problem by the polymerization of the hemes, obtaining a detoxification. As with other quinoline antimalarial drugs, quinine's mechanism of action has not been fully resolved. Therefore, the therapeutic effect of chloroquine is based on the fact that when it reaches the food vacuole of the parasite, it binds hemin (a toxic compound) and inhibits conversion of this compound to hemozoin (a nontoxic compound). Accumulation of hemin inside the food vacuole of the parasite results in the parasite's cell death.

In another interpretation, quinine may target malaria's purine nucleoside phosphorylase enzyme, subtracting energy from this key molecule. Thus, we are reporting two hypotheses: the one purely molecular based on the detoxification process and the other one focused on energy. In any case, the enzyme's key metabolic role must be considered. In consideration of the structural differences between quinine alkaloids and diterpene artemisinins, it has been necessary to explore other possible mechanisms of action.

For centuries, the complex of Cinchona alkaloids was the only available remedy to treat malaria and relieve its symptoms, in particular as an antipyretic. In Italy, before the Second World War, quinine was considered strategic for the health of the population and was therefore nationalized along with tobacco and salt. It was easy to find and buy quinine, since quinine tablets were sold by tobacconists or directly distributed by the state. In the same years, Agro Pontino was reclaimed by the Mussolini government, finally releasing Rome from malaria after thousands of years. Now it is a fertile land where valuable varieties of artichokes and tomatoes are produced.

Malaria and quinine were linked for centuries, and Cinchona trees were cultivated in many tropical sites to satisfy the enormous demand for quinine. Despite several reported syntheses, Cinchona trees remain the only economically practical source of quinine. Most natural products cannot be reproduced by chemical synthesis, due to the number of chiral centers usually present. A popular solution is hemisynthesis, consisting of the transformation of precursors present in the plant by a few steps. As a matter of fact, although many organic syntheses of quinine are reported and available, even if the ratio of any passage is high, more than five passages means a unsuitable cost of production. Another approach is when the structure of a natural product is used as a template for compounds with similar or increased activity.

However, on the basis of their quinoline structures, useful medical drugs were synthetized including chloroquine, which ensured adequate treatment for a long time. Until the early 1900s, quinine was used exclusively against malaria, until chloroquine and mefloquine were synthesized and introduced on a large scale. Chloroquine was the drug of choice because of its low cost and absence of quinine's side effects. In consideration of the subjects which should be treated by the drug, the cost of the remedy is a crucial factor. According to a 2002 study, prices for tablet treatments can be compared as follows: $0.09 chloroquine, $2.73 quinine, $5.04 mefloquine, and $5.34 artemisinin. The success of the chloroquine pill is second only to ASA (aspirin). Despite several collateral effects, chloroquine has been widely used by people living in or visiting places where malaria was endemic, confident in its power to prevent or limit malaria. The concomitant action of DDT on the vector and chloroquine on *Plasmodium* dealt an almost mortal blow to malaria—until the counteractions of *Plasmodium* and the mosquito.

The mechanism of action of antimalarial drugs is still largely unclear, since most of them were identified on the basis of their potent antimalarial properties in phenotypic screenings, without a rational design to interfere with a specific molecular target, but rather on the basis of efficacy. During the Second World War, the best cultivations of Cinchona trees were in Indonesia, thanks to the excellent work on hybridization carried out by the Dutch growers between *C. calyssaya* (yellow quine), and *C. succirubra* (red quine). During the Second World War, the United States was fighting in the Orient against Japan. The shortage of the Cinchona bark, due to the difficulties of obtaining the drug from the production places, became a serious problem, to treat the American soldiers, which came from the other part of the ocean and totally accustomed to other situations and life conditions. Under wartime pressure, research into synthetic production was undertaken. In 1944, the chemical synthesis of quinine was accomplished by the American chemists R.B. Woodward and W.E. Doering, two legendary pioneers in the story of organic chemistry. The synthetic approach to antimalarial drugs considered the need of the quinoline core along with amine groups, just like quinine. Quinine was used as a scaffold for production of synthetic antimalarial drugs, such as mefloquine, amodiaquine, and mepacrine. However, natural products are too complex and none of the syntheses can compete in economic terms with isolation of the alkaloids from natural sources. The synthesis must be stereo-controlled, in consideration of all four stereogenic centers of quinine, and needs an asymmetric catalytic method, as well as the difficult separation of the mixture of epimers derived by the synthesis. In any case, the best afforded solution was a production of synthetic quinine on 16 chemical steps.

Under the pressure of the war and the military command, English, American, and Australian researchers failed to follow scientific methods, hoping for a stroke of luck instead. They worked on the quinine structure, making a large number of derivatives and chemical similarities. They tested 16,000 compounds before finding a useful trace for a safe and efficient antimalarial drug. They knew that in 1856 a young chemist called William Henry Perkin, when attempting to synthesize quinine, had obtained mauveine, which was the world's first synthetic dye. The synthesis was accidental; Perkin was on his Easter holidays and performing some experiments in the laboratory in his apartment in East London. It was an example of pure serendipity, surely not the only one in science. Becoming interested in dye chemistry, Perkin obtained an important result with the synthesis of methylene blue, derived from mauveine (Fig. 5.8). Methylene blue was not efficient as an antimalarial, but it was useful to visualize the parasite. Nowadays we still use this reactive to evidence cells in students' exercitations and in particular to visualize microbes with a microscope in microbiology and plant cells in botany.

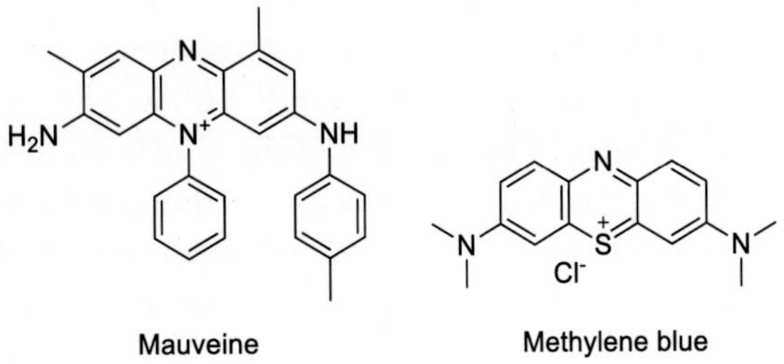

Mauveine Methylene blue

Fig. 5.8 Structure of mauveine and methylene blue.

The next story is important, wherein we can try to follow the fundamentals of modern medicine, the same ones that are still the basis of the current medical approach to treat insect-borne diseases such as malaria. Following the development of the theory of the "magic bullet," it is necessary to understand the importance and the limits of the "classic" medical approach in the fight against insect-borne diseases, as well as the central role of the immune system.

Among problems that medicine had to solve in the development of efficient drugs against insect-born diseases, the visualization and identification of parasites were decisive steps. Here we find another connection between malaria and blue methylene, and to explore it we must move to Germany between the end of the 19th century and the beginning of the 20th. It is a period of great importance for modern medicine, with the birth of chemotherapy, also known as the theory of the magic bullet. This term is a metaphor for a single chemically and physically definite molecule able to influence our physiology in order to restore homeostasis—in other words, good health—and seek out specific disease-causing agents. Direct consequences of this approach are modern-day pharmaceutical research and the relevance of the body's immune response. A vital protagonist in this radical change, and a pioneer of the main basis of our medical therapy, was the German biochemist Paul Ehrlich (1854–1915), who received the Nobel Prize in Physiology or Medicine in 1908. Ehrlich shared the Nobel Prize with Élie Metchnikoff for their separate researches on the body's immune response (Winau et al., 2004; Gensini et al., 2007). On that occasion, Ehrlich explained his chemical theory about the formation of antibodies, which are our best way to fight the toxins released by bacteria, and therefore the most effective way to obtain a medicine against vector-borne disease etiological agents by serum antitoxins and chemotherapy. Ehrlich was initially interested in a method to fight malaria by killing the evasive microscopic organism responsible of the disease. However, at that time bacteria were still practically invisible and mysterious, and it is much easier to fight an enemy if you can see it. To understand how Ehrlich reached the result of evidencing the parasites, we must consider his scientific formation. An example of a pathway full of serendipity, but also rich of geniality and capacity of treasuring any experience.

Ehrlich was born near Breslau, on the continually disputed border, at that time in Germany and now known as Wrocław, Poland. After becoming a medical doctor at the local university, and experiences in Strasbourg, Freiburg im Breisgau, and Leipzig, in Breslau he worked at the laboratory of his cousin, a pathologist called Carl Weigert, where he was fascinated by the pioneering use of aniline dyes as biological stains. Later, via his investigations at the Charité Hospital in Berlin he was able to demonstrate that selected dyes react specifically with various components of blood cells and the cells of other tissues. These studies were useful to utilize dyes for therapeutic properties to determine whether they could kill off disease-causing microbes. Among the utilized dyes were blue methylene and pyrrole blue, obtained by the condensation of pyrrole with the tetramethyl-diammido-benzophenol. First, he obtained promising results using methylene blue to kill the malaria parasite. Two other pioneers already have opened the pathway for immunology and parasitology: Louis Pasteur, with the development of the first vaccines, and Robert Koch, with his works on tuberculosis and his subsequent cure with tuberculin therapy. As consequence of the first obtained results, Ehrlich was invited to work at Koch's Institute for Infectious Diseases in Berlin. However, it is necessary to consider that Pasteur's vaccine and Koch's tuberculin were obtained from weakened bacteria, and this approach, albeit the expectative at that time were enormous, is not universally useful for microorganisms, as indicated by the difficulties to develop a vaccine against malaria. Ehrlich's strategy was based on new serum therapies, or cell-free blood liquid, extracted from the blood of naturally or artificially immunized animals to induce immunity in mankind, developing the of "antitoxins" to explain the immunizing properties of sera.

During his research, Ehrlich was in search of a chemical explanation for the immune response. In his theory, living cells and dye molecules are able to interact because both have side chains, consisting of a shorter chain or group of atoms attached to a principal chain in a molecule. In the case of dyes, the side chains are related to their coloring properties and can link with particular toxins. A cell under attack by foreign bodies produces more side chains, in the tentative of locking in the foreign bodies before they are in condition to damage the host cell. Later, the "extra" side chains break off and become the defense antibodies of the immune system, circulating throughout the body in search of toxins. It was these antibodies that Ehrlich first described as magic bullets. The philosophy of the "magic bullet" is to find the most powerful and effective molecule by synthesizing new chemicals, using as a template those already discovered. Ehrlich's serum therapy was considered the ideal approach to contend with infectious diseases—an ideal mixture of the developing medical approaches of its time.

The great chemical-pharmaceutical companies were sensitive to Ehrlich's work, first by Hoechst and later by Bayer. The researchers, now including an organic chemist, Alfred Bertheim, and a bacteriologist,

Sahashiro Hata, broadened the targeted microorganisms to include spirochetes, which had recently been identified as the cause of syphilis. Some spirochetes are well-known to cause disease, including syphilis and Lyme disease. Spirochetes are long and slender unusual bacteria with axial filaments, which run along the outside of the bacterial protoplasma, but inside the outer sheath. Most of them contain a distinctive double-membrane (diderm). Usually, they are tightly coiled, and their form looks like miniature springs or telephone cords. Members of this group are also unusual among bacteria for the arrangement of axial filaments, which are otherwise similar to bacterial flagella. These filaments run along the outside of the protoplasm, but inside an outer sheath; they enable the bacterium to move by rotating in place.

Beginning the study with an arsenic compound, named atoxyl, 3 years and 300 syntheses later (an amazing large number at that time), the Ehrlich's team discovered in 1909 the drug Salvarsan (Riethmiller, 2005). Salvarsan was first tried on rabbits that had been infected with syphilis and then on patients experiencing the dementia associated with the final stages of the disease. Astonishingly, several of these "terminal" patients recovered after treatment. More testing revealed that Salvarsan was actually more successful if administered during the early stages of the disease. Salvarsan and Neosalvarsan, introduced in 1912, retained their role as the most effective drugs for treating syphilis until the advent of antibiotics in the 1940s.

Strange relations between insect-borne diseases and some pathologies

We have seen the relation between *Plasmodium* and our red blood cells, and now we are ready to join together genetic, dyes, and cartoons with insect-borne diseases. You may know of the strange cartoon creatures named Smurfs (French: *Les Schtroumpfs*; Dutch: *De Smurfen*; Italian: *Puffi*), which in the 1980s were popular in many different countries and languages, appreciated by children and their parents. They form a fictional colony of small, blue, human-like creatures. They are very nice and amusing little creatures, but something in their living style is quite strange. They live in mushroom-shaped houses in the forest with the psychedelic fungus fly agaric (*Amanita muscaria*) all around. They are virtually all males; there is just one female. The Smurfs are a population genetically separated without any outside sexual genetic breeding and this is an explanation for their unique color, which however is not a unique case. In the Appalachian mountains (USA), there is a human community quite separated from the rest of the population, and sometimes a baby comes into the word colored blue, just like the Smurfs (Trost, 1982). Similar cases were observed in other populations subjected to a limited hereditary genetic condition, such as the classic case of "Blue Fugates" or Blue People of Kentucky. However, if treated with blue methylene, the blue disappears rapidly and completely. The strange case is related to a genetic shift inside red blood cells, named methemoglobin (Fox, 1982). Methemoglobin is a form of hemoglobin, where the central ferric (Fe^{3+}) form of iron is substituted by the ferrous form (Fe^{2+}), impairing the affinity for oxygen of ferric iron and affecting hemoglobin functionality (Fig. 5.9). The binding of oxygen to methemoglobin is increased and the hem site to release oxygen is overall decreased. This is a disadvantage considering that hemoglobin is involved in the oxygen transportation method; therefore, when methemoglobin concentration is elevated in red blood cells, tissue hypoxia may occur. However, the changes in heme are once again profitable in the case of insect-borne disease, confounding the parasite. Methemoglobinemia affects about 400 million people globally. In 2015, it is believed to have resulted in 33,000 deaths and it is more present in certain parts of Africa, Asia, the Mediterranean, and the Middle East..

Another hemolytic syndrome, favism, can be treated with methylene blue (and quinine) and is related to malaria. Favism is a hemolytic disease affecting some people in particular, and it can arise from ingestion of certain pharmaceutical toxins, or broad beans (*Vicia fava*). The term "favism" is used to indicate a severe reaction occurring on ingestion of foodstuffs consisting of or containing broad beans. Most of the time, people affected by favism do not know that they have a metabolic problem, derived by a glucose-6-posphate dehydrogenase (G6PDD), which is an inborne metabolic error that predisposes red blood cells to break down. G6PDD is an enzyme contained in red blood cells and therefore carriers of the G6PDD genetic trait may be partially protected against malaria.

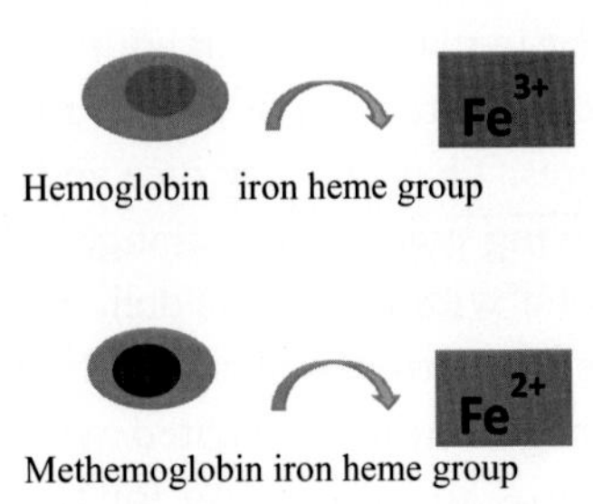

Fig. 5.9 The ionic change responsible for methemoglobin.

The symptoms, expressed within 6–24 h of the fava bean meal, consist of prostration, pallor, jaundice, and dark urine. These are the result of (sometimes massive) destruction of red cells (acute hemolytic anemia). The explanation is that fava beans are unique among other beans containing high concentrations of two glucosides, vicine and divicine, and their respective aglycones, convicine and isouramil. These compounds are powerful triggers of oxidative stress and are responsible for the characteristic hemolytic attacks.

Favism only occurs in people who have inherited the G6PD deficiency. However, awareness of the symptoms associated with favism was well-established long ago by experience, before the underlying mechanisms were understood with G6PD deficiency. The earliest reports are from Pythagoras forbidding his students to eat fava beans, and according to Diogene Laertius, even Pythagoras' death was caused by fava beans, when he was not able to cross a fava field, and in this way he was killed by his enemy. The strong aversion against Leguminosae by a rational mind, like that of Pythagoras, means that favism had already been recognized at that time as a dangerous disease with acute renal failure. This now confirmed by the common occurrence of G6PD deficiency in Greece.

Favism is more common and more life-threatening in children (usually boys) than in adults. Favism can be largely prevented by screening for G6PD deficiency and by education through the mass media or simply avoiding contact. For this reason, in supermarkets there is often advice concerning the presence and selling of fava beans. In a person who is G6PD deficient, favism can recur whenever fava beans are eaten, although whether this happens or not is greatly influenced by the amount of beans ingested and probably by many other factors. Once the attack is over, a full recovery is usually made.

It is 2000 years in Italy since we used to say "*repetita juvant*" (lit. "repeated things help" or "repeating does good"). Nobody speaks Latin any more (except in some cases in the Vatican), but some phrases are so true that they remain as roots to resist modernity asking for continuous changes. Some people affected by favism never have symptoms, and males are subjected to the disease more often than females. The reason for this difference is that favism is an X-linked recessive disorder, meaning that the responsible allele is located in chromosome X. Males possess only one X chromosome, inherited via the mother, and if the chromosome contains the gene that results in a defective G6PDD enzyme, the disease will be developed. In contrast, the presence of two X chromosomes in females allows a compensation that could mean they avoid the disease. The phenomenon is called X-linked recessive inheritance, meaning that the gene causing the trait or the disorder is located on the X chromosome. A mutation in a gene on the X chromosome causes the phenotype to be expressed in males (who are necessarily hemizygous for the gene mutation because they have one X and one Y chromosome) and in females who are homozygous for the gene mutation.

Several mutations of the allele that cause G6PD deficiency are found on the long arm of the X chromosome. Among them, two variants (G6PDA and G6PD) are the most common in human populations. G6PDA has an occurrence of 10% of Africans and African-Americans, while G6PD is prevalent in the Middle East. It is known as Mediterranean, being largely limited to people of Mediterranean origin (Spaniards, Italians, Greeks, Armenians, Sephardi Jews, and other Semitic peoples). Both variants are believed to stem a defense against malaria parasites. It is particularly frequent in the Kurdish Jewish population, wherein approximately one in two males have the condition, and the same rate of females are carriers.

The final consideration is that there are genetic solutions against malaria, but so far they are connected to important physiological dysfunctions. This is a biological rule: on the extreme sides of Boltzmann's curve, nature, through biological variation, provides lateral escapes, ready to be useful in case of future necessity. The study of these genetic situations is important and can be a starting point to develop new solutions.

However, methylene blue properties still have more to astonish us. Methylene blue was found also to be a potent antioxidant and nowadays there is a strong interest for its use in antiaging cosmetics due to its positive regenerating and protective action on skin cells. Coming back to the antimalarial drugs story, using methylene blue as a prototype, in 1934 chemists of the famous pharmaceutical company Bayer in Germany were able to synthetize resochin, which was effective but also quite toxic. Among the 1600 tested synthetic compounds during the Second World War, resochin, renamed chloroquine due to the presence of chlorine in its structure and its evident similarity with quinine, turned out to be the most effective, becoming a mainstay for malaria treatment despite its side effects. Chloroquine was therefore widely utilized from that time on, becoming the leading antimalarial drug. It was the template of a series of quinolone and quinoline antimalarial drugs of primary importance (Fig. 5.10).

Several current antimalarial drugs are utilized for the preventive treatment of malaria, named as prophylaxes, consisting of the ingestion of the appropriate dose before travel to the country where malaria is endemic.

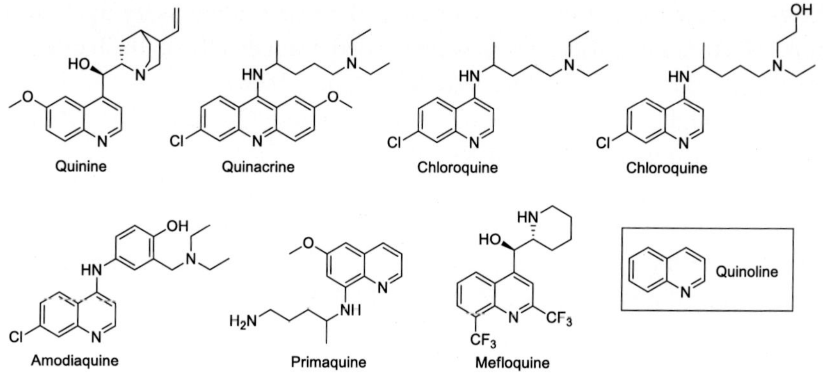

Fig. 5.10 Structures of chloroquine and related quinolone antimalarial drugs.

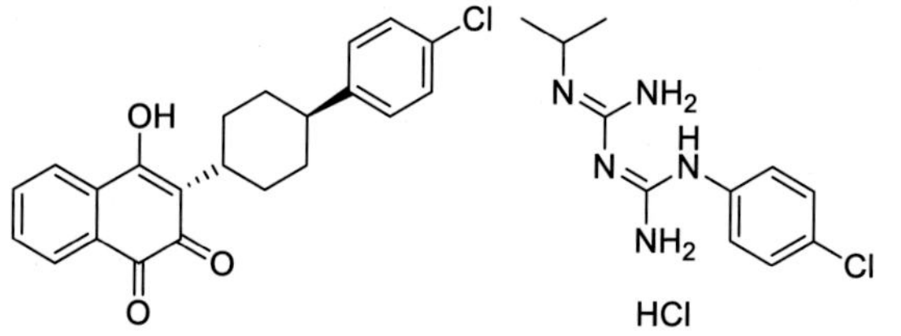

Malarone

Fig. 5.11 Structure of malarone (right) usually utilized with atavaqone (left).

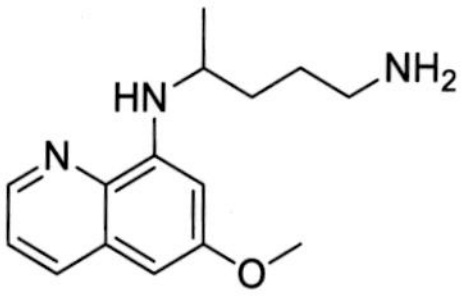

Primaquine

Fig. 5.12 Structure of primaquine.

Malarone (Fig. 5.11) merits a special mention, used alone or in association with other molecules, such as primaquine (Fig. 5.12) and mefluoquine. Malarone presents an ortho-quinone structure completed by an enol functional group, which can also be converted to chetone by a cheto-enolic conversion.

Ten years later, another war broke out, again in the East Orient: the Vietnam War. The North Vietnamese army was ravaged not only by the

USA military force, but also by malaria. Because soldiers were dying from this, effective antimalarial drugs were needed immediately on the battlefield. Again, quinine was difficult to find. With China being allied to North Vietnam, although not directly involved, the Chinese Chairman Mao Zedong ordered research about herbal remedies used in Traditional Chinese Medicine (TCM) against malaria. TCM is a treasure trove of medical indications, carefully described in many documents. The remedies of TCM need scientific validation, but they are considered a reliable starting basis for the production of pharmacological and medical drugs.

Many of them were utilized only because of the bitter taste, usually considered in antipyretic effects, but few contain active antimalarial constituents. The resulting list of plants was enormous, but among the hundreds of malaria treatments, the extract of sweet wormwood (*Artemisia annua*) emerged as very efficient in reducing fever. *A. annua* is a herbal species of the family Asteraceae. The term *annua* was assigned to this species since among the 400 species of the genus *Artemisia*, it is the only annual one. The genus is well-known for the species utilized in the production of liquors, such as *A. absynthium*, utilized for the production of absinthe; the same use also concerns *A. dracunculus* and *A. genepì*. The plant presents an erect brownish or violet brown stem, well-ramified with small leaves divided by deep cuts into two or three small leaflets, which are colored in lively green and characterized by an aromatic scent. The numerous tiny yellow-green flowers are arranged in loose panicles. Although the plant is common in China and temperate Asia, it can be found also in Europe, although its presence is discontinuous. In TCM, the infusion of the leaves is mainly reported as an antifever remedy, but the treatment of malaria is clearly evidenced in a text by the Chinese author Ge Hong (341 CE) and in a document found in a tomb in 168 CE. Later, it was time for science to take center stage. In 1972, the experiments on ether extract at low pressure of the leaves fully confirmed the toxic activity against *Plasmodium* and the Chinese researcher Tu Youyou isolated artemisinin also from two other *Artemisia* (*A. apiacea* and *A. leucea*) (Tu, 2011). One year later, thanks to the scientists at the Yunnan Institute of Materia Medica (China) and Shandong Institute of Traditional Medicine and Materia Medica (China), the active crystalline ingredient was obtained and the structure determined.

The compound was named "huanghaosu" or "huanghuahaosu," and later renamed "qinghaosu," but universally it is known as artemisinin (Fig. 5.13). Artemisinin was immediately considered an alternative to chloroquine, due to increasing resistance to chloroquine and related drugs.

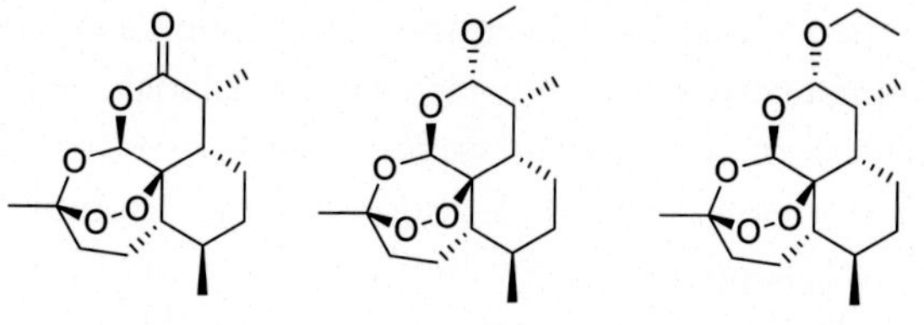

Artemisinin Artemether Arteether

Fig. 5.13 Structures of artemisinin and related compounds; note in particular the presence of the peroxide group and the number of carbons linked to two oxygens.

Animal trials on rodent malaria parasites with the crystals achieved excellent results in efficacy, toxicity, and safety, though the trials indicated problems concerning its solubility in water. In the plant extract, several other constituents help the solubilization, but the solubility changes for artemisinin, in case of its pure crystalline form, in consideration of the apolar basic structure and the presence of few active groups.

Artemisinin is a molecule completely different from quinine, chloroquine, and related compounds. The novelty is that this is not an alkaloid, completely debunking the theory that nitrogen is necessary for the activity. In fact, the bacteria do not possess a nervous system, although the neurologic side effects are of course related. Furthermore, whereas *Cinchona* is a subtropical and tropical genus, *Artemisia* is typical of temperate regions. Although most people consider the plant as Chinese, *A. annua* is commonly present in Europe, and other species of the genus are well-known for other reasons, such as *A. absinthium* for the famous liquor. The *Artemisia* genus belongs to the Asteraceae (formerly Compositae) family, wherein alkaloids are rare and terpene and phenols dominate in the secondary metabolism.

Artemisinin is a sesquiterpene lactone with unusual functionality, as tested by the presence of an extraordinary concentration of oxygenated functions, including a rare endoperoxide bridge, which is essential for its antimalarial activity. Chloroquine is mainly a flat etherocyclic molecule with a hyperconjugated center, based on two moieties. In contrast, artemisinin has a complex compact tridimensional structure full of chiral centers. However, its most interesting function, the peroxide, constitutes its Achilles heel. In a terpene, based on a hydrocarbon skeleton, the presence of oxygen is a sign of reactivity and instability at the same time. The oxygenation of the terpene skeleton must be considered a peril for the stability of the natural product, but it enhances its reactivity. Being very reactive, artemisinin is also unstable with a half-life of about 1 h, and this is a problem in treatments and precludes the use for malaria prevention. Artemisinin's discovery represented the advent of a completely new class of antimalarial agents, and, considering the success of artemisinin, several derivatives were proposed to improve its performances, like bio-disponibility and solubility in water or oil. The carbonyl group in artemisinin was reduced to obtain its derivatives, such as the water-soluble artesumate and the oil-soluble artemether in 1987, and dihydroartemisinin (DHA) in 1992. These derivatives showed a great antimalarial activity, and therefore they were introduced in the most important antimalarial drugs. However, again the natural source is easy and economic. Considering that the commercial source of artemisinin is still from the *Artemisia* plants, yields vary significantly depending on growth and habitat conditions, and various attempts to select high-producing cultivated plants or to obtain the drug using bioengineered microbes have been reported.

Considering the clear differences with quinolones, novelties were expected also regarding the mechanism of action of artemisinins. Also in this case, there are some aspects that are not clear. According to some researches, the drug acts by lowering the energetic cell source represented by the mitochondria. Metabolism of artemisinin and its derivatives is believed to be mediated primarily by the liver cytochrome P450 enzyme CYP2B6. This is in accordance with the particular oxygenated structures of artemisinins and the potentiality of carbon-centered generating free radicals or reactive oxygen species (ROS), since they are in situ activated to react.

Usually, free radical forms of oxygen are considered toxic, and several plant drugs containing phenols are used as antioxidants and antiROS. However, in this case the particular endoperoxide bridge characteristic of the trioxane pharmacophore of artemisinins is essential for their antimalarial activity. When the peroxidic oxygen is replaced with a carbon (e.g., 1-carba-10-deoxyartemisinin) the derivative is devoid of antimalarial activity. However, endoperoxides, being rich of oxygen, are dangerous for the stability of the compound. In fact, we are in presence of a molecular paradox: too activity is important to attack the parasite, but it is also a risk for the stability of the compound. This is an advanced version of the magic bullet theory, which is called by pharmacologists a *pro-drug*, meaning that it does not act in the original structure but becomes active by transformation after administration of the drug. The molecule is able to pass several passages of demolition maintaining intact its structure and only in presence of the target generate its active form. This is an aspect typical of several natural products, practically designed to obtain such results. Through this functionality, the compound should modulate parasite oxidative stress and reduce the levels of antioxidants and glutathione (GSH) in the parasite. As confirmation, artemisinins autoinduce P450 metabolizing enzymes, resulting in lower serum concentrations of the drugs in subsequent administrations. As peroxides are known sources of ROS, this finding has inspired the design of the next generation of antimalarial endoperoxides, including a number of derivatives such as the two bioactivation characters previously reported. The idea, on the basis of the synthesis of the derivatives, is join the oxygen-generating ring of artemisinins (Fig. 5.13) with the reductive scission models binding of low-valent transition irons (ferrous heme or nonheme, exogenous Fe^{2+}). The open peroxide model suggests that the ring opening of artemisinins may be driven by protonation of the peroxide or by complexation with Fe^{2+} electron transfer inducing reductive scission of the peroxide bridge to produce an O-centered radical, which self-arranges to generate a C-centered free radical.

Once formed, artemisinin-derived free radicals cause damage to cellular targets in their vicinity through alkylation. However, to obtain the oxidation of the erythrocyte membrane and consequent damage of the parasite, high concentrations of artemisinins (or chloroquine quinolones) are necessary and this is an open door for the insurgence of the resistance phenomenon. Using the membrane channels, the *Plasmodium*-resistant population of the erythrocytes is able to extrude part of the antimalarial drug, lowering its concentration to nontoxic levels. The mechanism is well-known, as experienced for several medicinal drugs, in particular in the case of anticancers. Furthermore, high doses of antimalarial drugs in the attempt to enrich the therapeutic doses increase the collateral effects. In the case of artemisinins, the limited tendency of these radicals to damage selectively cellular targets for intramolecular reactions, like typical alkyl agents, must be considered, and some data are in favor of artemisinin-derived radicals low toxicity.

A proposed mechanism concerns the classic heme hypothesis, the same already proposed in case of quinolone antimalaric drugs. Evidence supports the hypothesis that quinoline antimalarials act on the digestive vacuole of the parasite, interfering in polymerization of heme units to form hemozin by ferriprotoporphyrin IX. Quinine prevents the hemozoin crystals from growing by intercalating the quinolone rings between the aromatic groups of the ferriprotoporphyrine molecules. Hemes, being produced while breaking down hemoglobin in human red blood cells, are toxic to the parasite in high concentrations. Therefore, the mechanism of action considers that quinine binds to heme by the formation of an intramolecular hydrogen bond, interfering with hemozoin formation, thus leaving toxic ferriprotoporphyrin IX heme to thwart the parasite.

However, quinolones and artemisinins are completely different in their structures and the heme hypothesis has also been challenged. In fact, not all pharmacological data are congruent. For instance, the ability of mefloquine to affect hypnozoites is difficult to explain solely by prevention of heme formation. Mechanisms of action have been suggested, such as inhibition of vacuole-vesicle fusion or binding to essential proteins in the parasite. It was observed that during the development of parasites inside erythrocytes, the fusion of digestive vesicles occurs, giving rise to a large digestive vacuole. Instead of the mechanism preventing the heme's sequestration into hemozoin, another explanation of antimalarial drugs may be related to oxidative damage of digestive vacuole macromolecules and membranes, blocking its action. The digestive proteolysis is realized by proteases that contribute to hemoglobin breakdown, as well as other DV-associated proteins. Therefore, the digestive vacuole is important for the parasite, allowing the parasite to process 60%–80% of the erythrocyte hemoglobin. In this way, the *Plasmodium* cells can utilize the pool of amino acids, which are necessary for their survival and growth. Vacuoles act as degradative organelles, thanks to their low pH value. The membrane of the organelle works by ion pumps and transporters that maintain its low pH. The study of this organelle's functionality has been pivotal in the development of parasite resistance to several antimalarial drugs. In a similar way, the activation of artemisinins should be the result of accumulation of intraparasitic heme, which may also explain the selective toxicity of artemisinins and related trioxanes toward malaria parasites. In this regard, it is necessary to recall the deep structural differences between artemisinin and quinolone alkaloids. The gradient of pH values of the cytoplasm of red blood cells is around 7.4; instead, inside the parasite food vacuole the pH reaches 5.5, determining a further protonation of chloroquine. In the degradation process of hemin, as already described, the pH of the food vacuole plays a fundamental role. Most of the degradative enzymes are optimally active at pH 4.5–5.0, which probably allows the efficient proteolysis of hemoglobin, but also ensures the chloroquine's efficacy.

Before continuing the story of the mechanism, we must remember that chloroquine, as with other antimalarial quinolones, is an alkaloid. The word "alkaloid" derives from alkali, meaning a basic reactivity, able to change the pH of an aqueous solution in favor of oxydrilions (OH^-) against hydrogenions (H^+). The word originates from Arabic *al-qaly*, meaning "ashes from the saltwort," the ash that is obtained from plants living in the sands of salted waters, like Salicornia. True alkalis are the basis of the elements of I and II group of the Periodic Table, such as K, Na, and Ca, which on reacting with water give K(OH), Na(OH), $Ca(OH)_2$, etc., whose reaction produces directly OH^-. In organic compounds, the presence of these metallic elements in great quantities is forbidden, and basicity is obtained by a different reaction. The presence of N linked to carbon(s) inside the molecule allows the reaction of the lone pair with hydrogenions, which are therefore subtracted to the aqueous solution in favor of the OH^-. However, the basicity of such azotated compounds is not efficient like the inorganic ones, therefore they are called alkaloids, similar to humans and humanoids. Furthermore, the basicity is strongly influenced by the availability of the two electrons of the lone pair, which is a consequence of the remaining part of the molecule.

Chloroquine (Fig. 5.14) has three N, at neutral pH of the N of the side chain, with a free lone pair; it is easily protonated (CQ+), whereas at acid pH even the N integrated in the quinolone ring is protonated (CQ++). This change affects the membrane permeability of chloroquine and related compounds. The alkaloid chloroquine is a lipophilic weak base, and it is able to pass through biological membranes in the uncharged form (Fig. 5.15). Once inside the vacuole acidic compartment, chloroquine is protonated and

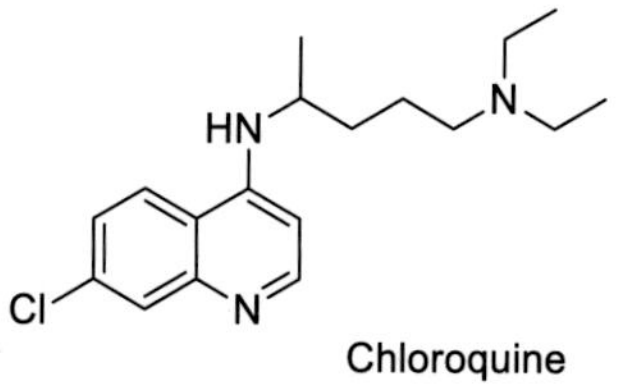

HN N Cl N Chloroquine

Fig. 5.14 Structure of chloroquine.

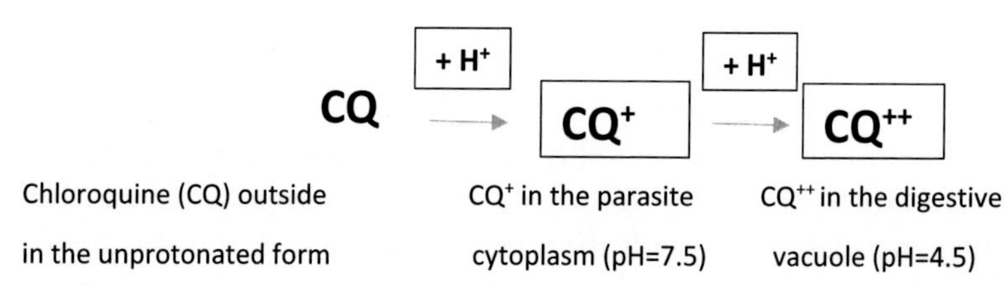

Fig. 5.15 The change of pH between the cytoplasma and the vacuole of the erythrocyte generate a double protonation of chloroquine and consequent trapping of the alkaloid.

trapped because the protonated base is relatively impermeable. The concentrative uptake of the drug means that the acid proteases of the parasite no longer function effectively. However, the parasite actives respond to restore the extrusion of chloroquine by the mechanism that we call Erythrocite membrane resistance.

The parasiticidal effects of artemisinins are effective on early ring-stage malaria parasites with little hemozoin, as well as observed in *Babesia* and *Toxoplasma* parasites that do not form hemozoin. Other models may be compatible on the basis of iron-dependent generation of ROS, calling for a role of the C- or O-centered radicals with the cellular redox systems or lipids. The explanation of the antimalarial activity is suggestive, being related to the peroxide bridge typical of artemisinins structures.

In this case, the high energy forms of oxygen, so far considered dangerous and negative, could exert positive effects in situ, revealing a reverse of the dominant axiom, but this is perfectly coherent with the Greek ancient term of a drug, *pharmakon*, meaning both medicine and poison. The meaning of this paradox is that something devoid of physiological effect is useless, but too much activity must be carefully utilized.

The future

The results of the fight against malaria are fluctuating. After a period of great success against the disease, by the global spread of DDT and antibiotic availability, malaria restarted when this pesticide was prohibited due to its environmental damage. The wide use of chloroquine has also been crucial in limiting malaria incidence, but over time resulted in high levels of drug resistance by the malaria parasite *Plasmodium*, particularly the most prevalent type, *P. falciparum*. Resistance increased every year, and nowadays it is considered that more than 50% of this parasite infecting people is resistant to treatment with chloroquine. The potential development of resistance to artemisinins by *P. falciparum* is a subject of close surveillance and extensive research. Studies at the Thai-Cambodian border, a historical epicenter of multidrug resistance, raised considerable concerns on resistance development, due to reduced susceptibility to artemisinins by prolonged parasite-clearance times. The future scenario is that all available antimalarial drugs will become ineffective, with the partial exception of quinine treatment, which has much less parasitic resistance around the world. However, quinine-resistant strains of *P. falciparum* were also reported, although in very limited numbers compared to other antimalarial drugs.

Although global incidence is still at high levels, in recent years malaria infection rates have been falling worldwide, with special reference to Sub-Saharan Africa, thanks to a general approach involving an increase of hygiene, environmental control, nutritional care, and other factors, in addition to drug treatment. However, several issues could reverse this positive tendency. Besides the already named resistance, other social factors such as human migration, social and political instability, shortage of food and other supplies, together with exponential increase of population must be considered. A number of malaria prevention and control tools currently available in advanced countries are becoming prohibitively expensive, and are thus not readily available for poor and marginalized populations in tropical and sub-tropical areas worldwide.

In conclusion, as in 2015, new chapters were added to the secular struggle against malaria, such as the Nobel Prize assigned to Y. Tu for the discovery of artemisinin and some advances in vaccine development. This was a recognition of the importance of the research against malaria and the need to explore new drugs, including as a priority natural products.

In any case, the number of people affected by malaria is still very high and there is a real need for new approaches to boost malaria prevention and control. On the drug side, after chloroquine, parasite strains resistant even to artemisinin have been detected, with quinine still remaining the only effective, but not decisive product. Furthermore, the RTS,S vaccine does not offer protection against *Plasmodium vivax* malaria, which predominates in many countries outside of Africa. Recently, a massive campaign of vaccination was undertaken in Mozambique, including also neighboring countries, and this will be an important test of the efficacy of this approach.

The fight against malaria

In summary, the malaria story could be at a sensitive point. Throughout history, it was possible to rely on several, but limited in number, efficient antimalarial medicines as powerful tools in malaria control. Drugs using *Plasmodium* as their target have developed different modes of action, developed to find the weak point of the microbe. Several successes were obtained, but only partially. Over the past 10 years, there has been a substantial regress of the enormous burden of malaria globally using a strategy involving several approaches in treatment of the disease: (a) improvement of living conditions and hygiene, including possible prevention measures; (b) introduction of new strategies of control, like insecticide-treated bed nets; (c) improved access to early diagnosis; (d) introduction of drugs combination therapy; (e) limitation of reproduction and diffusion of the vector; and (f) production and validation of vaccines.

The financial support for specific current programs on malaria highlights the production of effective vaccines, new antimalarial agents, and other novel control interventions. However, the key current challenge concerns the resistance phenomenon, based on the spread of parasite strains that are resistant to multiple antimalarial drugs. If this challenge is lost, the scenario of an exponential increase in morbidity and mortality in many malaria-endemic countries could be expected. After the development of resistance to medical drugs from the chloroquine family, artemisinin was considered a significant new hope. The key of any medical drug, even if positively validated for centuries in the perfect tests in thousands of human subjects, exactly as in traditional medicine, is the elucidation of the mechanism of action. The definition of the action of a drug is not only the necessary tool against empiricism, but i is indispensable to improve the medicament and avoid toxicity and side effects. However, it is not easy. In the knowledge of the dimension and characters of the concentric and interactive worlds, that still are not sufficiently explored, the most important one consists in the molecular level. It is also very difficult for us to understand. Nobody was able to see a molecule, or even indirectly perceive its presence. We need to utilize complicated devices to catch molecules and only after a complicated process of separation. It is necessary to imagine the pathway of the molecule after the entrance in our body, consisting in a myriad of dangerous and misleading molecular meetings to enter finally the right place and interact with the exact molecular target. It is like finding a needle in a haystack. However, it does work and is an irresistible challenge for human nature, probably the most important, as it is related to health. In this case, we have to face the multiform organism *Plasmodium*. There may be a lot to investigate about the mechanism of action of antimalarial drugs and consequent resistance.

Although reported cases of malaria are in decline and most of them evidence a static endemic situation, the disease cannot be underestimated. Firstly, although the general aspects are as described, the toxicity is highly variable, since the vector and parasite can be different. One percent of symptomatic infections may become complicated and develop into severe malaria. Severe malaria cases may manifest as anemia, hypoglycemia, metabolic acidosis, repeated seizures, and/or multiple organ failure. Other possible symptoms are jaundice, shock, renal failure, lactic acidosis, abnormal bleeding, pulmonary edema, and adult respiratory distress syndrome.

Cortical infarcts and cerebral venous or dural sinus thrombosis as part of disordered coagulation may also develop. Bacterial coinfection may be observed, particularly in those with shock, and accounts for the majority of late deaths, but the most delicate target is the CNS.

Malaria is still a huge threat for public health and socioeconomic development in resource-limited settings of the world. Data appear positive. Malaria mortality rates have fallen by 47% globally since 2000, and by 54% in Africa. Most deaths occur among children living in Africa, where a child dies every minute from malaria, but this is also decreasing. However, the trend could be reversed due to the resistance phenomenon.

To deal with the serious situation caused by resistance, scientists are exploring new strategies, first repudiating the pharmaceutical axiom based on a single active molecule, embracing the philosophy of the multicomponent/multitarget. In the case of artemisins, artemisinin-based combinatory therapies (ACTs) have been introduced and widely deployed in malaria-prone regions. An example is the utilization of armethere in combination with lumefantrine. From the comparison of the structures of the two compounds, we can argue that whereas armethere is a sesquiterpene derived from artemisinin, lumefantine (also known for its use as coarthemeter) is a very different compound (Fig. 5.16). Therefore, there is a mixture of antimalarials of first and second generation.

Current and future research strategies to eradicate malaria can follow different approaches, including the following:

(i) The classic pharmacological treatment based on the active molecule that, according to the "magic bullet" paradigm. This "magic bullet" is a chemically defined molecule with a pharmacologically known mechanism and target, the activity of which is confirmed by clinical trials. It can be of synthetic or natural origin, albeit the former is usually

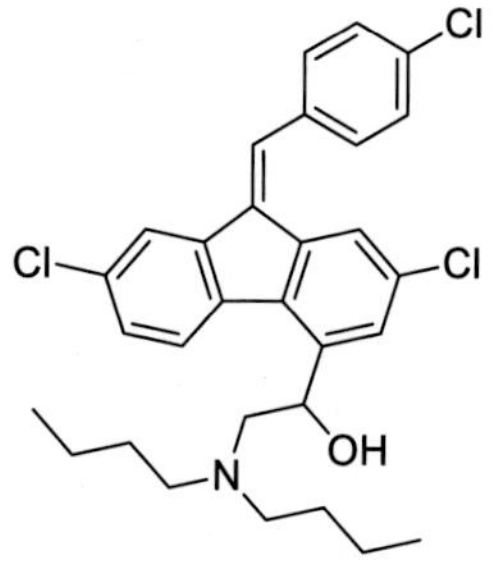

Lumefantrine (benflumetol or co-artemether)

Fig. 5.16 Structure of lumefantrine.

preferred by pharmaceutical industries. So far this approach, as for other diseases, is inevitably leading to resistance by the parasite.

(ii) The use of natural products, like quinine and artemisinin, as templates for production of a range of active molecules. The possibility of finding useful natural products is high, but there is no valid strategy to select the right molecule. So far, results have been obtained mainly using ethnopharmacology indications, but this requires careful scientific validation and could need tremendous screening.

(iii) The development of a vaccine, announced several times and so far still in progress. The vaccine hypothesis seems quite difficult, and so far insect-borne diseases have not been totally solved by vaccines.

(iv) Searching, or waiting, for natural mutations of the parasite into less aggressive forms, making the host tolerant. This involves more or less just hope, and the target is still argued, in terms of whether to try to change the parasite and/or the vector.

(v) The control of mosquito vectors, mainly based on insecticidal treatments against mosquito larvae, with serious impacts on the environment and human health.

So far, no single strategy can be considered preferable, and the mixing of more strategies is considered prohibitively expensive.

Exploring the future, some other considerations are necessary. Endemism is a typical characteristic of malaria. Thus, despite eradication in the developed parts of North America and Europe, if you are born where malaria has been endemic for at least hundreds of years, there are two possibilities: you will be a victim of the disease, like most patients 5 years or younger, or you will learn to live with the disease for the rest of your life.

Another consideration is also necessary, though it can be considered apathetic, since it is not completely respectful of the human life. The number of people regularly killed by the insect-borne diseases is so relevant that in case of sudden absence of such mortality, the quantity of new survivals will change drastically the number of inhabitants for several countries, causing enormous problems about the resources of water and food, with enormous pressure and instability. The inherent cause is that this epidemiologic situation is the consequence of a complex interplay of multiple factors. These include the mosquito vector, like density, ambient and environmental conditions, population movements, altitude, parasitemia rates among endemic populations, and the anopheline mosquito species. Others are related to the parasites, like resistance and virulence of strains.

Viral insect-borne diseases

In Table 5.2, we can see that insect-borne diseases caused by viruses are prevalent against the bacterial ones. This could be surprising considering the extreme simplicity of viruses structure against complexity and capacity of bacteria, but evolution and functionality must not always be in accordance, as evident in the current presence of old forms of organisms. As is well-known, viruses are very different from bacteria and are not considered as living organisms, though many aspects of their reactivity are similar, as evidenced by insect-vector diseases. Usually, in case of infection, people use to seek help in the antibiotics, ignoring the fact that they are not useful against viruses, also as consequence that the pool of efficient antiviral drugs is very limited.

This part of the book is focused on a special actor, very clever and able. It is simply a molecule but one that is able to undergo the processes already encountered, and find vectors and hosts to organize the superorganism strategy. This is an example of the absence of fundamental differences of behavior between organisms and other molecular matter. The most important difference is that the virus strategy is based on patience and capacity to wait, hidden deep inside the body of the host without any evidence of its presence, until the moment it strikes. Usually, the symptoms are not evident and the disease generally does not involve the death of the host, which would be negative to the infected agent and necessary for the agent proliferation (Table 5.5).

Viruses involved in insect-borne disease are a special category, coming from a long selection. They are classified as arboviruses, consisting of a heterogeneous group of viruses transmitted by a hematophagous arthropod vector. Ticks, sandflies, and mosquitoes are their vectors, and they predominate in tropical regions. The most well-known and severe arthropod-vectored diseases are caused by an arbovirus, like yellow fever, *togaviridae*

Table 5.5 Classification of *Flavivirus*.

Unranked	Virus
Realm	*Riboviria*
Phylum	*Incertae sedis*
Family	*Flavivirus*
Type species	Yellow fever virus

(equine encephalitis) chikungunya virus, and *bunyaviridae* (California encephalitis and Bunyaviral hemorrhagic fever). However, we must assign a special place to *flaviviridae* (dengue, yellow fever, St. Louis encephalitis, and West Nile encephalitis). The difference in viruses concerning vector-borne disease consists of the behavior inside the host. Human infections with most of these arboviruses are incidental, as humans are unable to replicate the virus to high enough titers. Humans are then a dead-end host, enabled to re-infect the arthropods needed to continue the virus life cycle. Yellow fever, dengue, and Zika viruses still require mosquito vectors, but are sufficiently well adapted to humans not to depend upon animals. Viruses are not considered organisms, but the arboviruses' behavior is very similar to that described in other vector-borne diseases. We may consider this as an example of evolution without selection.

Flavivirus is a genus of viruses in the family Flaviviridae. The name is derived from yellow fever, which in turn is named for its propensity to cause yellow jaundice in victims. *Flavi* or *Fulvi* (indicating fair hair) is the historical name of a dynasty of blonde Roman emperors, as reported by several historians. Suetonius described Augustus as having "*subflavum*" hair, and "his complexion was between dark and fair" (*Augustus*, 79). Nero also had "*subflavo capillo*," in contrast to the typical Mediterranean style. Wilhelm Sieglin compiled a list of 27 blond-haired Roman gods and goddesses, 10 blond Roman heroes and heroines, and 63 blond Roman historical figures, many of whom were Patricians. This could confirm some genetic theories based on frequent use by the Roman Patricians of names such as *Rufi*, *Flavi* or *Fulvi* (indicating yellow hair), and *Caesulla* or *Ravilia* (indicating light eyes), demonstrating strongly Nordic racial affinities and genomic influences.

Essentially, a virus is a molecule of nucleic acid inside an envelope. The general structure of the cell is the same, consisting of concentric spheres. The virus core is an RNA or DNA molecule, which is the active part, consisting of only 10,000–11,000 bases (compared to our 3 million pairs of bases). The virus is enclosed in a protein coat, named a capsid. The capsid is surrounded by a glycoprotein envelope, wherein surface proteins are inserted, helping the core to perceive the environmental changes and attach to its host cell. Dimensions are ultramicroscopic (20–300 nm in diameter, much smaller than a bacterium) and therefore the virus pertains to the molecular world. Its shape can be very varied and complex (icosahedral, helical, etc.). The main difference to bacteria and other organisms based on cells is the absence of growth: viruses only exist to make more viruses. A virus is a very efficient molecular mechanism, dedicated to the multiplication of its molecule. The virus particle attaches to the host cell's external envelope before penetrating it. The virus then uses the host cell's machinery to replicate its own genetic material, obtaining a copy of its genome. As the newly formed viral particle pushes against the host cell's plasma membrane, a portion adheres to it. The plasma membrane reacts and envelops the virus, but the cell membrane becomes the viral envelope. The virus core is then released into the cell. This process slowly uses up the host's cell membrane and metabolism, and usually leads to cell death. The virus particles burst out of the host cell into the extracellular space, resulting in the death of the host cell. Once the virus has escaped from the host cell, it is ready to enter a new cell and multiply, causing an increase of the infection. Viruses are not able to obtain replication alone, since they are metabolically inert and therefore all viruses are obliged to be infectious agents.

Therefore, in conclusion there are no good viruses, since they replicate only within the cells of living hosts, causing cell damage. Once replication has been completed, the virus particles leave the host by either budding or bursting out of the cell (lysis).Viruses attack any form of living organisms, including bacteria, and their subunits, like white blood cells.

The idea considering the viral insect-borne diseases limited by the simplicity of viral structure, mainly consisting of a nucleic acid material, can be misunderstood. This molecule is able to perform several activities that are considered peculiar to living organisms. Viruses in *Flavivirus* are linear positive-sense RNA enveloped in a capsule with icosahedral and spherical geometries. There are 32 Flavivirus species recorded as virulent for mankind, including West Nile virus, dengue virus, tick-borne encephalitis virus, yellow fever virus, Zika virus, and several other viruses which may cause encephalitis. They are classified as arboviruses, since transmission is a consequence of the bite from an infected arthropod (mosquito or tick). The viruses can infect several hosts, with important animal transmission routes, but humans are considered dead-end hosts, being unable to replicate the virus to high enough titers to re-infect the arthropods needed to continue the virus life cycle. However, there are important exceptions in yellow fever, dengue, and Zika viruses. These three viruses still require mosquito vectors, but are sufficiently well adapted to humans not to depend upon animal hosts. Therefore, again there is concomitant participation, potentially involving transmission from nonhuman vertebrates to humans with an intermediate vector arthropod. The known nonarboviruses of the *Flavivirus* family reproduce in either arthropods or vertebrates, but not both, with one odd member of the genus affecting a nematode. As a consequence, *Flavivirus* utilizes different strategies to infect the host and replicate. In addition to the codification of viral proteins, inside the genome virus consisting in its RNA, there are special signals that are able to activate counteractions regulating amplifications of viral components. In the adaptation to the type of host, special roles have been assigned to the parts of the RNA present in the non-codifying region of the *Flavivirus*. As a confirmation, dengue virus is able to adapt to different hosts, like mosquitoes or human cells, by selection of different viral variations. The adaptation to the host is performed by viral structures present in the 3'UTR region, which is associated to improvement of the so-called "viral health." Therefore, replication of the elements of the viral RNA changes according to the different involved cellular types. In some aspects, this can be considered an adaptation, like the metamorphoses described earlier. That means that there is no absolute need for the cell envelop to obtain some results by the nucleic acid. Other virus transmission routes for arboviruses include handling infected animal carcasses, blood transfusion, childbirth, and consumption of infected food, such as unpasteurized milk products.

We can now consider the characters of each viral insect-vectored diseases, always remembering that inside any infection there is a complicated mechanism.

Dengue

Dengue is another of the most important mosquito-borne diseases in the world (Hong-Juan et al., 2012). Again, it is spread in tropical and subtropical regions, but the pathogen is an arbovirus, which is the general name assigned to viruses affecting mankind. In this part, we shall encounter arboviruses several times. Arboviruses are the main agent of several arthropod-borne diseases. The term "arbovirus" is self-explicative of this intimate relation, since the "ar" refers to arthropod, and "bo" refers to borne. In the arthropods, we include mosquitoes, ticks, fleas, and gnats. In the main classifications, arboviruses are divided into *Flavivirus*, *Alphavirus*, and *Bunyavirus*. *Flavivirus* are single-stranded, enveloped RNA viruses. These viruses are all transmitted by either ticks or mosquitoes. West Nile virus, Zika virus, and yellow fever are examples of mosquito-borne flaviviruses. *Alphavirus* also includes single-stranded RNA viruses, but they have a coated virion, which is the infective active component of the virus. *Alphavirus* is transmitted through mosquito bites and is responsible for chikungunya, Ross River virus, and both western and eastern equine encephalitis diseases. *Bunyavirus* includes enveloped RNA viruses, which can be transmitted by arthropods or rodents. We shall focus on the ones that are transmitted by arthropods, including ticks, mosquitoes, and various types of flies. The diseases attributed to *Bunyavirus* include Jamestown Canyon virus and California encephalitis.

Arboviruses are enveloped. The virus envelopes protect the inner part, which contains the capsid with the viral genome. Envelopes are chemically constituted by phospholipids and proteins, derived from cell membranes, as well as some viral glycoproteins. Thanks to the envelope, viruses try to avoid the host immune system, and glycoproteins on the surface of the envelope serve to identify and bind to receptor sites on the host's membrane. In this way, the viral envelope fuses with the host's membrane and defeats the host cell's external defenses, and the capsid and viral genome can enter, causing infection. Once inside, the virus can locate the molecular resources necessary for its replication.

Therefore, most vector-borne diseases are consequences of the ability of a virus to infect and diffuse inside a host cell. Despite this astonishing capacity, the arbovirus is a very simple molecular machine. Simplifying, we have a protein envelope and a nucleic acid core, plus some enzymes. If you want to control a virus, you must mainly target on the external barrier, the envelope. Without this protection, the virus can be easily attacked by the immunity system or a vaccine. Current researches on viremia, such as Zika or Ebola, are focused on this aspect. The idea is that when deprived of its protection, consisting in the involucre, the virus can be defeated by our natural defenses, without any other external intervention. In such approach, sustain the general health state of the patient is fundamental.

In the case of dengue, the viruses belong to the genus *Flavivirus*, whereas the main vector is *Aedes aegypti*, followed in a lesser degree by *Ae. albopictus*. The incidence of dengue, in contrast to the situation of malaria, has grown dramatically around the world in recent decades. The main difference between dengue and malaria diseases is that a virus is responsible for the infection and the illness. Other differences must also be considered, however. Dengue has emerged as a worldwide problem since the 1960s, and is reported as endemic in more than 100 countries in Africa, America, and the Eastern Mediterranean, but South-East Asia and Western Pacific are the most affected by the disease. Although mainly hot countries are impacted, including tropical and subtropical areas, the disease is common in many popular tourist destinations in the Caribbean (including Puerto Rico), Central and South America, South-East Asia, and the Pacific Islands.

In the United States, local spread of dengue occurs infrequently. Dengue is the most common and important worldwide viral disease transmitted by mosquito vectors. No vaccines are available and therefore vector control with insecticides remains the most important tool for prevention and interruption of the transmission of this disease.

Anyone who lives in or travels to an area with risk of dengue is at risk of infection. Before you travel, you should locate country-specific travel information to help you plan and pack. Dengue is considered a typical disease for passive dispersion, meaning the spread of the vectors and dengue virus from one region to another through means of transport. The increase of tourism is one cause of diffusion of vector-borne diseases, and in such cases, most of the classic considerations about insect-vectors must be totally revised.

Until 2005, the reported data considered about two-fifths of the world's population to be at risk of acquiring dengue. However, the numbers of dengue cases, as for other insect-borne diseases, must be considered only an indication, since many situations are underreported and misclassified. Reported data tell us that more than 2.5 billion people, corresponding to more than 40% of the world's population, are now at risk of dengue infection and there are 50–100 million cases of acute febrile illness yearly, including about 500,000 cases of severe dengue. The incidence of dengue has grown dramatically around the world in recent decades. According to WHO current estimations, there may be 50–100 million dengue infections per year, but this number is the medium of an estimate interval 284–528 million, of which 67–136 million are manifested clinically, considering any severity of disease, 0.5 million are hospitalizations, and 22,000 deaths occur worldwide every year. Another study estimates that 3.9 billion people, in 128 countries, are at risk of infection. However, it is necessary to remember that the risk of infection means a potentiality, and the disease could not be expressed in evident symptoms.

We know that a recovery from infection provides a lifelong immunity for the subject against that particular serotype. Cross-immunity to other serotypes after recovery is only partial and temporary. Currently, there is no specific treatment for dengue, although a vaccine is currently being developed. Its prevention and control depends solely on effective vector control measures. However, again the overuse of insecticides has generated resistance to several insecticides by the vector.

The agent

The disease is caused by a single stranded RNA virus of genus *Flavivirus*, comprised of four closely related but antigenically distinct serotypes, named DENV-1, -2, -3, and -4. Viruses are not considered living organisms, but the rules of homeostasis are completely respected. They are most efficient in reproduction, this being the unique pursued aim of these molecular machines. In fact, we must consider that organic matter, inside an organism or not, is subjected to the same general physical and chemical factors acting in the environment. Dengue is classified in different illnesses, such as undifferentiated fever, dengue fever, dengue hemorrhagic fever, and dengue shock syndrome, with various degrees of symptoms and gravity of manifestations, including acute fever, symptoms of headache, retro-orbital pain, myalgia, arthralgia, and rash. Typically, dengue illnesses begin 5–7 days after the infection, as the consequence of a blood meal from an infected mosquito. Within the vector, the virus infects the mosquito mid-gut and subsequently spreads to the salivary glands over a period of 8–12 days. Therefore, after the bite of the infected mosquito, there is an incubation period of 3–14 days before symptoms appear. After this incubation period, the virus can be transmitted to humans during subsequent probing or feeding. Viremia in human peripheral blood peaks during the early days (first 2–3 days) of acute illness and then declines sharply, after the introduction of virus targets directly on antiinfective human defenses. Human cell types, i.e., monocytes, macrophages, and dendritic cells, acting as major phagocytic cells of the innate immune system and responsible for detecting and removing invasive pathogens, are permissible for infection. These cells are also antigen-presenting cells critical for the initiation, expansion, and polarization of adaptive cellular immunity.

Dengue virus genome is coated with a host cell-derived lipid bilayer. The genome is an 11 kb-longsingle positive-sensed RNA molecule encoding 10 viral proteins: three structural proteins, capsid, membrane, envelope, and seven nonstructural proteins. The envelope proteins on the viral surface are major players for viral attachment, endocytosis, uncoating, and fusion, binding to cellular receptors and direct virus particles to the host cell's endosomal compartments. Dengue virus enters host cells through receptor-mediated endocytosis. Once inside, the viral uncoating occurs in the cellular endosomal compartments and subsequently viral RNA releases into the cytoplasm of the host cell, near the ER (endoplasmic reticulum), which is populated by ribosomes. It is worth remembering that in this reticulum, which is part of the cell endomembrane system, the proteins' synthesis is performed, and further the proteins are glycosylated in the Golgi apparatus, to be finally encapsulated in microvesicles and in this form travel inside the cytoplasm, before entering the cell metabolism. The viruses mimic this pathway. Through a mechanism termed "budding," the nucleocapsids join with the endomembranes, forming progeny of viral particles in the rough ER cisternae. The budding continues with transport of the viral particles to the Golgi apparatus, wherein the initially attached glycan is further modified by a complex process of trimming or remodeling, resulting in varying oligosaccharide structures.

Glycosylation is fundamental, since it can promote proper protein folding required for protein functioning, and in this case affects interactions between virus and receptors, and alters antigenic structures. However, the modified viral structure can be recognized by host immune cells and antibody, thus our body impacts viral replication and infectivity. This is our best defense against the virus, but immune efficiency can be decreased by several factors opening the way to viral infection. This equilibrium beside attack and defense, the balance between inside and alien forces, is the key of any insect-borne disease, indeed of any disease. Once again, mankind is the main decisive factor, leaving the others in a second role.

The fight against viruses, including dengue virus, has been underway for a long time, based on different strategies and targets. Research against dengue is exploring the aforementioned aspects of the disease. Attention is focused on those who experienced dengue asymptomatic infections, to address protective immunity better. Another approach includes viral and host cellular molecules involved in receptor interaction and infection. Another front is focused on the abovementioned role of glycosylation with respect to receptor binding, viral tropism, and virulence of infectivity, and enhancement of innate immune mechanism, such as adaptive humoral and cellular immunity. However, so far several aspects of the dynamic process of the different stages of dengue virus infection are not well understood.

The vector

Ae. aegypti is the primary vector species responsible for transmitting the viruses that cause dengue, but also for chikungunya fever, urban yellow fever and other important diseases involving viruses. *Ae. aegypti* has white bands on its legs and spots on its abdomen and thorax. The viruses are passed on to humans through the bites of an infective female *Aedes* mosquito, which mainly acquires the virus while feeding on the blood of an infected person, but other possibilities can occur. Mosquitoes do not naturally carry viruses—we must remember that mosquitoes must acquire them from an infected person before they can transmit them to another person. As was reported for malaria, only female mosquitoes bite, in search of blood for egg production, and therefore only female mosquitoes are responsible for transmission of the above virus diseases. *Ae. aegypti* females are usually not good fliers, since they may spend their lifetime in or around the houses where they emerge as adults and they usually fly an average of 400 m. However, several forms of passive transportation are possible and this means that people, rather than mosquitoes, rapidly move the virus within and between communities and places.

Again and as always, we must consider the human role. The life cycle of the vector may be as short as 10 days or, in cool weather, as long as several months. Therefore, the mosquitoes must survive during winter, finding refuge in our warmed houses or other protected places. The indoor habitat is less susceptible to climatic variations and increases the mosquitoes' longevity.

Ae. aegypti (as evidenced by the name) originated from Africa but is now distributed globally in tropical and subtropical regions. Distribution in new areas was assisted by mass human forced migrations, like that to the New World associated with the massive slave trade between the 15th and 19th centuries from Africa to the Americas, and then to Asia as a result of trade during the 18th and 19th centuries. A further worldwide redistribution occurred as a consequence of the Second World War following troop movements. Therefore, once again the virulence and impact of an insect-borne disease is strictly linked to human activities. Furthermore, as for other insect-borne diseases, recently rapid human population growth and increased urbanization has led to an increase of incidence, which is favored in crowded and dense human habitats. Permanence and diffusion are helped by substandard housing, inadequate hygiene conditions, insufficient and/or polluted water supply, and inadequate or absent waste management. All these factors may cause an abundance of mosquito breeding sites, but we can add storage of drinking water and other urban water containers, including plant/flower-pot bases, rain pools, guttering, tires, and small quantities of still water. Any water reservoir provides an ideal habitat for reproduction and growth of *Ae. aegypti* larvae. As for any other mosquito, the immature stages need water-filled habitats, which can consist of artificial containers closely associated with human dwellings and often indoors. This means a large spread of the vectors and their control needs the collaboration of inhabitants, since usual massive treatment of areas by insecticides can be insufficient or inefficient.

Public enemy No. 1

It is evident that *Aedes* mosquitoes are favored in human settlements, being highly adapted to rural and urbanized areas, since they generally live indoors and near people. Inside houses, mosquitoes use to rest hidden in cool shaded places, such as in wardrobes, laundry areas, and under furniture. Therefore, if during the winter we heard the typical sound of the mosquito's flight, we should not consider this presence unusual.

If you ask people about the most aggressive and dangerous creature on the planet, they might suggest a lion, tiger, crocodile, or shark, or similarly aggressive and powerful creatures. I would reply—after mankind—Aedes albopictus. All the other creatures are relatively very rare and avoid contact with harmful human beings. I remember the episode of a Ph.D. visiting student of my laboratory coming from Central Africa, who went to the Zoological Park to see elephants and lions for the first time.

This is the final act of a long history. Where are the immense mammoths and the ferocious saber-toothed tigers? They were exterminated by the inexorable traps and poisoned arrows of our ancestors. In contrast, insect vectors of diseases are still everywhere, and some of them dominate several habitats. To win the struggle of survival, you do not need force and superpowers, but smartness and perfect knowledge of your habitat, including the strategies of natural enemies.

As already reported, also in developed countries, malaria and other diseases cannot be considered ended and peril could enter from abroad, as evident in the case of *Ae. albopictus* (syn. *Stegomyia albopicta*) of the Culicidae family. Among Culicidae, this mosquito is acknowledged as the most invasive mosquito species in the world. However, in terms of its ecological and physiological plasticity, its potentiality is much higher. Its environmental impact is expanding in several countries, sustained by climate changes and concentration of human population in huge cities. *Ae. albopictus* is a dangerous vector due to its aggressive daytime human-biting behavior and its ability to transmit many viruses, including dengue, yellow fever, West Nile, and chikungunya. It may also act as a vector of filariasis, with special reference to *Dirofilaria immitis*, *Dirofilaria repens*, and *Setaria labiatopapillosa*.

In comparison with other insect vectors, even in the same genus like *Ae. aegypti*, *Ae. albopictus* has evidenced a higher migration and adaptation capacity. In recent decades, *Ae. albopictus* has spread from Asia to areas of Africa, Europe, and the Americas. Therefore, although both have been implicated in large outbreaks of chikungunya and dengue, whereas *Ae. aegypti* is confined within the tropics and sub-tropics, recently *Ae. albopictus* became established also in temperate and even cold temperate regions, expanding its distribution as soon as climate changes favored its presence. Another important characteristic is the possibility to survive and reproduce in a large range of aquatic microhabitats, rural or artificial, in peri-urban and urban areas, from shady city parks to flower vases, water storage vessels, and concrete water tanks in bathrooms—practically everywhere humans live. This mosquito has demonstrated a special capacity of adaptation, as well as the possibility to feed on many targets and therefore be an efficient vector of several viral diseases.

Depending on the mosquito's species and age, bites can be relatively painless, so persons may not notice they are being bitten or may think they are being bitten by other biting insects, or bites may be very painful and irritating for several days. Usual targets are people's feet and ankles, which mosquitoes may bite repeatedly. Another recent novelty of this mosquito is biting during daylight hours. Biting activity is higher in the 2 h after sunrise or before sunset, but it will bite throughout the day, differently from mosquitoes endogenous of temperate zones. As a result, bed nets do not prevent disease transmission. These mosquitoes are hard to catch and move very quickly, darting back and forth.

This mosquito is commonly known as the (Asian) tiger mosquito or forest mosquito, being characterized by the evident and recognizable white bands on its legs and body, which resemble those of a tiger, but also to express its violence (Roiz, 2001). Males are very similar to females, but much smaller, and their antennas are clearly thicker. Again, this species is native to the tropical and subtropical areas of South-East Asia, and it was confined for a long time in these regions. In the past few decades, it has spread to many other countries thanks to climate changes, but again mankind should take central responsibility. This mosquito was probably introduced through the transport of goods in international travel. From Asia, the tiger mosquito passed to the eastern states of the USA and then in 1990–1991 to Italy, thanks to the resistance of its eggs, and later to other parts of Europe. Italian people did not have previous experience with this mosquito, but immediately perceived its arrival. The striped appearance of this species was perceived not only for the painful bites, generating a severe itch and large swellings, but also for the all-day disturbance, with no break. In Europe and the USA, *Ae. albopictus* is an alien species in a condition to dominate its habitat. Usually the dimensions of females are superior to those of autochthonous species. The new vector was rapidly able to become a significant pest, thanks to some additional characters respect to those of native mosquitoes, which are not able to deal with the invasion. Autochthonous mosquitoes are common in water pools of any type and shape, whereas *Ae. albopictus* needs smaller pools, preferably shaded, like those easily found in human sites, especially in towns. Another important difference is in its movement capacity. Common mosquitoes often move for a few hundred meters, while other species can reach more than 10 km. The tiger mosquito lives in a limited area (*c.*200 m) and therefore egg deposition sites can be found near where the insect can be observed. In addition to humans, this mosquito also bites other mammals and birds. The female mosquito uses to integrate the bool feeding with other source of food, like nectar and other plant-produced substances, of course the same behavior of males and many other insects. During the feeding, the female is particularly perseverant and careful. In a first step, it flies around looking for the best target. Once this is decided, it starts to feed but soon suspends the feeding, before the ingestion of the necessary quantity for the development of the eggs. From any bite, usually only 2 μL are assumed, but the bite is sufficient to transfer the parasites. The mosquito visits several targets, causing the infection of many hosts, and the transfer of the disease from species to species.

The tiger mosquito is an epidemiologically important vector for the transmission of many relevant viral parasites, including yellow fever virus, dengue fever, chikungunya fever, and several filarial nematodes such as Dirofilariasis immitis. Furthermore, *Ae. albopictus* is capable of hosting the Zika virus and is considered a potential vector for transmission among humans (Romi, 1995; O'Meara et al., 1995). In particular, in Italy the alert was immediately raised after the first reports about the mosquito's presence, followed by several programs to control the diffusion, increasing every year. In Italy, the mosquito's successful integration is considered a consequence of recent climatic changes, in particular the higher temperatures in winter.

As already reported, *Ae. albopictus* is native to tropical and subtropical regions with warm and humid climate; however, it has been adapting successfully to temperate regions, where it is now able to hibernate over winter (Roiz, 2001). The history of the Asian tiger mosquito is strange, since its origins are humble, and very different from the current widespread diffusion. It originally lived on the edges of forests in East and South-East Asia, breeding in tree holes and other natural reservoirs. Centuries ago, it spread to Madagascar and Indian Ocean islands, but only in the past 50 years has it spread to all inhabited continents, and this recent diffusion is considered in large part as a result of increased global air travel and seaborne trade. The worldwide trade in secondhand tires, probably from Georgia (USA), which often contain water and are an ideal place for resistant eggs, is considered the key factor of the arrival of *Ae. albopictus* in Europe. The eggs can survive in such a refuge until the tires reach their destination. The eggs can also survive cold winters because they go into a state of dormancy or "diapause," allowing the *Ae. albopictus* mosquito to persist in areas with a temperate climate. In newly infested areas, the mosquito has adapted easily to human settlements, where pots, vases, and buckets can act as breeding sites, provided that there is a bit of vegetation. Therefore, the case of the tiger mosquito can be considered emblematic of climatic change and human influences.

Although eggs from strains in temperate zones should be more tolerant to the cold than those from warmer regions, the distribution of this species was limited since the larvae and eggs of *Aedes* sp. transmitting mosquito-borne disease cannot survive snow and temperatures under freezing. Generally winter had been the most important front line blocking the tiger mosquito's invasion. Changing these conditions, the distribution areal can consequently move to northern territories. The climate change models indicate that by 2030, the temperatures will have risen sufficiently to enable *Anopheles* mosquitoes to conquer Europe until Wales, where the climate in becoming increasingly mild and wet (Bendict et al., 2007; Caminade et al., 2012). Tiger mosquitoes are also favored because they can survive throughout winter in suitable microhabitats, and typically fly and feed in daylight. In fact, in 2018 these mosquitoes were reported in the southern and central regions of France, and consequent alerts reported 333 cases of dengue, 12 of chikungunya, and 21 of Zika diseases. These are very low numbers in comparison with the overall population, but enough to trigger concern.

The argument is why has *Ae. albopictus* so far been absent in Europe when it is now so abundant and prominent? Again, we must consider the human influence. Beside the climate changes, in particular the rise of temperatures, especially in winter, which allowed survival during the year, the mosquito adapted perfectly to urban communities being associated with inhabitants' lifestyles in major cities, whereas *Anopheles* sp. and other ordinary mosquitoes preferred living in wetlands. Furthermore, it is more aggressive and virulent in comparison with other mosquitoes, affecting the environment equilibria in favor of its needs. The danger of introducing alien species, such as *Ae. albopictus*, is a current problem of high importance, as evident in the controls at airports, and it will be necessary to return to this aspect.

Aedes larvae and pupae are usually targeted using organophosphates and insect growth regulators. Indoors residual spraying and insecticide-treated bed nets are also employed to reduce transmission of malaria in tropical countries. However, these synthetic chemicals have strong negative effects on human health and the environment, and induce resistance in a number of mosquito species. The main problem concerns the spread of the mosquito in microhabitats inside villages and towns. Therefore, eradication is difficult and ordinary methods of disinfestation are inefficient without the active collaboration of the population. Capillary information about movements and activities of inhabitants are essential in the eradication campaigns. In this case, the computer simulation of flows, obtained by monitoring of the cellular activity, are useful and the data can be crossed with project like the Gridded Population of the World by NASA checking the geographic position of each individual in a selected area. Otherwise, after the insecticide treatment, the mosquitoes can re-emerge more active and dangerous than before.

The future of dengue is not predictable, but as with other vector-borne disease, it is re-emerging in areas that have been disease-free for relatively long periods of time. The episodes are so far limited and short-lived, but some in particular are interesting. Xiao et al. (2016) reported on a Dengue outbreak in a city in southern China that had been disease-free for more than two decades. According to the details of the report, the infection, due to Dengue serotype 1, was introduced by a traveler from South-East Asia and transmitted by *Ae. albopictus*. However, *Ae. aegypti* is the most important vector of dengue, since *Ae. albopictus* is a less competent vector of arboviruses, and causes milder epidemics. Therefore, the replacement of *Ae. aegypti* with the tiger mosquito could even result in public health benefits. However, there is no solid evidence for this, and the milder course of the outbreak could in part be explained by the relatively short duration of the hot season in some affected areas. *Ae. albopictus* is rapidly increasing its changing global distribution, and its expansion could create new opportunities for viruses to circulate in new areas, becoming a common cause of epidemics in countries so far fre of *Ae. egypti.* In particular, the worldwide trade in secondhand tires or other goods, which often contain water and are an ideal place for eggs and larvae, has been a key factor in the large-scale conquest of *Ae. albopictus*, which adapts easily to new environments, even in temperate climates. Similar episodes have been described in Mauritius and Hawaii. Therefore, airport controls are important and necessary, but it is unlikely that they will block the tiger mosquito.

Chikungunya fever

Chikungunya virus (CHIKV) is an alphavirus of the Togaviridae family vectored by *Aedes* sp. mosquitoes. Perhaps due to its difficult name, this is less famous among the diseases here considered, but this is not a reason to underestimate its value and importance. We shall consider epidemiology, biology, treatment, and vaccination strategies of CHIKV, which is a case study of insect-borne diseases.

"Chikungunya" is Maconde (a local Tanzanian language) for "he who bends" and describes the huddled-up body posture caused by the chronic and incapacitating arthralgia that normally accompanies acute infection in human populations. Arthralgia (from Greek *arthro-*, joint + – *algos*, pain) literally means joint pain. The term "arthralgia" should only be used when the condition is noninflammatory, and the term "arthritis" should be used when the condition is inflammatory. Since 1952, CHIKV has been responsible for sporadic and infrequent outbreaks. However, since 2005, global CHIKV outbreaks have occurred, inducing some fatalities and associated with severe and chronic morbidity. CHIKV is thus considered as an important re-emerging public health problem in both tropical and temperate countries, where the distribution of the *Aedes* mosquito vectors continues to expand. The virus is endemic to tropical regions, but the spread of *Ae. albopictus* into Europe and the Americas coupled with high viremia in infected travelers returning from endemic areas increases the risk that this virus could establish itself in new endemic regions.

It has already been reported that insect-borne diseases, like other infective pathologies, are typically subjected to explosion outbreaks followed by inexplicable rapid disappearance of the mass phenomenon. This is particularly important for CHIKV, characterized by an alternance of emergences and re-emergences in different parts of the world. It is important to follow the itinerary and time sequence of CHIKV outbreaks as they offer exemplar behavior of an insect-borne disease. Before 1999–2000, human infections in Africa had been at relatively low levels for a number of years, but in 2000 there was a large outbreak in the Democratic Republic of the Congo. In February 2005, a major epidemic occurred in islands of the Indian Ocean, reaching a peak in 2006. In 2006 and 2007, India was mainly affected, but later several countries in South-East Asia were also impacted. In 2007, there was a recrudescence in Africa, in Gabon. In 2005–2007, India, Indonesia, the Maldives, Myanmar, and Thailand reported more than 1.9 million cases. In 2007, transmission was reported for the first time in Europe, in a localized outbreak in northeast Italy, followed by several reports in France. However, vectors were able to land in America and spread. In December 2013, France reported two laboratory-confirmed autochthonous cases in the French part of the Caribbean island of St. Martin. Since then, local transmission has been confirmed in more than 43 countries and territories in the WHO region of the Americas. This is the first documented outbreak of CHIKV with autochthonous transmission in the Americas. As of April 2015, more than 1,379,788 suspected cases of CHIKV had been recorded in the Caribbean islands, Latin American countries, and the United States of America. A total of 191 deaths were also attributed to this disease during the same period. Canada, Mexico, and the USA have also recorded imported cases. On October 21, 2014, France confirmed four cases of local CHIKV infection in Montpellier, France. Also in late 2014, outbreaks were reported in the Pacific islands. Currently, a CHIKV outbreak is ongoing in the Cook Islands and Marshall Islands, while the numbers of cases in American Samoa, French Polynesia, Kiribati, and Samoa have reduced. The WHO responded to small outbreaks of chikungunya in late 2015 in the city of Dakar, Senegal, and the state of Punjab, India. In the Americas in 2015, 693,489 suspected cases and 37,480 confirmed cases of CHIKV were reported to the Pan American Health Organization (PAHO) regional office, of which Colombia bore the biggest burden, with 356,079 suspected cases. This was less than in 2014, when more than 1 million suspected cases were reported in the same region.

In 2016, there were totals of 349,936 suspected and 146,914 laboratory confirmed cases reported to the PAHO regional office, half the burden compared to the previous year. The countries, reporting most numbers of suspected cases, were 265,000 (Brazil), 18,000 (Bolivia), and 19,000 (Colombia). The year 2016 was the first one that autochthonous transmission of chikungunya was reported in Argentina, following an outbreak of more than 1000 suspected cases. In Africa, Kenya reported an outbreak of

CHIKV resulting in more than 1700 suspected cases. In 2017, Pakistan continued to respond to an outbreak which had started in 2016.

Zika virus

The reported cases of Zika disease are relatively few in comparison with other insect-borne diseases. To understand the phenomenon associated with the Zika disease, we must introduce the word "pandemic," more commonly associated with influenza. A pandemic is the worldwide spread of a new disease. This is the main difference to an epidemic, which is a term broadly used to describe any health problem that has grown out of control. The widespread occurrence of a disease is considered when defined in a community at a particular time. In other words, the term "epidemic" should be preferred in the case of a disease actively spreading, whereas pandemic relates a geographic spread affecting a whole country or, more appropriately, the entire world. A pandemic occurs when a new virus or other infecting agent emerges and spreads around the world, thanks to the absence of efficient immunity in most people. Fear and panic about pandemics are generated with the feeling of a rapid spread, facilitated by tourism and travel, and the possible general infection of everyone. Pandemic derives from the Greek *pan*, meaning "all," and *demos* meaning "people." However, if for influenza vaccines offer an efficient form of defense, for insect-borne diseases the situation is less favorable. Alarms and alerts about outbreaks are therefore the result of a mass emotive reaction by the population, due to the possibility of transmit immediately and everywhere information. Globalization is surely part of pandemic alerts. As clearly evidenced in the COVID-19 (SARS-CO2) outbreak, the emergence of a new strain of virus, as well as its consequences, is not predictable. Every measure is important and it is not possible to decide a priori if the counteractions are adequate or consequences of an exaggerated impulsive reaction.

The Zika virus is the cause of the Zika disease, and is a *Flavivirus*. The vector of Zika virus is an *Aedes* mosquito, which can bite monkeys and humans, with *Ae. africanus* and *Ae. aegypti* as the principal probable vectors (Benelli, 2016). Symptoms generally include fever, mild headache, aches, joint pains, maculopapular eruptions, and rashes in different parts of the body. Most of those affected not experience any symptoms at all, but in certain cases the consequences can be more significant. The Zika fever is common in West and Central Africa, but it also occurs in many Asian countries, and since 2014, its presence in American regions has been reported. The very few cases identified in the United States were considered travel-associated and in 2018 and 2019, there were no reports of Zika virus transmission by mosquitoes in the continental United States. It is possible that the Zika fever was overestimated, as in the case reported here, but this most recent intruder in insect-borne diseases rapidly acquired the status of a global threat, attaining higher virulence and causing complex clinical manifestations, including microcephaly in newborns, infants, and babies, and Guillain-Barré Syndrome. Owing to its rapid spread and the dissemination of information about possible consequences, a general alert raised public concern and generated panic and fear.

However, most of the world did not know about the existence of the Zika virus and its danger and potential impact until the 2016 Summer Olympic Games in Rio de Janeiro. The situation generated a debate about the danger for approximately 16,000 athletes and 600,000 visitors coming from abroad to the Games. Data confirmed the potential risk in the form of the state of Rio de Janeiro recording 26,000 suspected Zika cases, the highest of any state in Brazil, with an incident rate of 157 per 100,000, the fourth highest in the country. Everything started in December 2014 with the first reports of an unknown exanthematic disease outbreak in Brazil. Later, the disease was identified as the Zika infection. In May 2015, the spread of the disease to other parts of the country, first in the states of Pernambuco (PE), Rio Grande do Norte (RN), and Bahia (BA) in the northeast region, then in other states of the central-west and southeast regions. Again, geography gives us some help in interpretation. The state of Pernambuco is in the extreme eastern part of Brazil, projecting out to the Atlantic Ocean and practically in front of Africa. The state is part of the *sertão*, a very beautiful equatorial place with splendid *playas* full of *coqueiros* in the coast, but the internal part has needed to face the progressive desertification in the last decades (*sertão* in Portuguese means great desert). The rains could be enough, but they are concentrated in a short period. In recent years, this situation generated a massive migration toward the coast with the major towns Recife and Macejò, causing usual conditions for the development of infectious diseases and their progressive expansion. An outbreak is usually a combination of factors, like a massive susceptible population, climatic conditions conducive for diffusion of the mosquito vector, socioeconomic impact, and alternative nonvector transmission vehicles. However, this rapid spread was probably due to the special characters of the spread of this disease, in comparison with other ones, including sexual transmission. In fact, dengue virus took several decades to diffuse, bound by a climatic barrier in the south and low population density areas in the north, while the Zika virus epidemic rapidly spread across Brazil within months, and beyond the area of permanent dengue transmission. Thus, Zika spread rapidly across Brazil and to more than 50 other countries and territories on the American continent. Its speed was one of the key factors for the epidemic alert, leading Brazil to declare a national public health emergency in November 2015, confirmed by the World Health Organization three months later, and culminating in an explosive situation involving a decision about the Olympic Games. The debate was furious, being without any reliable results, as is common when science is mixed with economic and political interests. Again the WHO was asked and responded to the claims, evidencing the general low impact of the Zika virus, resulting in mild symptoms if at all in those affected, opening the door to the Games with its authority. In effect, the epidemic was progressively subsiding, thanks to Brazilian efforts to mitigate the risk to athletes and visitors by a series of indications to take precautions in protecting themselves from mosquito bites and practicing safer sex. This is another indication that the agent cannot be considered totally guilty of the epidemic, but the social situation, with absence of hygiene and adequate control, are necessary concurrent causes. The parasite does its homeostatic work, and humans amplify this. Therefore, something changed.

2016–2018: According to WHO data: "incidence of Zika virus infection in the Americas peaked in 2016 and declined substantially throughout 2017 and 2018." In fact, due to its pandemic potentiality, the Zika disease is the most carefully detected one by international and local heath institutions. This tendency can be confirmed in the following timetable of the Zika outbreak.

2018–2019: The Pan American Health Organization (PAHO) in Brazil reported 19,020 cases of Zika disease, of which 1379 were laboratory-confirmed. In 2019, as of week 9, the Brazilian Ministry of Health reported 2062 probable cases, compared to 1908 cases reported over the same period in 2018.

2018: PAHO, in addition to Brazil, reported data on other countries in South America, with the majority of cases in Peru (984), in Bolivia (1736 with 486 laboratory-confirmed), and Colombia (857, laboratory-confirmed). For the whole of 2018, Mexico reported 860 confirmed cases of Zika disease, an evident decrease in cases compared to 2017, at 3260. In 2018, Guatemala was one of the countries much affected with 2300 ZIKV disease cases being notified during 2018, compared to 703 cases reported in 2017. Of these cases, 106 were confirmed in 2018 and 164 in 2017.

2018–2019: In 2018, El Salvador reported 481 cases of ZIKV disease, against 450 cases in 2017 and 128 Zika disease cases, in 2019, as of week 11, compared to 66 cases for the same period in 2018. In March 2019, Peru and Colombia remained among the countries with the highest numbers of cases, with 275 and 110 cases, respectively. In Central America, Mexico reported 13 confirmed cases, compared with 39 for the same period in 2018.

2019: No vector-borne locally disease cases had been reported by EU/EEA countries in Europe as of week 12 in 2019, whereas in the period 2015–2019, 22 EU/EEA member states had reported 2398 travel-associated infections through the European Surveillance System.

The diffusion of Zika due to travel and tourism is carefully detected in the USA, but a significant decrease in travel-related cases returning to the continental United States has been observed for the same period, from 4897 cases in 2016 to 72 cases in 2018. In December 2017, it was possible to find the most recent mosquito-infected autochthonous case in the continental United States, as reported in Hidalgo County, Texas. Since then, and as of March 6, 2019, no autochthonous mosquito-borne transmission had been reported in the continental United States. Data from the other parts of the world confirm these trends.

Finally, Zika, despite general concern, is not presenting a current menace, but it can be considered an example of positive control of an insect-borne disease when appropriate attention is exercised, although the social effects should be considered and amplification by information limited if possible.

Unfortunately, considering previous experiences in other insect-borne diseases, this trend cannot be considered reliable. However, we can trust reliable diagnostics for detecting the Zika infection and several research advancements in drug/therapeutic targets and identification of vaccine candidates. Despite these research achievements, currently there is no effective drug or any vaccine available against the Zika fever.

Yellow fever

There are several similitudes with the dengue disease. Again, the agent belongs to the *Flavivirus* genus and it is present in tropical areas of Africa and South America. Remember that the word *flavus* means "yellow" in Latin and the yellow fever in turn is named for its propensity to cause yellow jaundice in infected people. Again, *Ae. aegypti* is the most important vector, at least in America. Monkeys are the main reservoir of infection; by transmission from monkey to monkey, the virus is transmitted and can infect humans, with a vector change. This disease is the cause of 200,000 clinical cases and 30,000 deaths every year. A total 90% of cases are reported from Africa, mainly the Sub-Saharan regions, and these predominantly affect young people. However, also in this case these data must be considered indicative, since most infections and deaths occur in rural areas, where surveillance and reporting can be inadequate. In the Americas, the disease is distributed from Philadelphia in the USA to Mendoza in Argentina. The disease can be fatal, acute, or mild, and even unapparent. There is no antiviral treatment, but it is possible to prevent yellow fever infection by vaccination.

West Nile fever

Again, this disease is caused by a virus of the genus *Flavivirus* and the vector is an *Aedes* mosquito, aided in this case by *Culex* (Aziz et al., 2015). The disease has been reported in regions of all populated continents, including the whole of America. The infection effects can be invisible, or cause mild fever, meningitis, encephalitis, acute symptoms similar to poliomyelitis syndrome, and ultimately death. The most interesting aspect of this disease is the range of hosts involved. West Nile fever can infect birds, horses, and humans, with different signs and symptoms, including none at all, until the point of death. The first cycle of the virus concerns several bird species and mosquitoes, supporting replication of the virus. The other targets of the virus are mammals, mainly horses and humans. Although in horses the cyclical disease is characterized by mild to severe symptoms of ataxia, involving weakness, muscle twitching, and nerve deficiency, including cranial damages, in humans most infections are asymptomatic. However, in the 1990s, the disease became an important human and veterinary challenge. The impact of the disease increased noticeably in human infections. People over the age of 50 are more at risk and can develop serious forms of this disease. The problem is that the incubation period of this disease can cover several weeks and approximately 80% of infected people are totally asymptomatic or symptoms can be confused with others in aged patients. In fact, in many developed countries, vector-borne tropical diseases are so rare that the experience of medical doctors to diagnose these diseases became inadequate in the absence of direct experiences. A vaccine is available, but only for horses.

A similar case concerns St. Louis encephalitis, a viral disease transmitted by *Cules* sp.; birds are usually the hosts, but it can also affect humans and other hosts. This disease is currently present in Canada, Mexico, and Central and South America. The clinical symptoms of this encephalitis are often not manifested, except during epidemics, wherein children and elderly are the most susceptible, with a mortality is 5%–20%. No vaccines or other effective treatments are available, including antibiotics, because they are not effective on the virus. Similar diseases are eastern equine encephalitis virus and western equine encephalitis, wherein again birds, mosquitoes, horses, and humans are involved. Again, mortality is between 5% and 15%, and infants, children, and elderly people are most affected. Around 50% of survivals result in severe permanent brain damage. There is no effective vaccine for humans so far.

Lyme disease

Lyme disease, also known as Lyme borreliosis, is probably less famous compared to West Nile fever, but it has been described by the *New York Times* as "the infectious disease able to diffuse most rapidly worldwide after AIDS." There are several reports of its increasing diffusion in Asia, Europe, and South America. Lyme disease is the most common disease spread by ticks in the Northern Hemisphere, with an estimated 300,000 infected people a year in the USA alone, and 65,000 people a year in Europe.

The name of this disease relates to he town of Old Lyme, Connecticut, USA, where in 1975 the first epidemic was reported. According to the reports, in Lyme a mysterious increase of arthritis cases occurred, especially in infants, causing at first apparently innocuous cutaneous erythema on the thorax and other parts of the body, which are considered symptoms of juvenile rheumatoid arthritis. However, in contrast to this disease, the symptoms persisted with increase of the erythema, associated with fever, headache, joint pains, and tiredness. The acute form of the disease includes loss of the ability to move one or both sides of the face, joint pains, neck stiffness, and heart palpitations, among others; shooting pains or tingling in the arms and legs, memory problems, and tiredness can also occasionally develop. Months to years later, repeated episodes may occur.

The infection is caused by a bacterium of the genus Borrelia, spread by ticks, but it is transmitted to humans by the bites of infected ticks of the genus Ixodes. Approximately 70%–80% of infected people develop a rash, which is typically neither itchy nor painful.

As for other arthropod-vectored diseases, infections are most common in the spring and early summer. Beside prevention and measures of control, research is ongoing to develop new vaccines. An effective vaccine for Lyme disease was marketed in the United States between 1998 and 2002, but it was withdrawn from the market due to poor sales.

Other vector-borne diseases

It is necessary to mention Japanese encephalitis (1E)-epidemics, whose incidence has been reported to be particularly high among pediatric groups with high mortality. Nearly 1.4 billion people in 73 countries worldwide are threatened by lymphatic filariasis, a noxious parasitic infection that leads to a disease commonly known as elephantiasis. Filariasis is vectored by mosquitoes, with special reference to the genus *Culex*. The main control is currently against mosquito larvae and consists in the treatment with organophosphates and insect growth regulators, but with negative effects on human health and the environment. Green-synthetized nanoparticles have been recently proposed as highly effective larvicidals against mosquito vectors, whose details can be found in Chapter 7, dedicated to new solutions.

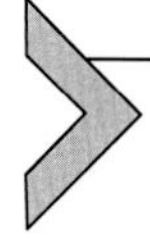

Insect-borne diseases affecting animals

Metamorphoses in the food environment

The food planetary situation is full of contradictions. On one side, the alimentary systems—consisting of production, transformation, distribution, and consumption—have the potential to nourish the global population, in terms of quantity and quality, as well obtain the respect of the environmental sustainability. However, despite this positive potentiality, food production currently still appears in the opposite direction. Although food world production in calories is proceeding in accordance with the growing population, at least 820 million persons do not have sufficient food, and even more are consuming food of low quality, insufficient in micronutrients and essential amino acids. The environmental impact is modeling the planet in the Anthropocene Age, with something like 40% of the soil designated for production of food and consumption of 70% of available fresh water. The conversion of natural ecosystems into pastures and arable lands is affecting the species equilibria in any habitat. About 33% of hydric stocks is over-exploited and more than 60% is used up in production. These and other congruent data were published in the EAT-Lancet Report, named "Food in the Anthropocene" and produced by a selected pool of experts. The report also contains a series of precise indications, expanding the concept of planetary boundaries to identify a useful compromise between production and sustainability in the coming years, to meet the challenge of an estimated 10 billion inhabitants by 2050. In other words, we know what we need to do as a species, but actually making this happen is another matter.

In this part of the chapter, we will focus on farm animals and their products, in consideration of the economic importance and the strategic relevance of agricultural and livestock production (Murugan et al., 2016). This argument must be based on two considerations, the quantity and the quality of food, which are usually not coincident. The first aspect is strictly related to insect-borne disease, due to the massive use of chemicals against pests to protect production in pre- and post-harvesting steps. The second aspect evidences several changes in the production and consumption of food, with massive orientation of consumers on certain types of nourishments.

In Italy (excluding several forms of allergy to one or more food components), something like 6% of the population are vegetarian or vegan. Even vegans can be taxonomically classified into relative and absolute vegans, on the basis of their ethical considerations, refusing not only foods like milk, eggs, and honey, but in the case of sexual vegans, which avoid physical contact with omnivorous, because they are considered impure being in contact with meat. The alimentary tribalism is considered a form of recent tendency, but it is an old story. The refusal to eat meat, in particular bloody meat, can boast ancient roots. In the hypothetical Golden Age, mankind was happy and naïve, eating everything nature spontaneously offered in great quantities; only in the Silver Age did meat enter mankind's menu. Later, discrimination among foods was seriously carried out and considered necessary. Practically every religion contains a detailed list of foods to be avoided. Several of these prohibitions have historical dietetic reasons, but warnings of different kinds survived for centuries and are still present in all current religions.

Nowadays, an alimentary tendency is not a selection of the right proteins, but adhesion to a living style, indicating not a physiological aspect but a philosophical choice. Food considered as identity of a selected group, in order to be safe and purified by obesity and fats, like a necessary deprivation against abundance and obtain a soul purification. Avoiding certain foods, it is possible to maintain an evidence of separation and identity against a world dominated by the globalism. The famous sentence of Feuerbach can reloaded. We are not what we eat, but what we do *not* eat. This phenomenon is important, but still marginal. There is still 96% of the population waiting for their meat input. The need for such high production leads to massive breeding and shifts the attention away from the quality of food. If we aspire to eat better food, we must consider the production steps. Most of the meat that we eat derives from intensive breeding, where the animals are allowed to live only for the duration of their growing period and in abnormal conditions. Only the quantitative result is considered. In chicken breeding, the conversion between feed and growing should reach more than 50%, which means that for every 100 g of food, at least 50 g remains inside the animal. After 4 weeks, the chickens are ready to be converted into meat and consumed in different forms. The main problem is that chickens are confined in a small overcrowded space, where the light is always on to avoid continuous mortal attacks inside the population. In such challenging conditions, these organic machines of conversion of vegetable matter into meat need continuous human assistance because diseases and epidermics are unavoidable.

The health of animals in industrial breeding, like chickens, fish, cattle, pigs, and others, needs to be continuously medically supported, in particular in the earliest stages of growth. The result is that the meat that we eat is full of estrogens, hormones, and antibiotics, usually significantly above the suggested limits. The problem is that the antibiotics pass through the alimentary chain, are transferred in the marketed products, and eventually reach us and are accumulated by us. This connection between food and feed is already evident in many sectors. Fish consumption is nowadays deeply dependent on aquaculture. Again, production is supported by heavy utilization of antibiotics, but this will not be possible in the future.

After a long period of total confidence in antibiotics' power and consequent enormous benefits for mankind, things are changing. The continuous appearance of new bacterial resistant strains is anticipating the postantibiotic era, when antibacterial drugs that currently work will be inefficient. Many fish producers, supported by legislation, are already partially replacing antibiotics with selected natural products introduced in food, in order to increase natural immune defense, and important results are being obtained.

Other breeding sectors are following their example. This could be the compulsory future of breeding and meat production, not considering the already present and increasing market of ecologically produced products. However, it will be a long road before things really change.

The situation concerning animal farming and derived products is every day worse. In industrial farming, animals are considered not as living organisms, respecting their lifestyle and environmental needs, but simply a method of conversion of food into other foods, generally food for animals into food for humans. Not considering the extreme conditions, here again there is a continuous input of antibiotics, but we must consider the epigenetic factor. When we change radically the environmental conditions of animals, we are producing animals very different from the starting ones, because of the accumulation in DNA of small but significant changes. The changes are translated in the composition of the derived products. Therefore, the present nutritional input is very different from previous input, and this must be considered as the cause of several consequences in the alimentation chain and at the end in our physiological equilibrium, and the source of many modern diseases (Kant et al., 2015; Donin et al., 2018).

Food for a global population of 7 billion humans

Livestock production is growing rapidly as a result of the increasing demand for animal products. Global meat production and consumption was predicted to rise from 233 million tonnes (2000) to 300 million tonnes (2020), and milk from 568 million tonnes to 700 million tonnes over the same period. Egg production was also predicted to increase by 30%.

This forecast shows a massive increase in animal protein demand, required to satisfy the increased human population. The considerable and growing demand for animal protein is focusing attention on the sources of feed protein and their suitability, quality, and safety for future supply.

Significant increases in global demand for livestock products will clearly require increasing amounts of feed protein supplies, and sources and alternatives will need to be reviewed continually. There would seem to be strong justification for research and development investment into a number of very promising new sources. What is certain is that there will need to be considerable increases in feed manufacture, requiring a thriving, successful, and modern feed industry.

Safety issues will remain paramount in the minds of consumers following recent food crises, and continuing investment is needed into quality assurance programs to gain market access for animal products and retain consumer confidence.

How we feed and treat animals intended for human consumption raises important public health concerns. These concerns arise not only from what is fed to animals but also from gaps in regulations and systems intended to ensure the safety of feed and the food supply. A variety of substances, including many waste materials from the agriculture, food, and rendering industries, are "recycled" into feed for food-producing animals. Some of these ingredients used in feed, particularly those from animal and mixed sources, may result in unwanted feed contaminants or have other unintended consequences. Animal feed plays an important role in the cycling of dioxin, arsenic, pathogens, antibiotic-resistant bacteria, prions, and other substances of public health concern. These problems have drawn attention to feeding practices within the livestock industry and have prompted health professionals and the feed industry to scrutinize closely feed quality and safety problems that can arise in foods of animal origin, as a result of animal feeding systems. Given the direct links between feed safety and safety of foods of animal origin, it is essential that feed production and manufacture be considered as integral parts of the food production chain. Feed production must therefore be subject, in the same way as food production, to quality assurance including feed safety systems.

However, assuring feed/food quality and safety is only part of the problem, because ensuring animal health and welfare and reduction of environmental impact are also fundamental. It is estimated that there are 5×10^6 EU farmers, 130 billion animals for food production, and 450×10^6 tons of feed yearly. Since the BSE-related 2000 ban of animal proteins in the EU, soy meal represents the world's best protein. However, self-sufficiency of EU country members is low (2%), resulting in a 10% increase of soy consumption and the addition of pure amino acids. Sources of protein for animal feeds are many and varied, with considerable opportunities for further diversification and substitutions. More research is required on alternative sources before many of the opportunities can be exploited in practice.

Quality protein can be provided sometimes from various crop residues and by-products of food and drink manufacture, such as brewers' grain and maize gluten meal. These by-products are many and varied, and differ considerably in the value and significance for animal feed protein supply. Today's technological capacity for reusing and further processing of these waste materials is limited due to poor knowledge about potentially useful primary and secondary plant products that are still contained in the residues. Food is lost or wasted throughout the supply chain, from initial agricultural production down to final household consumption, often only partially considering the quantity of daily produced organic waste for families. In medium- and high-income countries, food is wasted significantly at the consumption stage, as it is often discarded even if it is still suitable for human consumption. Significant losses also occur early in the food supply chains in industrialized regions. In low-income countries, food is lost mostly during the early and middle stages of the food supply chain due to deterioration, whereas much less food is wasted at a consumer level.

However, some of these by-products provide a valuable local source of protein, which can be inexpensive, accessible, and continuously available from the local food industry. Their use can also be regarded as a significant recycling opportunity, and more of a closed system for waste disposal. Food safety considerations may still dominate this protein supply route, with restrictions on certain by-product materials or their treatment before use in animal feed. Production of meat and other livestock-derived products can be highly affected by insect-borne diseases. Medical treatments are crucial for meal production and animal welfare, and these are strictly related to antibiotics and insect-borne diseases, but preventive counteractions are once again necessary and decisive. Regarding the insect-borne diseases affecting animals, a single report is evidenced here and this case utilized as a model of current and possible devastating diseases interesting livestock. The sequence of reported events, as well as interpretations and evaluations, may be a starting point to predict and successfully face the emergence of devastating vector-borne pathogens.

Bluetongue disease

Chronology of bluetongue virus (BTV) spread in Europe

1969: First isolation in Greece of strain BTV-9 and BTV-1.

1998: Isolation in Tunisia of the strain BTV-2, endemic of Sub-Saharan Africa, and belonging to strains from South Africa, Nigeria, Sudan, and the USA.

2000–2001: Isolation in Greece and Turkey of the European strain BTV-1, similar to viruses that have been isolated in India. First outbreak in Bulgaria.

2003: A new strain BTV-4 type, different from that of Greece and Turkey, is isolated in Corsica, Sardinia, and the Balearics.

2004: Detection of strain BTV-4 in Sicily and France.

2005–2014: Distinct strains are still entering in Europe, affecting at least 12 countries and more than 800km further north than before.

2014: In Sardinia, BTV caused the deaths of 13,000 sheep and damage costing 42 million euro. A total of 5772 infection sites were detected.

Bluetongue (BT) is a vector-borne disease of domestic and wild ruminants caused by viruses of the genus *Orbivirus* transmitted by *Culicoides* midges (Bowne, 1971). BT is a complex multihost, multivector, and pathogen disease, whose occurrence is fluctuating, re-emerging after long periods of absence, when most transmission occurs silently in disease-resistant animals, and short devastating phenomena affecting in preference certain breeds of sheep, which are economically very important for production of high-quality wool (Saegerman et al., 2008). Cattle are the most common reservoir, being able to be subclinically infected also for prolonged infection. The disease's name stems from the most evident manifestation of the disease, a typical blue coloration of the tongues of infected animals.

The disease manifests a pre-patent period of 3–8 days. The symptoms in sheep consist of transient fever (up to 106°F or 41°C), edema of the face, lips, muzzle, and ears, excessive salivation, and hyperemic oral mucosa. Later, at 7–12 days, several lesions may progress with petechial hemorrhages, erosions, and ulcers; a marked pulmonary edema is also often seen. Wool became fragile and no longer useful for the market. As a consequence of severe oral lesions, animals are reluctant to eat and suffer acute diarrhea, and muscle and coronary band pain limits movements. Finally, many affected animals become depressed and die, although others fully recover. Although the target of this disease is limited to sheep, cattle, and some species of deer and camelids, the effects can be epidemic and devastating, causing alerts for immense economic damages. While BTV first appeared in 1998, a transmission and maintenance network has recently emerged in Europe, causing the deaths of more than 1 million sheep, which are considered victims of viremias caused by BTV, with annual losses of US$125 million in the USA alone in trade of animals and animal products. In recent years, the livestock industry in the USA suffered he deaths of approximately 179,000 sheep within 4months caused by epizootics of BTVs. Decreased trade associated with BT outbreaks has become an even greater threat to the livestock industry than that of the disease itself, becoming a major obstacle to exports of U.S. ruminants and ruminant products.

The virus

BTV includes viruses of the genus *Orbivirus* pertaining to the family Reoviridae, which is a large group of double-stranded RNA viruses with 10–12 genome segments (Howell and Verwoerd, 1971; Verwoerd et al., 1972). The group is complex and varied, as consequence of a strong capacity of adaptation to the environmental changes. BTV cross-reacts with many antigenically related viruses including Palyam virus and the viruses that cause epizootic hemorrhagic disease and African horse sickness. BTV replicates in both arthropod and mammalian host cells. The virulence of BTV varies markedly; even strains with matching serotypes have variable virulence. A total of 26 serotypes of BTV have been identified worldwide with testing through viral antigens. The tests indicated great differences in the distribution since only five have been recognized within the USA and at least nine in Europe.

BTV presents interesting peculiarities. Its structure was found to be the same as that found in reovirus, consisting of a similar segmented, double-stranded RNA, with even polypeptides present in the capsid, four of which are major and three are minor components. However, Verwoerd et al. (1972), suggesting a direct coding relationship between the genome of the two taxa, also evidenced differences in genome segments, which may be utilized to code for certain capsid proteins. In fact, two of the BTV polypeptides are present as an outer diffuse protein layer surrounding the capsid. The authors speculated that "this outer layer probably has some of the functions of the reovirus outer capsid, one being the 'masking' of a viral transcriptase which could be demonstrated after its removal."

The vector

The virus is transmitted by certain biting species of *Culicoides* midges, pertaining to Diptera, Ceratopogonidae. *Culicoides* species total more than 1400, which are divided into many subgenera, although the classification and taxonomy of the genus are controversial and confused. Several species are known to be vectors of various diseases and parasites, which can affect animals, but less than 1% are considered responsible for transmission of BTV (Table 5.6).

C. imicola is considered the most common species vector in Southern Europe and has been recorded in Africa, Asia, and Europe so far. However,

Table 5.6 Scientific classification of the main vector of BTV in Southern Europe.

Kingdom	Animalia
Phylum	Arthropoda
Class	Insecta
Order	Diptera
Family	Ceratopogonidae
Subfamily	Ceratopogoninae
Genus	*Culicoides* Latreille, 1809
Species	*Culicoides imicola* Kieffer (former name *C. pallidipennis* Carter)

there are reports of other suspected BTV vectors, such as *C. pulicaris* and species in the *C.* (*avaritia*) *obsoletus* complex. There are also differences concerning distribution and habitat, in particular the environmental factors. In Sicily, the species *C. imicola* resulted not present in undulated areas of high altitude. This was related to the very hot summers of the island, in particular in inner areas, because in highly angled undulating topographic areas, rapid desiccation leads to drying of soil, which prevents proper larvae development. Further examinations have indicated that a suitable climate and nutritious soil must be considered the biggest determining factors in distribution, rather than altitude. A likely reason is that *C. imicola* pupa are especially prone to drowning, so their eggs are often laid in surfaces free of running water. Therefore, as already reported the larvae, as with other early stages of the insect, need moist soil, so there tends to be a trade-off between dry and wet areas.

BTV was first described in South Africa, where it has probably been endemic in wild ruminants since antiquity. Since its discovery, BTV has had a major impact on sheep breeders in the country and has therefore been a key focus of research at the Onderstepoort Veterinary Research Institute in Pretoria, South Africa. Several key discoveries were made at this Institute, including the demonstration that the etiological agent of BT was a dsRNA virus that is transmitted by Culicoides midges and that multiple BTV serotypes circulate in nature. Multiple serotypes circulate each vector season with the occurrence of different serotypes depending largely on herd immunity. Indigenous sheep breeds, cattle, and wild ruminants are frequently infected, but rarely demonstrate clinical signs, whereas improved European sheep breeds are most susceptible.

After a long period, characterized by sporadic incursions in some areas of Europe, resulting in the vector being confined to tropical and subtropical areas, the situation changed in 1998. The emergences in Europe of several new virus strains were reported and carefully monitored. Sero-surveys showed that virus strains were closely related to those that were previously circulating in the same regions, but more virulent. So far, we were not able to determinate the factors responsible of such genetic changes in this RNA arbovirus; however, it is clear that some small changed occurred somewhere in the Middle West selectively, probably independently and/or simultaneously. Otherwise, the changes could be manifested in limited strains and transmitted to others by the mechanisms already reported, causing a spectacular migration of BTV. The migration was carefully monitored by sequence analyses and it is considered a consequence of the climatic changes of the last decades in Europe. It is noteworthy as in this case the monitoring and the relative map of pathways were fully determined and clearly reported, whereas in other cases of insect-borne diseases affecting plants, like that further reported, there is total confusion. In any case, monitoring is fundamental to understand the developing of epidemic episodes and respond, possibly by prevention. This is possible nowadays thanks to powerful and appropriate techniques and devices, but their use is often limited or even absent. Research institutes and universities should be heavy involved in such monitoring. The current regime of large freedom in investigation items should be changed in favor of studies dedicated to the general interests, with the necessary capacity of prevision of outbreak and of the sanitary system to react. Therefore, the basic research should be combined with the current needs to obtain information and data about the going on of the phenomenons interesting the public concern.

From some point on the geographic map in the Orient, the migration of strains of BTV started to conquer new territories following two independent pathways (Pioz et al., 2012; Purse et al., 2005). Analyses followed and evidenced the routes of arbovirus in the conquest of southern regions of Europe. An eastern group, which resulted in viruses similar to those isolated in India, moved laterally toward the Occident and was able to cross North Africa through the African Mediterranean coasts. In this way, it arrived on the coast of Tunisia and finally crossed the sea and landed in Europe, proceeding by three directions of expansion: the Balearics (Spain), Sardinia/Corsica, and Sicily/Calabria. In particular, the BTV-4 strain was involved, but before leaving Africa it joined with another serotype, BTV-2, coming from Central Africa. It has been reported that BT is endemic throughout most of South Africa, since 22 of the 26 known serotypes have been detected in this region. Later, parts of southern regions of Italy were also involved.

The adult *Culicoides* insects are not valid flyers, but they can be easily dispersed by the wind in passive journeys for hundreds of kilometers, in just a few nights. Otherwise, they can be transported by the routes of ruminants, as results of animal trade. These routes are like consolidated "corridors" from South Asia to Europe, formed by the connected ruminant population, usually passing through Pakistan, Afghanistan, Iran, Turkey, and beyond.

Therefore, distance is not a problem, and so far the diffusion was mainly not allowed by adverse climatic conditions. A second route, involving the spread of the "western" BTV-1, −2, −4, and − 9, passed through Anatolian Turkey, Greece, the South Balkans, then Albania, crossing Mediterranean Sea and reaching the Apulian peninsula (Melior and Wittiman, 2002). It is interesting to note how these routes are exactly the same used over the years by human migrants to reach Europe, landing in Italy after a desperate and dangerous navigation with any kind of craft or through a long terrestrial trip without any help or support.

Changes in climate conditions in Mediterranean countries could made possible stable introductions, driving the spread of vectors and pathologic agents (Mann et al., 2009). The environmental effects must be included in programs of control completed by data to understand vector behavior and distribution, seasonal abundance, and infection rates, with the aim to predict disease incidence and spread across coastal and inland areas, in order finally to define their role in virus overwintering, as reported in the following case.

In this regard, it is important to connect Culicoides vector's life cycle and the environmental factors. After the three initial stages (eggs, larvae, and pupae) highly affected by the availability of moisture, since semiaquatic habitats are necessary locations for larvae, pupae, and early adults, environment changes and adults are able to look for better places. Later, the adult females begin egg maturation to obtain the oviposition and restart the cycle. During the maturation of the eggs, named the gonadotropic cycle, the female looks for blood-feeding and, in the case of infected blood, starts the cycle involving the host's participation. Therefore, the first cycle is concentrated on homeostatic development and survival of the vector, whereas the second one is a consequence of the need for reproduction, consisting of the BTV transmission cycle. Let us start the cycle with a susceptible ruminant, which develops a viremia and is infective to the vector during the so-called latency period. After 2–4 days, an uninfected adult bites the infective host in search of blood. In this stage, called the extrinsic incubation period, the virus can enter the midgut of the Culicoides, with dissemination through the hemocoel. This is a large body cavity that is formed from an expanded "blood" system. The result is the consequent infection of the salivary glands of the host. After 4–20 days of incubation, the infected insect will bite a susceptible target, causing infection and restarting the cycle. The real effects of the disease occur during the extrinsic incubation period. Attention must be focused on the efficiency of the superorganism, as a complex system, and the possible consequences in the realization of some crucial passages. The efficiency of the virus's dissemination is dependent on temperature, influencing vector competence and rate of virogenesis, as well as other stages of the vector's life cycle. There are many sensitive steps, but the superorganism generally works, albeit we experience only its positive achievements. Every times, survival of the precedent steps is in peril.

Larvae can have enormous difficulty surviving winter's cold months (or the contrary if temperatures are gradable), and adult activity rates can also be affected. The same consideration can be extended for the availability of moisture and the persistence of the BTV between seasonal vector transmission in the overwintering period.

Bluetongue epidemic in Sardinia

Since 2000, the Mediterranean island of Sardinia island has been affected several times by BT epidemics (Calistri and Caporale, 2003). Sardinia is the second largest island in Italy and in the Mediterranean Sea, and it is positioned exactly in the middle of the sea. However, despite its position, in contrast to Sicily, Sardinia has rarely been subjected to invasion since Roman times. The reason for this is mainly related to the pride of its people, always defiant in the face of any kind of submission. This characteristic is often related to the harshness of Sardinia's central mountains. Sardinia has its own language (not a dialect) and is famous for its splendid seaside, but also for the production of a special cheese, named pecorino, since it is made by the sheep milk (*pecora* is sheep in Italian). Therefore, sheep are from ancient times one of the most important economic resources in the economy of the island. In 2018, the number of sheep and goats in Sardinia was estimated at 3.7 million: 93% sheep, meaning more sheep that inhabitants, and the remaining 7% goats. In the same year, the presence of serotype BTV3 was reported again after a period of absence. The animals affected were very few, but sufficient to create an alert. In that year, the sheep deaths accounted for 29,302 and 2500 farms were interested, accounting for 824,620 involved animals. Therefore, BT is a nightmare for Sardinian shepherds despite the compensation received for the damages suffered.

The immunization of susceptible sheep remains the most effective and practical control measure against BT. In order to protect sheep against multiple circulating serotypes, three pentavalent attenuated vaccines have been developed. Despite the proven efficacy of these vaccines in protecting sheep against the disease, several disadvantages are associated with their use in the field. In Sardinia, live-attenuated (modified live virus, or MLV) vaccines against BTV, first developed in the USA, are available and are used to control BTV. In small ruminant livestock, MLV vaccines provide good protection and limit the damages, although they were not able to solve the problem totally against clinical disease following infection with the homologous virus serotype. These MLV vaccines must continue to be used in the immediate future for protective immunization of sheep and goats against BT. Two doses of MLV (in some instances, just one dose) can provide substantial immunity to the epizootic serotypes of either BTV. Again, monitoring is fundamental, because vaccination should be provided to the field within 2 years of the initial incursion. However, it is recognized that this is not the final solution, and further research and development are warranted to provide more efficacious and effective vaccines for control viruses transmitted by biting midges, like BTV and epizootic hemorrhagic disease virus (EHDV) infections. EHDV is another economically important virus that can cause severe clinical disease in deer and, to a lesser extent, cattle. Until the recent climate changes in Southern Mediterranean, about the control strategies in fringe regions, the utilization of EHDV was limited to Turkey, Cyprus, and Israel, which are the first countries interested in the migration routes. However, vaccination did not solve the problem completely, and other actions should be taken.

Meanwhile, other solutions have been proposed. A research project was performed to deal with BTV diffusion by the utilization of natural products, in particular the neem. Although vaccination can give results, the problem of controlling the disease remains. Once again the target to obtain preventive results is the vector, and in particular the stage of the insect when it is easier to limit proliferation. *Culicoides*, like many other mosquito vectors, prefer to populate moist microhabitats, like irrigation channels, drainage pipes, and dung heaps. In particular, they can often be found in drinking troughs or water pools, where sheep and cattle drink water during the day, especially in summer (Fig. 5.17–5.19).

Fig. 5.17 Typical drinking trough for sheep and cattle.

Fig. 5.18 Irrigation channel frequently used by cattle to drink water.

Fig. 5.19 Another source of water.

The research of a new solution started several years ago, a long period of gestation being necessary for the experiments. My group of research is integrated by entomologists of the University of Cagliari, Sardinia, with long experience of BT. A previous research, involving virus detection in insect body pools of the prevailing captured midge species and supplemented by specific body region analyses, evidenced the serotype BTV-1 and BTV-2 in Sardinia. However, although *C. imicola* represents the main recorded BTV vector, other European *Culicoides* biting midges turned out to be possibly implicated in virus transmission in Sardinia. The *Newsteadi* complex, based on *C. newsteadi* species A and species B, prevailed for 47.7% of the biting midge species, followed by *C. imicola* (27.8%) and the *Obsoletus* complex (*C. obsoletus* and *C. scoticus*) (17.6%), with a territorial distribution that confirmed and completed the French consideration. While *C. imicola* was more abundant along coastal areas, the *Newsteadi* complex was frequently collected at higher altitudes and the *Obsoletus* complex was notably associated to cattle farms. Furthermore, the analyses of thorax and head, containing salivary glands, of the vectors, evidenced higher infection rates associated with *C. scoticus*, *Newsteadi* complex, in comparison to *C. imicola*. The virus was detected in *Newsteadi* complex and *C. obsoletus* in winter and spring, whereas it was mainly found in summer and autumn in *C. imicola*. Therefore, at least in Sardinia, BTV is transmitted by multiple *Culicoides* vectors. These data are very interesting, showing the complexity of the BT disease, and any program of control on insect-borne disease should be primarily based on an accurate analysis of the phenomenon at the local level, since the climatic envelope, consisting of the range of climatic variation in which a species can persist, is the main factor to consider.

Field experiments were performed and the results published in 2017 in *Research Veterinary Science*. The focus of the research was the definition of an innovative program of vector control and started from the knowledge of the cycle vector and the necessity of moist sites, consisting of breeding muddy sites close to livestock grazing areas utilized by the livestock, essential for the larvae's survival. Aqueous formulations neem cake from *Azadirachta indica* (see Chapter 7) turned out to be effective to control young instar populations of *Culicoides* biting midges (Foxi and Delrio, 2010, 2013). The field experiments were preceded by laboratory bioessays, which showed a larval lethal concentration value (LC_{50}) of 0.37 g/L of neem cake after 7 days. Furthermore, the same solution was applied directly at a dose of 100 g/m^2 on the water's edge of a pond margin of a livestock farm in Sardinia and other similar locations. The resulting data 28 days after the treatments evidenced a significant reduction of in *Culicoides* emergences. The treatment proved to be more effective in the case of *C. imicula* compared to other vector species, such as *C. catanei*, *C. circumscriptus*, and *C. festivipennis*. Therefore, neem cake can be considered useful in programs for prevention of BT outbreaks, outlining the current lack of effective tools in the fight against *Culicoides* vectors.

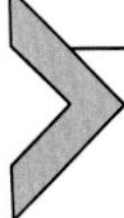

Insect-borne diseases affecting plants

The centrality of the nutritional environment

In all the arguments present in this book, it is necessary to realize that also in the case of insect-borne diseases, competition for food is the central aim of all activities of any living organism, and the disease is a collateral effect. In the last decades, several evolutions of the nutritional environment have changed radically the quality of food, including production, storage, distribution, and consumption. The changes in action indicate the key role of infection of plants that must be considered fundamental for human alimentation, even survival of mankind.

Nutrigenetics and nutrigenomics have demonstrated that the quality of aliments is important in prevention and on the impact of novel epidemic diseases, like globesity and related illnesses. The global nutrition environment evidences wracking contrasts and precarious equilibria, which can be derailed by environmental changes. Again, the problem of competition for food sources is fundamental. Mankind's world is divided into two parts: 520 million malnourished persons are on one side and the other side is overproduction with 20%–30% of the produced food lost or wasted, making a contribution of 8%–10% CO_2 increase only in 2010–2016. The food environmental scenario is in swidt evolution with several factors influencing the quantity and quality of food. The problem of desertification is well-known, involving the loss of precious cultivable lands, but another type of desertification concerns our everyday food (Kelly et al., 2016). Most of the millions of people living in developed countries suffer serious limited access to affordable fresh food (Karthika et al., 2016). This population, crowded into huge urban centers, creates devoted customers at hypermarkets in the weekend and consumers of ready-prepared food. They are subjected to the so-called "food desert." This desert concerns limited access to fresh whole food, like just-produced fruits and vegetables, which are considered very important for health maintenance and illness prevention. The Social Market Foundation reported that in 2018, 10.2 million people in the UK were interested in the food desert phenomenon, with 1.2 million living in deprived areas. The counterpart is an increasing consumption of ultra-processed foods, rich in preservatives, fats, salt, and sugars, but lacking in vitamins, antioxidants, and other natural products. Insect-borne diseases are involved in the storage and conservation of ultra-processed foods, since their organic materials are the ideal medium for larvae and microorganisms, which have sufficient time to work. Packaging is usually deficient, even wasteful, although recent solutions like intelligent packaging are now available.

The living conditions in towns and cities push us to the utilization of already prepared foods, determining the empires of takeaways and fast food restaurants, where the quality of food is an afterthought. The only reference is the taste of the ingested food, but nothing is known about the food's production and the origin, as well as other similar "details." The correlation between health and this kind of food has been well-evidenced by a recent cross-sectional study, indicating the risk in takeaway meals for coronary heart disease, type 2 diabetes, and obesity in children aged 9–10 years. Takeaway meals and fast-foods are known to be energy-dense and of poor diet quality. This kind of food should carry the same warnings used for tobacco products: "Fast food can cause a slow and painful death." The association between fast food and biomarkers of chronic disease and dietary intake has been clearly reported by many studies. In particular, Kant et al. stated that:

"American adults reported a mean of 3.9 (95% confidence interval 3.7, 4.0) AFH and 1.8 (1.6, 1.9) fast-food meals/week. Over 50% of adults reported ⩾3 away from home meals and >35% reported ⩾2 fast-food meals/week. The mean BMI of more frequent away from home meals or fast-food meal reporters was higher (Ptrend ⩽ 0.0004). Serum concentrations of total, LDL and HDL-cholesterol were related inversely with frequency of away from home meals ($P < 0.05$). Frequencies of fast-food meals and serum HDL-cholesterol were also related inversely ($P = 0.0001$). Serum concentrations of all examined micronutrients (except vitamin A and lycopene) declined with increasing frequency of away from home meals ($P < 0.05$); women and ⩾50-year olds were at higher risk."

The food desert is in practice a loss of precious molecular variability in alimentation, in favor of massive production of food consisting of a mixture of fats, sugars, and proteins, virtually repeated in every meal. Most of the current nutritional system is reduced in the composition to the basic nutrients, cooked in a simple and practical way, sustained by an aggressive marketing, ready to be quickly and easily consumed, without any idea about the story and the importance of what is going to be eaten.

The packaging with a beautiful picture of the food inside is the only appealing information, usually with few relation to the ingredients. This kind of food is ready to be cooked and eaten. No change or transformation is allowed, just eating. The lack of information and knowledge, including the correct scientific sensibility, is a general problem. I remember my difficulties explaining, at a professional meeting of an important U.S. government agency, that olive oil is not only a mixture of fatty acids and triglycerides, and the importance of its complex variable composition, being the result of the bio-transformation of environmental different inputs of the territory. That is the fundamental importance of the minor components, making the Mediterranean diet one of those most highly rated for health and classified by UNESCO, in 2010, as an Intangible Cultural Heritage of Humanity.

Fundamental competition for food

Although social attention is generally focused on insect-borne diseases affecting humans and the consequent medical treatments, parasites are spread all around and most economic damage concerns crops and other cultivated plants of strategic importance for alimentation of humanity and environment sustainability. Now we have clear the consequences of a human pandemia, but let us imagine the consequences of a pandemia affecting livestock or cereals, reducing totally the food's supply. In such case, there are not hospitals or medical supports and the carestia for billions of humans is the possible scenario. No available enough food means health loss and decrease of our immunity defenses. The damage and loss caused by these diseases are increasing and affecting strategic crops. However, once again the explanation of the causes of the phenomenon are quite complicated, and solutions are not as easy as might be expected.

The explanation of these diseases that we are going to use will not be different from the cases already reported. A forced alliance between very different organisms, based on the optimization of the trophic necessities of the parasite. Mankind in this contest is backstage; the negative consequences for us are simple collateral effects for the system, but enormous for the economy.

Plants are continuously subjected to several diseases wherein different organisms can be involved. As expected, crops and other economic valuable plants are mainly under investigation and considered in control strategies.

Crop and other agricultural production is expected to increase in relation to the needs of the world's population. As already reported, the global population is roughly divided between people in advanced countries facing excess of food and others experiencing difficulties in terms of limited availability and quantity of nutrients. The gap is rapidly increasing, because the second group is increasing in number and the first one is decreasing, as consequence of the globalization and concentration of production and distribution in multinational trusts. The global population is predicted to rise to 10–11 billion in the next 20–30 years, concentrated mainly in developing countries. To maintain these billions of people, so far the solution has been intensive utilization of the soil and natural sources to generate enough food, but unsustainable global situations in progress are also a consequence. Waiting for a desired equal distribution of life conditions, several aspects require immediate changes in favor of a sustainable and ecofriendly approach: new fertilizers and pest controls based on integrated activity systems.

Nowadays, agriculture also needs radical changes and brave decisions. Attention is usually focused on increasing the production, but postharvesting conditions, conservation, and waste are also very important, and these are the factors of loss of food, which are going to increase, in accordance with the production and the request. Losses due to arthropod pests currently account for around 20%–30% of the world's food production. Organic non-living matter is a preferred target for necrophagous, parasites, and demolish organisms, in search of nutrients.

Insect damage is important in the field and in stored products, besides microbial and fungus attacks. In general, counteractions are preventive if possible, but also based on any possible way to prevent economic damage. Although it has been demonstrated that excessive and inappropriate use of pesticides must frequently be considered counter-productive, the result of the increasing needs is that the use of insecticides has increased even more than necessary.

Agricultural production resorts to the use of a large quantity of insecticides and antimicrobials to increase production and preservation of foodstuff (Govindachari et al., 1992). Effects are not limited to the treated field, involving undesirable consequences for public health and the environment. Current pesticide pollution of Adige Valley in northern Italy, due to continuous heavy treatments of apple monoculture, is a clear example. Organism genetically modified introduction was presented as an innovative and definitive solution, but its utilization induced additional problems for farmers and the market. In any case, although scientists' opinions are mostly in favor of the use of OGMs, the position of consumers is exactly opposite. Either way, the expected drastic reduction of pesticides is still unachievable.

Most insecticides are usually utilized to reduce the damage caused by insects that destroy crops or transmit diseases. Already, agriculture accounts for 59% of the resistant insect species and veterinary pests for 41%. It is important to consider that, like antibiotics in the USA and UK mainly being utilized to treat livestock being considered immunostimulants, most insecticides are used in agricultural practice to improve production and preservation of foodstuffs. In both cases, use is now widespread, excessive, and arguably inappropriate. Furthermore, we may have to face new epidemic emergencies, due to several factors, including climate changes, concerning crops and livestock. Insects are vectors of important diseases involving nonhuman targets, causing important effects on plants and animals of strategic economic relevance. Recently, some of these diseases are of increasing concern to the general population, attracting a level of attention never experienced before, and generating great alarm in terms of the consequences of their rapid diffusion. The potential economic negative effects are enormous and the damage on the local economic system could be dramatic.

So far, much attention of research and public concern has been focused on vector-borne human diseases, in order to eradicate their presence and save as many lives as possible. However, if the resistance will affect and limit the life supports, like food and water, our surviving struggle will be in balance, due to two key factors: resistance to many pesticides and clear damage to the habitat. The story already reported for resistance to antibiotics can be repeated for insecticides, adding clear damage to the habitat. Even pyrethroids are now considered dangerous, since they can impair memory and movement in nontarget animals. It is well-known that without the continuous work of pollinating animals, humanity has no future. In a new hypothetical Ark, beneficial insects should take the first available places.

All the appeals and declarations for limited and selective utilization have so far had a limited impact. Mosquitoes are the most critical group of insects in the context of public health, because they transmit numerous diseases, causing millions of deaths annually. The frequent use of systemic insecticides to manage insect pests leads to the destabilization of ecosystems and enhanced resistance to insecticides by pests, suggesting a clear need for alternatives. Most plant-based products are not as effective as their synthetic counterparts, and to apply a mosquito control in a large-scale program under epidemic conditions may be unacceptable, also considering the relative cost. Natural products of plant origin with insecticidal properties have been implemented in the past for controlling various insect pests and vectors. Many studies have reported the effectiveness of plant extracts against mosquito larvae. Plant extracts are generally preferred because of their less harmful nature to nontarget organisms, due to their innate biodegradability.

Some of these diseases could in future involve the most productive and strategic plants have rapidly raised in significance and generated great alarm about the consequences of their diffusion. The potential economic negative effects are enormous and the damage to the local living system may be dramatic. Several epidemic emergencies are in action, and the emergency is going to become a normal trend, as a consequence of the permanence of several factors, including in first place climate changes.

On the basis of novel knowledge, some new approaches are emerging, changing the aspect of insect control. Integrated pest management is an important approach, developed in recent years to control disease vectors and limit economic agricultural damage, improving crop yield with minimal cost. Its main goals are to: (a) increase the knowledge of the relationship between insect pests and the other organisms involved, including the target one; (b) reduce pesticide application and quantity developing biological controls, farming practices, farmers' collaboration, and mechanical and physical controls; and (c) build new models of integrated managements on the basis of laboratory and field experiments, including the research of new active compounds. Novel pesticides, to be suitable, must be low cost, eco-friendly, from renewable and sustainable raw materials, nontoxic to nontarget organisms, of rapid degradation, and not accumulate in the environment.

This chapter will be mainly dedicated to the impact of migration by alien species as an element of habitat perturbation. In this case, environmental factors play a fundamental role in the comprehension of the phenomenon and its development. The effect on alien species is based on the potential impact of invasive species on the habitat wherein they are introduced. A habitat can be regarded as a complex dynamic system wherein, under the influences of biotic and abiotic factors, species tend to get their homeostatic equilibrium, and persist in the face of competitors, predators, and diseases. On the basis of knowledge of prior invasion history, emerging alien species can be considered as those never before encountered as alien or those already known but so far considered not invasive. Therefore, the key argument to understand the phenomenon is that the habitat we are considering is alien first for the invader, and that means a series of unsuccessful attempts until a final successful adaptive sequence of steps, determined by some changes in the habitat, which allow the invasion. In other words, the barriers that usually block the invasion can be removed by perturbations changing the habitat. Perturbations due to abiotic and biotic influences include those by the migrating species finding conditions to live even better than where they were native, also as a consequence of important climate changes in action or the possibility of finding food and resources easily. The influence of human activity is an important factor.

In the past, exchange of species between continents was considered an occasion, as demonstrated in the events consequent to the discovery of the New World. Christopher Columbus traveled to unknown lands in search of gold. After three journeys, the quantity of obtained gold was risible, but he found a much more important and economic valuable green treasure. Hundreds of species have been introduced to regions outside their native ranges, from and to America, and many have become permanent additions to local fauna and flora, also favored by agriculture care. Limiting the list to plant cases, examples of this interchange are tomato, cocoa, red pepper, and tobacco on one side and sugarcane, cotton, and mint on the other. In some cases, the introduction was initially complicated by prejudice, such as for potato, due to the general negative attitude against European *Solanaceae*, like henbane, jimsonweed, and deadly nightshade, in consideration of the content in tropane alkaloids and their effects on the CNS. However, American *Solanaceae*, like potato, tomato, sweet, and red pepper, contain alkaloids only in the juvenile stage, and therefore the positive alimentary results overcome any resistance and prejudice. On the other side, the success of cotton and sugarcane cultivations generated heavy impacts on the new habitats, due to monoculture, as well as forced human migration from Africa. One can only imagine nowadays Columbus coming back from a voyage to a new land and facing all the bureaucracy necessary to introduce just one new species from America.

Cultivation and alimentation, and also social equilibria, were radically changed by these introductions, though several other negative aspects, such as the abandonment of native sources of food or medicine, should also be noted. Therefore, devastating changes to local situations were, and often still are, considered to be a necessary secondary effect. Later, this approach changed in favor of maintenance of the current situation, considered the best obtainable. Nowadays, there are severe alerts concerning introduction of alien species, in order to avoid the risks and consequences of their introduction in other habitats and the unexpected damage they can produce. This means that a conservative tendency is becoming dominant in human approaches to the future, although history is full of examples of positive breeding and exchange. The problem is that the immediate effects of a change by introduction must be perturbative, just as in a chemical reaction to obtain a new molecule, it is necessary to break the bonds between the elements of the reagents using the activation energy. The final exergonic energy obtained is the positive result generated by the necessary destructive loss of the original starting point, forced to remain in the past by incoming modernity. Fear about the future, worrying about even minimal changes, before considering all the possible consequences, is a current dominant tendency, and a cause of social and political concern. Any resistance or restoration or attempts, to maintain the obsolete equilibria, are destined to be frustrated or outdated. In any case, besides the obvious initial negative effects, future consequences of an alien introduction are very difficult to predict, as is the assigning of victims and winners. The only smart reaction possible must be based on an understanding of the reasons and causes of the proceeding new order. Again, the analysis of the negative effects on the environment accuses mankind's tendency and exaggeration in influencing and diverting the natural flow of the events. In any case, destruction of the past is a natural consequence of ongoing changes in any single minimal part of the universe.

Exchange of living biomaterials poses a significant challenge to biosecurity interventions worldwide, as demonstrated by several reports and experience in any airport. Alien species can deeply affect current ecosystems, influencing ordinary human health and activities, thereby redefining the classical boundaries of biogeography, although the story of our planet is full of such phenomena, and they can be considered part of the driving forces of evolution and breeding. Invasion of alien species is a consequence of the physical forces governing movements of matter, but the same forces can provoke acceleration or stagnation. In this period, the biodiversity is a phase of rapid change with loss or danger of extinction of a number of species augmenting the extent of migrations.

The research reacted in good accordance to the increase of the alien phenomenon. The number of papers on alien or invasive species published per year (1980–2010), according to ISI Web of Science (accessed July 1, 2011), started from *c.*100 papers in 2000 to reach 1000 in 2005 and 1500 in 2010. The entity of the phenomenon can be deduced by a global database of the first regional records of alien species covering the years 1500–2005. In a database of considered 45,984 species, a screening established a first record of 16,019 established alien species, in order to investigate the temporal dynamics of the occurrence of emerging alien species worldwide. The study evidenced that 1%–16% of all species on Earth, with differences in taxonomic groups, could be qualified as potential alien species and suggested that a high proportion of emerging alien species remain to be encountered, whose future impacts may be negligible or revolutionary. The surprisingly high proportion of species in recent records, which have never been recorded as alien before, means that the potential risk may be high. The high proportion of these emerging alien species mainly resulted from the increased accessibility of new source species pools in the native range. Our ability to predict the identity and impact of future invasive alien species is pivotal, and it is largely dependent upon knowledge of prior invasions and understanding temporal trends, origins, their interactions inside new habitats, and the drivers of their spread, improving prevention and risk assessment tools. Risk assessment approaches that rely on invasion history will need to be prioritized, with the awareness that physical boundaries and governmental declarations cannot halt or prevent biological invasions completely.

Migrations and invasions are a constant phenomenon of biological activity, therefore attempts to control the situations and predict the consequences are probably useless. Despite all controls, the phenomenon is relevant and the rate of emergence of new alien species is still considered high. One-quarter of first records during 2000–2005 were of species that had not been previously recorded anywhere as alien and during the past two centuries, a strong worldwide increase of established alien species was registered.

The current state of the phenomenon shows serious driving forces in favor of an increase of migrations. Although a large variation across taxa was also evident as part of the phenomenon, model results show that the high proportion of emerging alien species cannot be explained solely by increases in well-known drivers, such as the amount of imported commodities from historically important source regions or global effects. Instead, these dynamics reflect the continuous incorporation of possible new regions into the pool of sources of potential alien species, likely as a consequence of expanding trade networks and environmental change. Increases of temperature and climate change, causing land degradation and sources spoilage, are among the prominent driving forces in action and this trend will probably be confirmed in future. Species, so far confined in a traditional territory, are practically forced to move by environment changes. These changes contemporaneously involve the starting territory and the new ones. This process compensates for the depletion of the historically important source species pool through successive invasions, opening the door to new phenomena.

These results suggest the importance of studying carefully selected cases in act. The study of the phenomenon of the migrations in act and their consequences should be released by alerts and any a priori protective reaction, and on the contrary based on the study of dynamics of past invasions and on adequate and better-informed predictions of future trajectories of alien species movements and accumulation. Considering that any alien species was once an emerging species, there must be conditions for its improvement. The dynamics of emerging alien species accumulation can provide a direct measure of ongoing invasion dynamics without the confounding effect of subsequent introductions either from the native range or from already occupied regions in the alien range. Later, some considerations were reported as model cases of potential alien species and the risk in their diffusion in restricted territories. In particular, we considered important the case of "emerging alien species," followed in its delicate transient status to alien species causing interest for environmental impact and diffusion of related diseases.

Invasive alien species

Somewhere else at some time, an emerging alien species can possess a high chance of becoming an alien species arising from a native or already known species. The pool of species involved does not encompass all native species, but is limited to those with a high potential of being introduced and establishing in a new region. Therefore, the flora and fauna of the relevant habitats must be compared. Knowledge about the proportion of emerging alien species will also be important for biosecurity, which often relies on information of known alien species, including species not yet recorded but suspected to have a high risk of arrival and impact.

Again, the past can give us some key information. The Aral Lake was once called the Aral Sea and used to be the fourth largest lake in the world, with an area of 68,000 km^2 and water 16–24 m deep. The original name means "Sea of Islands," referring to the more than 1100 islands that had dotted its waters, but now they are hills emerging from a desert. The Aral Sea drainage basin encompasses Uzbekistan and parts of Tajikistan,

Turkmenistan, Kyrgyzstan, Kazakhstan, Afghanistan, and Iran. Until the 1960s, the inflows, consisting of several rivers that fed the Aral Sea, were diverted by Soviet irrigation projects. Since that time, the Aral Sea has declined, the water disappearing and leaving space to the desert. In August 2014, satellite images revealed that the eastern basin of the Aral Sea had completely dried up, and desertification is continuing. The same situation was caused by Mongols, who diverted its two main southern effluents to the Caspian Sea. The eastern basin is now called the Aralkum Desert. Among the sand of the desert, there are still corroded carcasses of boats, once fishing in hundreds on the sea and now destinations for tourists. Most of remaining water is salaried, and the zoocenosis radically changed consequently. The alterations in the biodiversity of the lake represent not only a natural response to a decrease in water level and a subsequent increase in salinity, but also effects of nonnative species introduction. The only organisms able to survive are small crustaceans and microscopic colonies of *Artemia* sp., similar to the *A. salina* that we use in our laboratories to test toxicity. *Artemia* crustaceans were not present until the 1950s; *Artemia* is an alien species since its eggs were introduced by migratory birds. However, fossils reveal that there had been four similar desertifications during the last 20,000 years. Therefore, we have the possibility of a congruence of natural and human causes of the phenomenon, or simply it must be considered the further case of a nonsense action persevered by mankind, to force the natural equilibrium to the own temporary interest. Many similar examples can be considered and analyzed to understand other situations in progress.

Several selected examples of the environmental effects of invasive alien species in different parts of the planet can be reported as case studies. These are focused mainly on native species extinction, causing changes in species richness and abundance, alterations to food web interactions, monitoring of steps of invasion, routes of transfer, economic relevance, social and cultural consequences, etc. The economic impacts usually generate attention and alerts, but in some cases, exotic species have minimal demonstrated impacts in their new environments, or have immediate direct impacts or have indirect impacts that may not be immediately apparent. To distinguish, understand, and consider correctly each case, two aspects can be considered fundamental: environmental changes and the role of native species versus alien ones. So far, there are local cases, interesting limited parts of the planet. They are important case studies to understand the phenomenon and its driving forces.

The first case, dedicated to the Argentine stem weevil, reports on a consolidated classic invasion by an alien species and its catastrophic impact in New Zealand (Barker et al., 1989; Barker and Pottinger, 1986). However, counteractions so far have been unsuccessful. The second study case will focus on a developing insect-borne disease affecting olive trees in a southern region of Italy. Due to its developing and progressive damage, it can be considered a model of this kind of phenomenon. In the third case, fennel and its potential impact in pastures of some USA states are considered.

First case: Impact of alien species over decades

An important and well-reported case concerns the impact of invasive Coleoptera species *Listronotus bonariensis* Kuschel (Argentine stem weevil) to New Zealand natural grassland ecosystems. As already considered, the impacts of changes in isolated territories, such as islands, are very interesting as a case study. In past times, exotic invertebrates were generally considered of low impact on plants in New Zealand's natural ecosystems. This was possibly due to the high level of endemism of New Zealand native plants and their phylogenetic distance from host plants of many invasive plant pests. However, the case of this herbivorous invasive species indicates that geographic and/or taxonomic distances may have limited importance, and in some cases this can result in a clue about the environmental system. The adult of this weevil presents limited dimensions, up to 3 mm long. The body is characterized by a gray color with a waxy covering. This wax plays an important role in camouflage, since soil particles adhere to the wax, and when the insect is stationary, it is difficult to spot. Therefore, the adults, thanks to their mimetic capacities, small size, and ability to remain immobile for long times, are practically impossible to detect in the soil, and virtually invisible to predators or other enemies. Their larvae are tiny, cream-colored grubs with brownish heads, with egg incubation periods of 7–15 days, larval periods of 30 days and pupae periods of 13 days. The presence of the insect is usually obtained through the detection of the eggs, which can be observed in the inferior part of the sheaths, where they accumulate near the soil surface. Eggs present an enlarged form with round apexes. Initially their color is light yellow, but this changes to black and therefore they became visible and can be easily localized on the plant's surface, in particular on leaves. An important part of the battle against pests in agriculture concerns identifying the preference of each species in feeding. Each insect selects plants for feeding, which is part of its behavior in the ecosystem, and often this can suggest strategies for its control. Feeding in the case of herbal species, like cereals (*Graminaceae*) living for a short time, requires an exact correlation between the lifetime of the plant and the growing stages of the insect. Oviposition starts in coincidence with the growing of cereals, like mays and wheat, but damage is produced later, when the larvae are active and the plants are mature and grown. The larva needs large quantity of food to increase rapidly and it feeds insatiably on the crown, the stem apices, and the radicular meristems of the weeds. The eggs hatch and the larvae begin to feed inside the plant stem until they reach the second or third instar, at which point they burrow out of the plant and drop to the ground. Once outside the plant, the larva begins to feed at the base of the turf plant, ingesting plant material from the stems and crown. After the fifth instar is reached, the larva pupates within the upper soil profile and the adult soon emerges. Adults then mate and lay the next generation of eggs over winter to resume the life cycle in the following season. The favored production of eggs causes a massive feeding and the consequent destruction of the host plants, exactly when they should reach their final stage of maturity, just before the harvest. Therefore, the consequence of the presence of this insect is the damage of the crobs, with reduction of the number of plants and loss of production, diminution of the roots' volumes and radical apparatus, until the final complete plant breakdown. It is possible that other aspects are involved, such as the consequent prevalence of some native *Graminaceae* weeds, which pollute the field and probably facilitate pest diffusion.

The species is native to South America, and in particular widely distributed in the central regions of Argentina, Uruguay, Chile, Bolivia, and Patagonia. However, the main damage is reported for Australia and New Zealand. The status as a primary pest in an invaded habitat, not being a notable problem elsewhere in particular in the native regions, is similar to several other species. In this case, Goldson et al. (2001) attributed such eruptions of relatively minor pests to a lack of natural enemies in New Zealand's highly modified pastoral ecosystem. Again, the ability of human activity to change and reshape the territory, destroying natural biodiversity, is the main cause of the emergence of pests. In New Zealand, weeds are affected with consequent serious economic damage in several agriculture sectors and livestock. New Zealand's economy is largely based on highly productive pastoral farming, embracing extensive sheep grazing and large-scale milk production, requiring the availability of enormous quantity of grasses. Grasses are the main pasture for sheep as for other herbivorous creatures, like the Argentine weevil. Pasture covers more than 10 million hectares (38%) of New Zealand

and is the largest single land use in the country. This is made possible by a temperate climate and large territories, modified with heavy investment in land improvement, including the introduction of European grasses and regular application of imported fertilizers. Highly skilled farm management by owner-occupiers constitutes one of the highest ratios of capital to labor in farming anywhere in the world. It is important to consider that lamb, mutton, and beef represent a fundamental part of New Zealand's economy, including their globally precious wool. Wool was New Zealand's main export earner from the 1850s until the start of the 20th century. In 1920, wool contributed 26% of New Zealand's total value of exports, but by 2011 its contribution had fallen to 1.6%.

However, we must always remember that human activities are interconnected, as well as the consequences, and therefore we must refer to history and to the most impactful episodes, like wars. The New Zealand wool boom, which started in 1951, was a direct consequence of the U.S. policy in the Korean war (1950–1953). At the beginning of the war, the USA sought to buy large quantities of wool, necessary for soldier uniforms and to maintain stockpiles. As soon as the New Zealand wool was chosen, the prices tripled overnight and farms were asked for record production. After the war ended, the export price of wool declined, until it fell to 40% in 1960. However, the biological machine had already been started and New Zealand's sheep population continued to rise and produce. The sheep numbers rose from 34.8 million in 1951 to 70.5 million in 1982. A series of measures by the government caused a decline of this number, until it reached 39.3 million in 2004, and a slight increment of the sheep population put it at 40.1 million in 2006.

Nowadays, New Zealand is the world's top dairy exporter, accounting for a third of the world's dairy trade. Dairy alone now accounts for 35% of New Zealand's total commodity export value. Total sheep numbers as of June 30, 2018 totaled 27.3 million. However, this is down 0.8% on the previous June, whereas total beef cattle numbers on June 30, 2018 totaled 3.68 million, up 1.9% from the year before. Furthermore, total dairy cattle numbers rose slightly to 6.60 million at June 30, 2018 due to a decrease in the numbers of cows and heifers. Total beef and veal receipts were expected to total $3.42 billion FOB in 2018–2019. However, New Zealand beef and veal exports were forecast to decrease by 3.1% to 415,000t shipped weight, despite the more favorable exchange rate. For 2018–2019, the number of cattle processed for export was forecast to decline by 3.0% to 2.51 million. This forecast decrease followed from a 9.6% increase in

2017–2018, due to a large increase of cows and bulls. The mix of cattle classes was expected to remain relatively steady in 2018–2019; however, some alien species have been causing a decline of production, including the famous Merino wool.

Let us now return to the alien species with the last part of this story. The high quality and production of products derived from pasture fueled a high rate of export of meat and wool, until the invasion of New Zealand's native grassland flora by the Argentine stem weevil. The presence of this species in New Zealand was first reported in late 1927 by Marshall (1937). However, its introduction is likely to have been earlier in the 20th century and there may have been more than one introduction. In any case, in the next 30 years the occurrence of this weevil became an abundant and massive agricultural pest. It was first recognized as a pest to wheat in 1933, but it was not until the late 1950s that the weevil's potential for damaging pasture was recognized. From that period, the impact of this pest increased and major damage was evidenced in crop grasses, and in some of New Zealand's introduced Graminaceae including common cereals and pasture grasses, like perennial ryegrass (*Lotium perenne*) and Italian ryegrass (*Lotium multiflorum*). *L. bonariensis* is also a major pest of cool climate turf grasses in Australia, and can frequently be found in association with these plants. The consequence is that this weevil has been highly adventive and has acquired host plants that it did not evolve with, at least at a species level. This is the main reason for its diffusion and consequent agricultural damages, which however were not equal in whole New Zealand.

Several other exotic species of Curculionidae have been studied in New Zealand native grasslands, but few have been recorded to be feeding or breeding on New Zealand native plants. Their presence in native grasslands may often be simply a case of vagrancy, for example, in the case of the lucerne weevil, *Sitona discoideus* Gyllenhal, but its hosts' preferences are restricted to species of *Medicago* spp. and *Trifolium* spp., and it is unlikely to have host plants in New Zealand's native flora. In contrast, a flightless, polyphagous, European weevil, *Otiorhynchus ovatus* L., which occurs in tussock grasslands in central Otago, might feed on some New Zealand native plants. Furthermore, this species, and three other *Otiorhynchus* spp. that are established in New Zealand, were not recorded on native plants sampled. That means that each case must be considered and studied as singular, and general considerations or extrapolations may be misleading.

Having become established throughout New Zealand, by the early 1990s *L. bonariensis* was estimated to be causing damage to the intensive pastoral sector amounting to NZ$78–251 million (US$50–170 million) annually. Responses were focused on the use of pesticides and biological control, and have been considerably offset by the successful introduction of the parasitoid *Microctonus hyperodae* Loan (Hymenoptera: Braconidae) in 1992, but recent research has indicated that *M. hyperodae* is becoming less effective as a biological control agent for reasons that are not entirely understood. This is a further indication of the difficulty of biological control to produce the expected results, or probably that the natural clock needs different times for production of visible effects.

Second case: Novel insect-borne disease in action

The second case was selected due to its potential impact. Therefore, in this case, we shall try to understand the conditions for development of a dangerous alien species.

We shall focus on an insect-borne disease, though the scenario is far larger. Pathogens are distributed everywhere and their targets are any organic matter, including living organisms. Their effects on cultivated plants are under examination, with some cases of particular interest, with insect-borne diseases first in line. The interest is strictly related to the economic damage. Therefore, there is no real interest in environmental impact; effects on crops and cultivated plants are only considered for the loss of production.

We must recall that insect-borne diseases are the results of a complex multiorganism interaction. The network of several different collaborating organisms is on the basis of diffusion, effectiveness, and metabolism of insect vectors, including the resistance phenomenon. The integrated network acts like a "superorganism," integrating functions of all the different types of involved organisms. A disease affecting the host is the result of a determined and useful collaboration between totally different organisms, from bacteria to multicellular organisms, and an integrated system is key to survival and proliferation. This is an important lesson, based on the consideration of the existence of several levels of eco-friendly interactions in the environment. The consequences, which we consider as negative, are only the collateral effects of competitive struggle. Here, an important recent episode of insect-borne disease affecting olive trees is reported (Figs. 5.20 and 5.21). In addition to the economic impact, the Xylella affair is an emblematic case evidencing the difficulties of facing such emergencies with ordinary responses (Benelli, 2018; EPPO/OEPP, 2016; Nicoletti et al., 2016; Purcell, 1997).

Fig. 5.20 The effects of Xylella attacks on olive trees in Apulia.

Fig. 5.21 All the olive trees in this field were devastated by Xylella.

- *Xylella fastidiosa* and the olive destiny
- Chronology of *Xylella fastidiosa* outbreaks

1870: Reports in California of grape wine "mysterious disease" causing the deaths of plants.
1890: The disease practically disappeared or was not recorded.
1892: Newton B. Pierce reports on the disease in California and on the damage to grape plants.
1920: New epidemic diseases appeared in California, apparently not linked to the previous episode.
1920: Alfa-alfa disease (AD), no other cases reported.
1930: Hewitt names the rediscovered grape wine disease the Pierce disaster (PD).
1930: Reports on disease in peach and other trees.
1933: PD spreads in southern USA and is considered relevant from an economic point of view.
1940: Major epidemic disaster; vectors come from alpha-alpha (AD) through a "virus"; the research is related to xylem sap-feeders and considered xylem-limited.
1970: Almonds and oaks also affected; symptomless plant host discovered.
1972: PD is considered important, but classified as one of several diseases affecting grape wine and other cultivated species.
2011: First cases of dehydrated olive trees near Lecce town in Puglia; the disease is named Olive Quick Decline Syndrome (OQDS) on the basis of its effects.
2012–2014: During these years, in most of Salento, which is the peninsula of Apulia region, the presence of *Xylella* spread affecting the centenary olive trees. The analyses report an increasing presence of OQDS.
2014: The disease spreads, affecting more than 9000 ha. The alert is raising public concern as well as farmers' warnings.
2014: An International Symposium on the European outbreak of *Xylella fastidiosa* in olive trees is held in Gallipoli (October 21–22, 2014), and followed by technical laboratory workshops at the CRSFA, Locorotondo (October 23–24, 2014). This was the first meeting about OQDS, with the participation of more than 200 experts. The Proceedings of the Symposium are published in the *Journal of Plant Pathology* (2014), 96 (4, Supplement).
2015: The counteractions to limit OQDS demonstrate inefficacy; the only adopted measure is the eradication campaign, in accordance with EU protocols and financial supports. Eradication of the infected trees

starts with the destruction of dozens of trees, and continues with the removal of any plants near the affected trees. The population of the region demonstrates against the eradication campaign.

2015: The regional court accepts the considerations against the eradication, but the campaign goes on although slowing down respect to the original plan. In July, the French authorities notify the Commission of the first outbreak of *X. fastidiosa* subspecies *multiplex* in Corsica and in PACA (departments: Var and Alpes-Maritimes), 25 outbreaks in PACA and some 350 in Corsica. The presence of *X. fastidiosa* subsp. *pauca* is also detected in one outbreak in PACA (Menton). Ornamental plants, such as *Polygala myrtifolia*, cause the main host plants to be affected, although new plant species are detected as investigations progress.

2016: A line of containment on the northern boundary of Salento is adopted. The line consists in 2km of eradication of any plant, situated coast to coast between the Ionic Sea and the Adriatic Sea.

2016: Sporadic news of the disease's presence is reported in Corsica and other parts of Europe.

2017: By the beginning of the disease up to 2017, Xylella has infected up to 1 million trees, causing the death of most of them. Reports about the presence of disease outside of the containment zone are presented, whereas in Salento the massive ecological disaster is evident.

2018: In March, the Italian authorities notify the presence of *Xylella* in different parts of the buffer zones already established with a large number of outbreaks in the zone adjacent to the northern zones of the buffer zone. On June 27, the EU Commission extends the demarcated area in Apulia by around 20km toward the north of the region. The EU Commission starts a campaign of controls about Xylella to entire territories of state members (EPPO/OEPP, 2004). In 2016–18, 40,600 inspections on 20,000 samples were analyzed in garden centers, nurseries, and other sites across the EU territory, with the exclusion of the demarcated areas. In the demarcated areas established in the EU territory, instead, more than 110,000 samples were analyzed in 2016. On April 10, the Spanish authorities also notify the Commission of the first presence of *X. fastidiosa* susp. *Multiplex* on one olive plant in an open field, in Spain mainland, in the autonomous region of Madrid. The area is demarcated and eradication measures taken, and movement of specified plants out of that area is blocked.

2019: The number of continuous outbreaks leads to the conclusion that eradication of the pest in the buffer zone is no longer possible and, because of the significant delays in the removal of those infected plants, the risk of further spread toward the north of the Apulia region is possible, since the range of host plants increases as investigations progress. In Apulia, Xylella has mainly infected olive trees. No infections have been confirmed so far on *Vitis* and *Citrus*. Movement of specified plants out of demarcated area is currently not authorized, except for grapevine nursery material subject to hot water treatment. Olive cultivars resistant to Xylella are started in several areas of Salento where trees had been eradicated. As *X. fastidiosa* is considered to be established, the entire territory of Corsica is declared as an area under containment. At the present time, the EU territory, with the exception of the officially demarcated areas, is considered free from *X. fastidiosa* based on official surveys.

2019: Several analyses report and confirm the presence of Xylella out of the zone so far considered infected, causing a general alert about the expansion of the disease.

2019: OQDS is currently considered as one of the greatest threats to European agriculture and landscape, with economically relevant damage and difficulties in stopping its further diffusion. Olive oil producers in Apulia react to the OQDS damage and consequent decrease of olive oil production with massive imports of oil from Spain and Greece. The different quality and cost of the imported oil cause an invasion of low-cost oil in the market.

The second case here reported concerns the olive tree and the production of olive oil in a region of Italy, once the most productive. Olive tree (*Olea europea* L.) is a wood species endemic of Mediterranean regions, wherein Greece, Spain, and Italy in particular are contending for primacy in olive oil production. The olive tree has been cultivated from ancient times and was probably one of the first plants to be selected for production of edible oil of high quality using olive oil mill (Fig. 5.22) totally dedicated only to this kind of product. Although the species is highly resistant and able to live for more than a century without particular treatments, recently a great alert changed the scenario completely and the very survival of most olive trees is in danger. The current numbers evidence a situation that cannot be underestimated. In 2018, the Italian production of olive oil registered a decrease of more than 43% in comparison to 2017 (185,000 t against 428,922 t). The Italian production was inferior to that of Greece, which also reported a decrease. Spain confirmed its first place with about 1.5 million tonnes, with an increase of 24%. However, the Spanish production is the result of enormous imports from Northern Africa and of production methods with

Fig. 5.22 The modern system of extraction of olive oil in Italy, useful to obtain extra virgin olive oil of high quality.

devices that are not utilized in Italy. The cost is lowered, but the quality cannot be compared. In practices, the trade name "extra virgin" is the same but the oils are very different. The important nutritional properties of olive oil, as a fundamental ingredient of the famous Mediterranean diet, are at risk, as well as the presence of precious antioxidant polyphenols and unsaponifiable oily compounds, but there is much more. In Italy, food like olive oil means tradition and culture, including a great variety of produced oils. Each region is proud of many cultivars selected from a long time ago and people consider as part of its character the taste and the quality obtained with care and traditional methods. The same goes for wine, pasta, cheese, and many other food products, appreciated worldwide for their unique characters. Traditional and distinct foods are considered a necessary mark for each town and village. Food made in Italy, synonymous with high quality and precious tastes, is made by a mixture of multiple differences and highly variable local production, generated by the fusion of ethnical different traditions both near and far, in part autochthonous and in part consequences of the influences of many invasions affecting the Italian peninsula and its main islands. Italian food products are desired by many producers in the world, but their quality cannot be reproduced.

In 2017, the per capita consumption of the olive oil in Italy was 8.9 kg yearly (+4.7% vs 2017), but the same year registered also an increase of the prize up to 40%. The reasons for these changes in a single year were considered a consequence of the weather in the 2017 winter, which had an exceptional freezing week, affecting seriously the olive trees of several regions, but in one part of south Italy the problems were very different and related to an insect-borne disease.

Why in Apulia?

First, why in Apulia? Italy is a peninsula, whose geographic form is considered similar to a shoot, or more often a boot. Islands and peninsulas are very interesting habitats, because of their endemisms, caused by their isolation, and capacity to preserve and produce living varieties. Apulia is the heel of the boot, and therefore a peninsula of a peninsula. Its territory pushes out into the Adriatic Sea, in great part virtually isolated from the rest of Italy. It is a plane territory, mainly surrendered by the Adriatic Sea, in the ideal position for the exchange with the Orient, characterized in the past by the import of species and silk from the Orient and the export of food of high quality (Fig. 5.23).

Before any consideration about the recent disease that destroyed the olive trees in this part of Italy, named Salento, we must consider the environmental factors and the sequence of events already reported. Salento is a sub-peninsula of the Italian peninsula, full of history. Because of its position and its arc form, it is described as the "heel" of the Italian "boot." The peninsula is also known as "Terra d'Otranto" and in the past was named Salentina. In ancient times, it was known as Messapia, because during the Bronze Age the Salento peninsula was inhabited by Indo-European populations. In fact, the first reports about the inhabitants of this land, around the 5th century BCE, concern the Messaps population, who were dedicated to agriculture, horse breeding, and pottery. Testimonies of this period are the 10 dolmens and menhirs in the lower part of the land, as well as the construction of cities with imposing walls. Salento is in the middle of the Mediterranean Sea and, being in front of Greece, looks like a bridge between central continental Europe and the Orient.

Fig. 5.23 The particular geographical configuration of the Apulian peninsula.

Due to its strategic position, Salento was the theater of several civilizations. Later, Salento history met with Oriental history due to its Mediterranean facing, and according to legend, inhabitants of Crete founded the town Lecce. In fact, during the 8th century, Greek settlers founded along the coast several cities that are still very important, such as Gallipoli, Otranto, and Taranto, which would become landmarks of Magna Grecia, becoming small capital cities more prestigious than those of the fatherland. Romans and later Byzantines dominated the region, until the Norman rule that, with Federico II, brought the region to be a very important cultural center. Under the Swabians, a long period of decadence started and the peninsula went through a long series of invasions, sacks, and destruction. Another dramatic period for Salento was the Turks' invasion that in 1480 attacked and sacked Otranto, whose resistance was punished by the killing of 800 inhabitants.

Under Spanish rule, starting from the 16th century, towns were fortified and Lecce became one of the most beautiful and important cities for cultural and artistic activities. The beauty of the inland parts with their baroque aspect attracted nobles and scholars, and nowadays the magnificent architecture is considered unequaled in the world. The beautiful buildings of that period, known as the "Apulia's Baroque," are made by stones of blank color, typical of the territory.

All these civilizations sacred the olive trees and the olive oil, which from antiquity is extracted from the drupes with sapient art. Salento's landscape is dominated by magnificent olive trees, which state as silent giants stating from centuries to protect the territory and signing the boundaries between the blue of the sea, the gold of the coastal sands, and the green of the vegetation. The habitat, dependent on the olive trees for thousands of years, without their protection now risks desertification as a possible future.

In 2013, this beautiful part of south Italy, the Salento Peninsula, well-known for its production of olive oil and wines, was troubled by a dramatic phenomenon, never reported in human memory. The olive trees started to lose their leaves and were rapidly reduced to skeletons. It all started somewhere near the town of Gallipoli, in the south of Apulia. The diffusion of the decline was rapid and in April 2015 the whole province of Lecce and other zones of Apulia were highly affected, always focused on the Salento Peninsula. Almond and oleander plants in the region also tested positive for the same disease, but without epidemic episodes. The disease has been called Olive Quick Decline Syndrome (OQDS) and its characteristics had never been recorded, in particular on olive trees.

The disease first caused withering and desiccation of terminal and lateral shoots, which are usually produced by the tree in great quantities and distributed randomly. OQDS results in the collapse of the rest of the canopy and ultimately the death of the trees. The giant trees, considered an emblem of nature's puissance and vitality, which survived for centuries despite any kind of offense and difficulties, in a few months were reduced to pitiful skeletons. The scale of the phenomenon was underestimated and information largely lacking until, at the beginning of 2015, the epidemic feature of the disease was evident to everyone, with thousands of centenary olive trees completely dehydrated by the disease. Up to a million olive trees were clearly affected or resulted infected in the southern part of Apulia. Only two evident responses were performed: a careful detection of the diffusion of the disease, and the eradication of affected trees and the nearest ones to these trees. After the eradication, the treated areas appeared totally changed, practically like a desert. In addition to the heavy economic damage, loss of the olive trees means a tremendous cultural and environmental impact on this territory, where they are the symbol of region's identity and provide important support for tourism. Nowadays, the situation reports damage totaling 1.2 billion euros and 10 million plants damaged in a territory of 720,000 ha, practically most of the Apulia territory. This could be only the beginning.

The disease

The bacterium responsible for the disease is considered to be *Xylella fastidiosa*. In this regard, OQDS arose due to a well-known plant disease, which affects several economically important plants including grapevines, almond, pear, peach, coffee, and citrus, but ornamental plants such as oleander and *Prunus* spp., forestry crops (sycamore, mulberry, red maple, elm, oak), several weeds, and native plants are also infected (Romi, 2010; Reiter, 1998, 2001). In the latter cases, the bacterium often does not cause any visible symptoms, but the plants act as efficient healthy carriers. The disease is called PD (Pierce's disease, or even PD disaster), from the name of the scientist who first reported the effects. However, although the presence of *X. fastidiosa* in Italy had already been reported, OQDS was totally unexpected, although the conditions for the infection had already been reported by scientists. Plants infected by *X. fastidiosa* mostly grow in parts of the American continent (North, Central, and South America), that have mild winters and long growing seasons, in particular in southern California, whose climate is similar to that of the Mediterranean (Chatterjee et al., 2008). The disease models have long predicted that the pathogen could establish in Mediterranean regions, owing to similar climatic and environmental conditions. For instance, Greece, southern Spain, and Italy have been identified as particularly at risk from *X. fastidiosa* if vectors of the bacterium occur.

Indeed, in southern California, both the host range and spread of *X. fastidiosa* have expanded further as a consequence of the introduction and spread of an alien new vector, the glassy-winged sharpshooter (*Homalodisca vitripennis*). Wong et al. (2005) from the University of California-Riverside reported olives as a host of *X. fastidiosa*, examining more than 500 samples from plants located in five cities (Fillmore, San Diego, Redlands, Riverside, and Tustin) in southern California counties. After increasing incidences of olive tree mortality occurring in the Los Angeles area in 2008, surveys were conducted by Krugner et al. (2014) to evaluate the association of *X. fastidiosa* with scorch and dieback symptoms in olive trees in southern California and the southern San Joaquin Valley. However, until the arise of ODDS, no significant alert was reported, and *X. fastidios*a was considered among the 150 pests affecting olive trees (Janse and Obradovic, 2010; Krugner et al., 2014). In fact, PD had previously been considered to affect mainly grape wine and sporadically other plants, such as oleander, almond, cherry tree, *Polygala myrtiflora*, *Spartium junceum*, and a high number of other targets, with different effects (Costa et al., 2000). PD is known to be prevalent within the USA from Florida to California and outside the USA in Central and South America, with varying effects and impacts (Hernandez-Martinez et al., 2006, 2007).

Therefore, this is more or less the story of a unique phenomenon, whose nature and causes must be considered with great attention, at least for the dramatic consequences and the possible scenario. Owing to the climate and soil condition, as well as a strong tradition of cultivation, olive trees are the dominant plants in Apulia, real living monuments to the spirit of this territory. Cultivation of olive trees has been a fundamental occupation of the local population as well as a key component of alimentation. In 2015, the olive oil produced in Apulia accounted for 60.4% of total Italian production, followed by Calabria and Sicily with 17% each. In 2018, oil production in Apulia decreased by 58%, with damage costing 1 billion euros. To face the increasing requirements of the Italian market, the result was an import of 43.3 million euros' worth of oil from Greece and Tunisia. The imported oil was converted and marketed as Italian oil, despite its low quality, and caused confusion and disappointment in consumers. However, if adverse weather conditions are considered a normal possibility and producers have faith in the next year, the ODQS disease may be a different matter entirely. People, and especially the scientific community, were totally unprepared to face the situation and the first response (or lack of one) was to wait for a natural positive solution by the capacity of olive trees to solve the problem independently. Even information supplied to the national population and press was limited until the damage from the disease was too evident. The first data of 2019 registered a decrease in the local production in the region, but a great quantity of olive oil in the market at low prices. A liter of extra virgin olive would usually cost at least 8–10 euros, but in the supermarkets it was going for as little as 2–3 euros per liter. This olive oil was imported and very different in composition from what was locally produced. The consequence is that in Apulia, oil companies have plenty of stocked imported oil, ready to introduce into the market as locally produced, thanks to the absence of any legal protection by the EU.

Once again, the correct interpretation of a vector-borne disease must rely on the past. As already reported, roots of ODQS can be found in the previous story of the similar disease, PD. Everything started unexpectedly in the 1880s, when a "mysterious and unknown" disease spread in the Los Angeles area in California, affecting deeply the precious cultivation of grapes (*Vitis* spp.). Rapidly, the new disease assumed an epidemic state and destroyed *c.*14,000 ha of grapes. Approximately 50 wineries had to close down because of shortage of materials. In 1887, N.B. Pierce (1856–1916) studied in detail the disease and reported the diffusion methods and the affected plants. He described the disease on grapes in California near Anaheim, and therefore it was known as "Anaheim disease." Later, the disease was named after him: Pierce's disease (PD) of grapevines. Growers, administrators, policy makers, and researchers worked hard on finding a solution, but no cure has been identified. However, the research was successful in identifying the Gram-negative bacterium *Xylella fastidiosa* as responsible for the disease. The mechanism of the pest was explained by the introduction on the scene of the vector, identified as a glassy-winged sharpshooter. A related disease was recorded in peach (*Prunus persica*) in 1890 in the USA, with outbreaks (mainly in Georgia) in 1929, 1951, and 1976; this was named phony peach disease (PPD). The causal agent of PD was isolated from grapes in a pure culture for the first time in 1978 (Davis et al., 1978, Turner and Pollard, 1955, 1959). Now, 125 years later, PD is still a significant concern for grape and wine producers in the southern USA (especially California, Texas, and Florida), and the affected plants account for 150 species and innumerable cultivars. Since 2013, the centenary olive trees, considered the heart of the natural landscape of the region, started to die. The cause was deemed to be a vector-borne disease, known as PD, although with several differences compared to past episodes (Almeida et al., 2005). The bacterium responsible was *X. fastidiosa*.

In Italy, the disease's progress is in accordance with the epidemic model. Everything started from a little area in Gallipoli, near the town Lecce, practically at the end of the peninsula. Slowly expanding in the first year and during the subsequent 2 years, most of the olive trees of the southern part of the Apulia region were totally destroyed. Thereafter, the diffusion was very rapid, epidemic, and devastating. Concerning olive trees, PD had previously been considered one of 100 diseases affecting the species, without any report of epidemic virulence. The disease menaces the surviving olive trees, at least in southern Italy. Therefore, affected countries, like France and Greece, are on alert and asking for rapid control of the discase before pandemic diffusion.

The first key consideration is that such virulence of a PD disease never experienced before presented several relevant coexisting novelties, concerning the vector and the parasite. Therefore, traditional methods of control must be considered obsolete or inadequate. New problems ask for new solutions and other points of view.

The parasite

Preferred common name:

Known as the pathogen of Pierce's disease of grapevines, which is spread by xylem feeding leafhoppers, known as sharpshooters.

International common names:

English names: alfalfa dwarf; almond leaf scorch; citrus variegated chlorosis; dwarf lucerne; leaf scorch disease; oleander leaf scorch; pear leaf scorch; pecan fungal leaf scorch; pecan leaf scorch; periwinkle wilt; phony disease of peach; plum leaf scald (Table 5.7).

Table 5.7 Scientific classification of the parasite.

Domain	Bacteria
Kingdom	Eubacteria
Phylum	Proteobacteria
Class	Gammaproteobacteria
Order	Xanthomodales
Family	Xanthomodaceae
Genus	*Xylella*
Species	*Xylella fastidiosa* (Wells et al., 1987)

The pathogen of ODQS disease was precisely identified as *Xylella fastidiosa* subsp. *pauca*, strain CoDiRo. *X. fastidiosa* is an aerobic, xylem-inhabiting, vector transmitted, Gram-negative bacterium of the monophyletic genus *Xylella*. The microorganism form is typically rod-shaped, with a diameter of 0.2–0.5 μm and a length of 1–4 μm. Their cells are covered by a thick cell wall, with grooves and ripples. There are several intrinsic problems when studying and trying to control Xylella. The name *fastidiosa* was assigned in consideration of the difficulty to culture the causal agent of PD on an artificial medium, and in any case very slow growing organisms can be obtained. The attempts to reveal the parasite by Pierce stated in the 1880s, but the causal agent of PD remained a mystery for long time. It is noteworthy in the above classification of the parasite in 1987, indicating the relatively recent scientific determination of the bacterium. In fact, for a long time the microorganism was able to dodge the analysis of scientists, many of whom considered that it could not be cultured at all outside the host and was generally presumed to be a virus or a nonculturable bacterium. In only a few cases was it possible to culture the bacterium successfully on an appropriate agar media. Now, considering that cultivation in laboratory conditions is still practically too difficult, genomic methods of detection must be used. Again, we are facing a galaxy of organisms, complicated by any other microorganism modulated by the need of adherence to the micro-habitat. *X. fastidiosa* can be divided into four subspecies (and several strains) that affect different plants and have separate origins. *X. fastidiosa* subsp. *fastidiosa* is the most studied subspecies, as it is the causal agent of PD; it is thought to have originated in southern regions of Central America and is able to affect a great number of different plants. *X. fastidiosa* subsp. *multiplex* is thought to originate in temperate and southern areas of North America and can affect many trees, including stone fruits such as peaches and plums. In particular, *X. fastidiosa* subsp. *pauca* is thought to originate in South America and can affect South American coffee crops in the form of coffee leaf scorch. *X. fastidiosa* subsp. *sandyi* is believed to originate in the southern part of the USA and cause oleander leaf scorch disease. The subspecies causing PD in Salento is *X. fastidiosa* subsp. *pauca*, and this was a source of speculation about its introduction in Salento.

X. fastidiosa, once installed, infects the conduction system of the host plant. The proliferation of the parasite works by blocking the xylem, which conducts the water and minerals around the plant. Within susceptible plant hosts, *X. fastidiosa* forms a biofilm-like layer within xylem cells and tracheary elements that can completely block the water transport in affected vessels.

Symptoms include chlorosis and scorching of leaves, and the entire plant can die after 1–5 years. Unusual features of the bacterium virulence in Salento were the speed and scale of the damage to olive trees. Pierce's disease is considered less prevalent where winter temperatures are cold—that is, at high altitudes and in inland northern areas. In these conditions, the bacteria are not in a simple situation. They do not have locomotion organs, like flagella, and, therefore have limited active mobility. However, at least one subspecies has two types of pili of different size and on only one pole; longer pili are used for locomotion while shorter pili assist in biofilm formation inside their hosts. These details are important since the bacterium needs to move with a characteristic twitching motion, traveling upstream against the heavy flow of the sap inside the xylem vessels. The bacterium has a two-part life cycle: inside an insect vector, and inside a susceptible plant. While the bacterium has been found across the globe, only once the bacterium reaches systemic levels do symptoms present themselves.

The vector

The vector question is very important although so far underestimated, since among the hypothesis of OQDS explosion, it is possible that *X. fastidiosa* acted as a quarantine agent in Europe that had been previously reported in the Mediterranean region, but did not spread, probably because of the lack of a suitable and efficient vector (Table 5.8). The vector could be an alien species, which found a suitable environment, or an already present insect, induced to change its plant host. In any case, the climate and habitat change effects can be key factors.

Table 5.8 Classification.

Kingdom	Animalia (animals)
Phylum	Arthropoda (arthropods)
Subphylum	Hexapoda (hexapods)
Class	Insecta (insects)
Order	Hemiptera (true bugs, cicadas, hoppers, aphids, and allies)
Suborder	Auchenorrhyncha (free-living hemipterans)
Superfamily	Cercopoidea (spittlebugs)
Family	Aphrophoridae (spittlebugs)
Genus	*Philaenus*
Species	*Philaenus spumarius* (meadow spittlebug) (Linnaeus 1758)

Several vectors of *X. fastidiosa* have been proposed in the case of Pierce's disease, in accordance with insects present in the affected territory. Among them is the glassy-winged sharpshooter, *Homalodisca vitripennis* (Germar), which is a large leafhopper, species native to the Southeastern United States. It is an alien species introduced into southern California, where it has become a serious threat to viticulture. Glassy-winged sharpshooters have large smoky-brown wings with red markings and are very good flyers, making them able to transmit plant diseases further than other vector leafhoppers, such as the blue-green leafhopper, *Graphocephala atropunctata* (Signoret).

Xylella is an obligatory insect vector, transmitted by the bite of xylem-feeding insect feeding into the xylem of a host plant, but infected plant material for vegetative propagation (i.e., grafting) can produce mature plants, that also have the same disease. In the wild, infections tend to occur during warmer seasons, when insect vector populations are at their highest.

The bacterium relies on insect vectors. Known vectors of *X. fastidiosa* are xylem-sap feeder insects belonging to the families Cicadellidae, Aphrorophoridae, Cercopidae, and Cicadidae within the Cicadomorpha order, generally known as spittlebugs. The latter are named due to their ability to produce a large quantity of spittle in the form of a white foam. The insect feeds on the xylem, the water-conducting tissue of both herbaceous and woody plants. Thanks to the xylematic conductor system, the water and mineral salts contained in the soil can reach the active photosynthesizing tissues, mainly present in the leaves. The known vegetal host range of the bacterium is vast, including more than 100 plant species (Turner and Pollard, 1955, 1959). Preferred plants depend on the season and locality, but, in general, the preferred species include crape myrtle, citrus, and holly. Glassy-winged sharpshooters tend to feed on last year's growth and meristematic growth, and excrete copious amounts of liquid as they feed. They ingest 100–300 times their dry body weight in xylem fluid per day, and in large populations, their high volume of excreta ("leafhopper rain") can become a problem, leaving white residue on leaves. Due to the dilute nutrient content of xylem fluid, glassy-winged sharpshooters must have special adaptations to obtain the proper balance of nutrients. Feeding times are thus orchestrated to coincide with the period of peak nutrient content in the host plant. In addition, a specialized structure of the digestive system known as the filter chamber is necessary to recycle the ingested fluid and improve nutrient absorption (Fig. 5.24). The final resulting excreta, as expected, is 99.9% water, inorganic ions, and ammonia, the sharpshooter's excretory form of nitrogen.

Fig. 5.24 Typical foam produced by the vector.

If there is no secretion, you can look to the movement of the adults. These insects are also known as froghoppers, and this name derives from both the resemblance of their body shape to that of a frog and their prodigious jumping ability. If you are near a plant with this kind of insect, you may see them jumping 1 m or more in every direction, despite their tiny size of 5–7 mm. *Philaenus spumarius* holds the world record for insect jumps. When leaping at an angle of 58 degrees above the horizontal, some have reached a maximum height of 58.7 cm above the ground. This means that this insect is able to move efficiently also vertically; starting from the ground level of herbaceous plants, it can easily reach the high level of trees.

If you have entomologic experience, you may notice two tiny black spots at the tip of the head and raised veins in the wings in the adults, whereas nymphs are green with black antennae. Bites of the adult cause discoloration and formation of the attacked plants caused by nutrition, but the insect did not represent a serious cause of depreciation or loss of agricultural production until *P. spumarius* was identified as the main vector of *Xylella fastidiosa*, one of the most dangerous bacterial plant pathogens worldwide and considered responsible for ODQS (Carlucci et al., 2013). The meadow spittlebug, *P. spumarius*, is one of the most abundant field insects in the Apulia region, although other species are probably involved (Rodriguez et al., 2019). The spittlebug feeding on the xylem-sap possesses a piercing-sucking beak, named a rostrum. Digging the rostrum into the tree for feeding, the insect causes the infection and the bacterium propagation closes the xylem vessels, causing dehydration of the plant. Therefore, once introduced the bacterium is diffused inside the plant by the xylem transport system. During this transportation, some populations of the bacterium proliferate inside the vessel, causing occlusion and blockage of the lymph sap and a drastic deficiency of water in plant tissues.

Leafhoppers are medium-sized plant-sucking insects comprising around 2400 species worldwide. They tolerate a wide variety of hosts and are associated with nitrogen-fixing legumes, actinorhizal plants, and some other plants which have a high xylem amino acid concentration. The meadow spittlebug is a homopteran insect very successful due to its polymorphism and adaptive capacity (Dongiovanni et al., 2018, 2019). Therefore, it is probably one of the extensively studied species in ecology and genetics, since it was studied with great attention by biologists for decades, fascinated by all these natural deviations from a reference model. Adults exhibit a heritable mimetic changes of color and pattern on the dorsal surface throughout its range, which also occurs in certain ventral parts. Another interesting aspect concerns its polyandrous nature, since females may mate several times with different males, meaning that the offspring of a single female may be fathered by several males. The result of an increased heterogeneity and high fitness is the present of the vector in habitats of great diversity, and therefore a potential of wide global distribution. However, probably the most important character is that this insect is highly polyphagous, since it feeds on wild herbaceous and woody plants, including some species of agrarian interest, accounting for more than 400 species, and reports mainly concern damage to alfalfa, red clover, wheat, oats, corn, and strawberries. However, some researchers have already noted relationships between spittlebugs and exotic plant species, in studies concerning the influence of alien species in evaluating whether introduced plants favor native or exotic and generalist or specialist herbivorous insects in newly evolving plant-herbivore networks. The results suggest a careful interaction between the alien species and the invaded habitat, wherein plants, insects, and other organisms must be included. Therefore, generalist herbivores can be favored by the invasion of the introduced plants. In contrast, exotic insects can be supported by introduced plants, increasing the local occurrence and range of insect pests. Herbivore accumulation on invasive alien plants increases the distribution range of generalist herbivorous insects and supports proliferation of nonnative insect pests.

If you want to feel the presence of the meadow spittlebug, you can take a walk in late summer and these little insects, when disturbed, will jump everywhere. Otherwise, in early summer you may see them at the stems of herbaceous plants. In summer, females deposit up to 400 eggs on the bark of the plant. In springtime, the eggs open and release the larvae which rapidly move to the nearest herbaceous plants. Later, the larva feeds on the stem of the host plant, and it is specialized in feeding on the lymph inside the tracheid vessels. The xylematic sap is low in nutrients, being constituted mainly of water and minerals. Therefore, the nymphs are forced to ingest large quantities of the sap, and their presence can be detected by the typical lather caused by the sticky foam, resembling the guttation in some plants (*spumarius* means "frothy"). The larva secretes this froth, known colloquially in different countries as "cuckoo-," "witches-," or "frog-spit." The foam is thus no more than the waste of the insect intestine, but it may afford protection from desiccation and predation. These insects feed the substances in the xylem, the ascendant conducting system of both herbaceous and woody plants. Its known host range is vast, including more than 100 plant species. Excreta are therefore produced in high number (leafhopper rain) and become a white residue, which can damage the leaves of crops.

However, there is another explanation for the feeding behavior of the vector, leading to reasons for the disease. Xylematic nutrition should not be limited to the ascending lymph, ascending from the reserve organs, but it could interest also the phloematic conductive system. In fact, the active conducting tissues, xylematic and phloematic, are collateral and the two kinds of lymph flow very near. Furthermore, through xylematic feeding these insects can have access to the assumption of nitrogen in the form of amino acids coming from the symbiotic nitrogen fixation. This endosymbiotic input can compensate the nutritional imbalance by the assumption of large quantities of liquids and by the microorganisms' nitrogen fixation. The result is a physiologic-anatomic system of filtering and selection of the nutrients, and these insects can rely on a specialized structure of the digestive system, the filter chamber. However, the system recycles the ingested fluid and improves nutrient absorption, but it is not selective in the possible intrusion of bacteria, with physiopathologic effects. In fact, Cercopoideceous insects, including *X. fastidiosa*, are cited in the literature as vectors of phytopathogen microorganisms living in the xylem.

Feeding times and growing of the adults are thus orchestrated to coincide with the period of peak nutrient content in the host plant. That means an absolute need to study this coherence in the habitat, based on the knowledge of the preferred host plants and the behavior of the vector. Once again, we are looking to a complex system of interconnected microhabitats, including the acquisition of *X. fastidiosa* while feeding from an infected plant. As already reported, the bacterium can survive only in adapted special environments. It establishes itself perfectly inside the insect, attaching itself to the surfaces of the mouthparts of the vector, on the floor of the cibarium, the apodemal groove of the diaphragm, and in the walls of the precibarial area both above and below the precibarial valve. In this way, the bacterium can persist and be ready to transfer itself in the new host through the biting of the vector. Nymphs of *X. fastidiosa* lose their ability to transmit the disease after molting. Transmission to other plants takes place during the seasonal flights of the vectors, when they are in large numbers.

Considering the previous similar cases of insect-borne diseases concerning plants, there are two possible main future scenarios. The epidemic could stop by inner or outer mechanisms, or the infection could spread out of Salento. In the second case, some models predict a meridional expansion, with a rapid transit through Calabria and the invasion of Sicily and Sardinia, but later Corsica and any Mediterranean region could affected. At this stage, any neighboring territory is in danger. Nowadays, there are no reasons to consider the Xylella outbreak already finished, since there are several reports about the presence of the paradise outside the zone of restriction imposed by EU. The vectors are not able to fly for long distance, but they can easily overcome the 2 km line of separation in several ways. When I was in Salento at the beginning of the infection, the windows of our cars were full of these vectors, trying to achieve transportation. The jump of these insects, although remarkable in consideration of the dimension, could not be enough to rich distant places, but they are light and can be easily transported by the wind for long distances. If you are in Salento, you may be struck by the changes. The previous green has been totally substituted by the dirty yellow of a desert sandy land where only the enormous carcasses of destroyed trees are present, as silent testimonials to a catastrophic event. Meanwhile, the situation in Italy alarmed all countries where the olive trees are present. The main measure so far was a general plan of detection of the current presence of the bacterium. The detection involved also the other side of the Atlantic Ocean, even where probably all the story started with Pierce's disease of grapes and almond leaf scorch.

In California, the incidence of *X. fastidiosa* in field populations of other species of leafhoppers has been observed by Frietag and Frazier to be as high as 18.6% for most of the year. In Florida, in contrast, natural infectivity of the glassy-winged leafhopper is very low. As to the vector efficiency of the glassy-winged sharpshooter, Costa et al. (2000) found that, under ideal conditions, 83% of oleander plants each exposed to a single leafhopper carrying *X. fastidiosa* became infected with the bacterium.

The situation in the United States was monitored in 2014 and reported in *Plant Disease* by Krugner et al. (2014). In the USA, California is the sole relevant producer of olives, with ~17,800 ha planted and a production value estimated at US$130 million per year, although olive trees are also sporadically present in Florida. The study was promoted by increasing occurrences of dieback and leaf scorching, which were considered as symptoms in California olive trees, with *X. fastidiosa* (Xf) as the causal agent. In total, 198 samples of olive trees showing branch dieback and leaf scorch symptoms were collected. Analyses by polymerase chain reaction (PCR) were performed to determine the association of Xylella with the disease. In detail: "Laboratory tests detected Xf in only the ~17% of the samples, from which six strains of the bacterium were isolated. DNA analysis and laboratory tests using grapevines and almonds indicated that Xylella recovered from olive belong to a subgroup known to cause almond leaf scorch but not Pierce's disease." Bacterial cells from laboratory cultures were inoculated into healthy olive plants and monitored over 1 year for symptom development and presence of bacteria, which tended to be self-limiting. Tested olive plants did not show any symptoms of disease, and Xf infection tended to be self-limiting. Results indicate that although Xf is unlikely to be the causal agent of olive leaf scorch/branch dieback, the infected olive may serve as a reservoir for Xf and insect vectors, thereby contributing to the diffusion of Xf. These results were partially confirmed also for the analyzed affected plants in Apulia, where the presence of Xylella was not always confirmed.

However, the paper contains also another clear indication, even more important:

"Collectively, the data indicate that *X. fastidiosa* is unlikely to be the etiological agent of olive leaf scorch/branch dieback, but olive may contribute to the epidemiology of Xylellae-caused diseases in California. Olive may serve as an alternative, albeit suboptimal, host of *X. fastidiosa*. Olive also may be a refuge where sharpshooter vectors might escape intensive area-wide insecticide treatment of citrus, the primary control method used in California to limit glassy-winged sharpshooter populations and, indirectly, epidemics of Pierce's disease of grapevine."

This is in perfect accordance with our experiences in the laboratory and in the field. Let us imagine the life of the vector during the feeding period. Herbal plants are surely preferred for several reasons, considering that to feed on the olive tree it is necessary to penetrate the thick and resistant bark of the trunk, relying on a delicate and minuscule stylet.

Attempts to find solutions among the biological control give some results, consisting in a parasitic tiny wasp, *Gonatocerus triguttatus*, which could control the early spring generation of the sharpshooter. As is usual in these cases, the wasp from Texas and Northern Mexico parasites the eggs of the sharpshooter with its eggs. An alternative could be the utilization of entomopathogens, such as *Hirsutella* sp., a fungus that is known to affect sharpshooters. It is important to underline that these organisms are already present in the southeastern United States, although improving their spread could be very difficult.

There are interesting similarities but also some important differences in the comparison between the Californian and the Italian cases (Overall and Rebek, 2017). The six strains of *X. fastidiosa* isolated from olive trees in southern California pertained to the subspecies *multiplex*. The glassy-winged sharpshooter, *Homalodisca vitripennis* (Hemiptera, Cicadellidae), resulted among the vector involved by transmission assays. However, it must be underlined that it has been demonstrated that this insect could transmit strains of both subspecies *multiplex* and *fastidiosa* to olive trees at low efficiency. Using insect trapping data, other vectors were proposed, like *Draeculacephala minerva*. Therefore, so far there are not evidences about a general widespread of the OQDS outside Salento, but this is not a reliable alibi to diminish the attention.

If the soil is totally deprived of any herbal species because it is cleaned by a tractor, as in the usual tradition in olive field, the only way to survive for the vector, and the parasite inside, is the transferred to the woody plants. Cleaning of the soil is a usual treatment in order to facilitate collection of the olives and prevent the *Bactrocera olea* attack on olives, but even more effective again the survival of herbs could be a prolonged period of months without any rain and high temperatures, as registered in the last years in the south of Italy. A combination of these factors could be the reason for a change of feeding behavior by the vector. This hypothesis needs confirmation. Therefore, in the last 2 years in collaboration with the team at ENEA, as part of my research group, attention was dedicated to the determination of the preferred feeding species for *P. spumarius* in central Italy. The aims of the study were not only the determination of the preferred species, but also the period of feeding, the relation between insect and plants, the possible presence of alien species, and the environmental factors influencing the vector behavior. The last argument is crucial: if the neotropical climate will dominate in the Mediterranean basin, the adaptation of plants can be influenced. Now, most plants are autochthonous or were introduced a long time ago, but other types of adaptations are possible in future. For instance, the baobab presents leaves only for 3 months during the year, a period coincident with the rains, but in addition, the trunk under the bark possesses protected green tissues that are able to perform the chlorophyll synthetic process throughout the year.

The investigation of the host plants influencing the distribution of vector of *Xylella fastidiosa* and *Phylenus fastidiosa* was tracked on long-term field observation, monitoring the presence of nymphs on 144 different target herbaceous species by visual count of the insect feeding foams (Benelli et al., 2018). In addition to providing new insights about the feeding behavior of the vector, the study was performed to verify if the spatial distribution of the nymphs was influenced by olive tree proximity, in order to obtain information about a possible relation between the olive tree and *P. spumarius* (Dongiovanni et al., 2018, 2019; Bodino et al., 2019; Latini et al., 2019; Krugner et al., 2014). The result of the research was a predominant preference of the nymphs at IV and V larval stages on a very limited number of herbal species in comparison with the plethora of herbs present in the field, but with differences during the year. Each year in March, the fennel, *Foeniculum vulgare* (Apiaceae), and the white bedstraw, *Galium album* (Rubiaceae), were the most visited, whereas from April the preference changed according to the seasonal development, shifting mainly to the sticky-willie, *Galium aparine*, followed by the hawkweed oxtongue, *Picris hieracioides* (Asteraceae), *Silene latifolia* (Cariophillaceae), and the common sowthistle, *Sonchus oleraceus* (Asteraceae); however, infestation on fennel resulted relevant in the May. In general, although nymphs feed on the same herbaceous plants of eggs deposition, adults have polyphagous feeding behavior. Thanks to their improved flying and jumping abilities, they could feed also on other plants, like shrubs and woody trees. *P. spumarius* certainly displays a good adaptability and therefore data on the biology and ecology of its local populations are of pivotal importance in any control program. This consideration is necessary to continue this narration, and to join the data from U.S. researches about Xylella occurrence with the Italian observations on vector feeding behavior.

We have seen that fennel is a key herbal species for diffusion of Xylella vector. It is evident that when we are talking of sweet fennel, we are referring to the wild plants very different in shape and constituents from those that we use for culinary purposes. The wild plants are present in the fields in summer, often as a highly invasive weed. In fact, from the yellow-green umbrella flowers, typical of the Apiaceae (Umbelliferae) family, at the top of the plant from late winter to summer, a single plant can produce thousands of seeds during its first growing season and hundreds of thousands of seeds in its second year of growth, multiplying its presence easily. Therefore, thanks to these characteristics and the long fruiting period, it is a strong candidate to become invasive and dominant, disturbing plant communities and habitats. As confirmation, *F. vulgare* is included in the CABI *Invasive Species Compendium*.

Fennel originated from South Europe, in particular with an areal distribution similar to that of the olive tree, but it was introduced in several temperate regions around in the world, including North America, and it is now known as wild anise in California. Owing to its typical strong smell, it is often mislabeled as anise in American supermarkets. In the USA, it can be found growing in San Francisco and on the Pacific coast, where it is considered a weed. After its introduction and naturalization in the 1880s in California, fennel escaped from cultivation, becoming a seed contaminant and an invasive plant. An alien invasive plant competes with other native plants for resources, making it harder for other plants to obtain sunlight, water, and nutrients, because the alien plant tends to take over.

In 2014, *F. vulgare* was included within the EU list of plant genera and species, as regards the need of measures to prevent the spread of the fastidious bacterium, and the relation between the bacterium and wild sweet fennel has been confirmed in other studies. Therefore, *F. vulgare* must be considered a European native emerging alien invasive species, potentially dangerous, also considering that so far the adopted measures to limit its spread have been unsuccessful. In other words, considering the proven existing relationship between the plant and the vector, confirmed by studies in 2018 (Morente et al. (2018) in Spain and Dongiovanni et al. (2018, 2019) in Apulia), the wide spread of wild sweet fennel colonizing new territories could contribute to the spread of diseases carried by the insect.

Finally, two additional considerations were included in the work of ENEA researchers. The above data outline the relevance of specific wild herbaceous plants for survival and growth of the most vulnerable stages of the vector. In olive cultivations of south Italy, the consolidate habitude of clean the soil from weeds in olive fields may be favorable to the vector, as already suggested. This treatment can favor the re-growing of wild fennel with the laying of eggs in autumn and cause ailments in nymphs in the hatching period.

In the last decades, many olive orchards in Apulia were abandoned, and the long tradition of careful cultivation was interrupted, favoring the weeding of uncultivated green areas. This is an additional potential factor of the OQSD outbreak, favoring the presence of invasive herbal species and alien vectors. The establishment of smart treatment to limit the plants,

Fig. 5.25 An olive tree significantly damaged by Xylella. The abundant production of shoots in the basal part is not useful to restart the tree's metabolism.

attracting the vector for feeding or egg laying, could make it possible to limit the utilization of insecticide (Fig. 5.25).

The future of olive trees

The Xylella alert and related events demonstrate the difficulties of governments and international agencies in adapting their strategies in the case of novel and unexpected alerts. It recalls in some ways the beginning of the movie *Godzilla*. When the Leviathan monster emerges from the sea and attacks Tokyo, the authorities spend a lot of time in interminable meetings, looking for a paragraph in the protocols where they might find written what to do in such cases. Godzilla is a special case and there nothing about an emergence like this in the protocols. Meanwhile, heedless of the bureaucratic impasse Godzilla continues its systematic destroy of the town. Against

Fig. 5.26 This is the end of an olive tree, as has happened in Apulia so far.

the unexpected attack of Xylella, the only adopted and strictly supported measures were ordinary ones foreseen in any kind of infection emergency, consisting of the removal of the infected organisms and acting to limit the zone of contamination.

So far, the only real performed action was the application of the EU protocol consisting of the eradication of any olive tree and creating a defensive line of 2 km in extension in the northern part of the Salento Peninsula, where any plant must be eradicated, in order to isolate the disease. This measure should stop the diffusion of *X. fastidiosa*. The blocking of plant imports from America is also in action. In addition, in this case, total control is quite impossible. It is very difficult that the EU's protective approach will have significant effects. The vectors are not good flyers; they usually move by jumping, but they can be efficiently transported by the wind and occasionally human or animal transportation, traveling kilometers in a single day.

Several aspects of the Xylella affair are waiting appropriate responses. We have seen the particular phytogeographic situation of the Salento Peninsula. It is possible that its partial isolation facilitated the diffusion of OQDS. Another agent could be assigned to the olive monoculture. Furthermore, as in many other cases, the environmental changes in action can be considered decisive for invasion or the increase of parasite virulence. Another influence deserving to be considered is the abandonment of agricultural care in many fields and the consequent degradation of the soil. Therefore, the absence of traditional treatment of the soil and the trees, due to urbanization of the population, could have interfered. We are probably facing a cooccurrence of all these factors and others as yet unknown. There are some unclear aspects of OQDS that were and are objects of speculation, such as the starting episode of the disease, and why olive trees were the selected target. In summary, OQDS must be considered a further edition of the PD pathology or a novel disease, which means an evolution or something never experimented. This is very important to understand the future of olive species and the possible effective measures.

In conclusion, there are two main hypotheses about future scenarios:

(a) a natural stress-induced dieback: this is consistent with widespread groves of various ages all suffering to different degrees and slowly declining rather than a virulent point infection that can be seen to spread. In other words, the disease is due to a "normal" increasing of virulence coupled with the "stress effect," derived from climate change and agricultural loss, that will affect mainly old trees, causing a turnover in favor of the new stronger generations;

(b) a modified, more virulent pathogen appeared, and plants defenses will not be able to face the new challenge, with devastating consequences.

The bacterium cannot be controlled by the use of chemical antibiotics, either because they are banned in agriculture in EU or because they are very costly and complicated. The insect vector could be controlled by appropriate insecticide, considering the environmental impact. Insecticides usually select as their target adults, but the larval stage is the best point at which to act on the insects, before they are able to move and fly away. In any case, it is fundamental to know exactly the vector's activities and plans in terms of where it feeds and reproduces.

Therefore, to obtain any real result, we must learn nature's lesson. Olive tree disease shows three main actors: the bacterium, the vector, and the plant (and probably a symbiotic fungus). Three fungal species were associated with the symptoms: *Phaeoacremonium aleophilum*, *Neofusicoccum parvum*, and *Pleurostomophora richardsiae*, although their effective roles are under discussion.

The associated organisms work together in the insect-borne disease, acting like a "superorganism." It is a very complex system, but in some ways also very efficient. The only way to face the *X. fastidiosa* challenge is through integrated pest management. It is necessary to operate considering the several aspects involved together: a treatment of soil to sustain the plant and its capacity to react to the disease; an insecticidal agent to control selectively the insect; and a natural, low-cost, and eco-friendly antibiotic treatment of new generation. Some may argue that this strategy is not so different from that suggested in the case of insect-borne diseases affecting mankind.

A preliminary key step is the reply to the question: how did *X. fastidiosa* become so dangerous in the last 2 years? We have known about the presence of bacterium in Italy for at least 30 years, and previously it was considered just one of the several diseases involving olive oil. Something happened in recent years changing completely the equilibrium between the microorganism and the host. There are several hypotheses about the causes of the change and a consequent relevant debate.

The second hypothesis is that a change in the *X. fastidiosa* genome occurred, giving rise to more aggressive and dangerous strains not present in the past (Steinbiss et al., 2016). In this case, there are two possibilities: the change is derived from some experiment or a biological cause. In the first hypothesis, a genetic laboratory generated a mutant population, which later escaped its creators. In the second hypothesis, due to climate change, some virulent strains from hotter countries were able to survive and proliferate against local populations. A conjecture is based on the possibility that the parasite was introduced through the import of infected plants. In fact, the hypothesis of the import of infected ornamental plants, like oleander from Costa Rica, has been seriously considered. However, it is complicated to imagine that all the imports of the infected plants were concentrated in Gallipoli and that they contained a special highly aggressive strain of the bacterium never previously reported. In any case, the cooccurrence of several improving basic factors must be considered, including climate changes.

The climate hypothesis

There is a public debate about the potential impact of climate change, in particular global warming, on human health and the future of the planet. However, there is often confusion about the usual terms, like climate (from the Greek *klima*, meaning inclination). Besides the influences due to the human activity, other factors can be considered as causes of the climatic variation: natural factors like the sun's radiant energy; alteration of the hearth's axis causing variation of its orbit. The effects of these factors on the climatic elements of the environment, such as temperature, humidity, wind, and rainfall, can cause significant local changes, but climate tendencies must be considered in averages for decades, centuries, and even millennia. The problem is that often the climate is confused with the weather, which actually refers to short-term climatic situations. This is the consequence of the weather's impact on our ordinary life conditions. In other words, people focus on updated forecasts, like the temperature in the next days or the possibility of rain. These are local and transitory situations, but more consideration should be dedicated to long-term aspects. This confusion is increased by continuous catastrophic warnings online and so on about extreme consequences in our lives every summer or winter, just to encourage us to click. It is pointless to insist that any forecast on the next week is not trustworthy, whereas the tendency of the global climate can be considered more reliable. In any case, meanwhile we discuss if the range of years is too short to decide if the warning current trend is a minor fluctuation or a first-order variation, we can observe the consequences of climate change on our habitat, including on insect-borne diseases. The debate on effects on climate changes is ongoing, increasing on one side the interest of ordinary people, but also the consciousness that the measures so far claimed are not adequate to the situation.

There is general concern about the recent increases of the global temperature and of carbon dioxide in the atmosphere. Data evidencing this reality and the relative impacts of the climate changes can be easily found in the reports of international agencies. Thus, United Nations (UN) at its annual 2019 convention reported fundamentally a global increase of the temperature of *c.* 1.5°C in comparison with pre-industrial times and an exponential increase of the percentage of CO_2 in the atmosphere. These increases are still considered acceptable, but the preoccupation concerns the effects of a confirmation of the tendency in the next decades. In 1901–2012, the increase was 0.89°C and a further increase of 0.3–0.7°C is expected in the period 2016–2035. A total increase of *c.* 3.7°C is expected. The population directly involved in the phenomenon is evaluated: 178 million for +1.5°C, 200 million for +2°C and 277 million for +3°C. We can assume that all these people will have only two possible futures: succumb or migrate, unless something changes the scenario.

The increases of temperature and CO_2, as well as the different distribution of precipitations, are considered the results of abiotic and biotic factors. About abiotic factors, like a change of magnetic field of the planet, we can do nothing and therefore attention is concentrated on biotic effects, although these are difficult to establish. Among the human influences, industrial and domestic pollution have been considered as mainly responsible for climatic change, and it is believed that a decrease of industrial activity could influence significantly the increasing tendencies. Governments are requested to act on pollution caused by urbanization, i.e., cars and houses heating, usually derived by changes of fuel. However, other effects are in action and these can be decisive as well. Larger populations need more food and cultivable lands are under increasing pressure. Intensive agriculture and the utilization of soil are responsible for at least 23% of biotic contribution to the greenhouse effect. Soil is continuously impoverished by intensive agricultural systems and the increase of CO_2 affects the nutritional value of food (−5.9%–12.7% in proteins, −3.7%–6.5% in zinc, −5.2%–7.5% in iron). Insect-borne diseases are mainly considered in terms of their effects on human health, but they are key factors on the production of food and its quality. Among the factors, influencing the evolution of food, the insurgence of organism genetically modified and the concentration of production and distribution under the control of multinational agencies. The consequences of these factors were the decrease of biodiversity and of number of varieties, as well the loss of local production, which are the result of a long and accurate selection.

The problem is the speed and the entity, not the nature of the phenomenon of climate change. Turnover of hot and glacial periods is a normal trend for our planet, mainly due to the quantity of CO_2 in the atmosphere. We know the occurrence of several marked changes in climatic conditions in Europe during the last 1000 years, mainly attested by the alternant records of temperatures. These changes deeply influenced human activities, like in the cold medieval period, when it was difficult to survive because of the low temperatures, which affected agriculture and crop production. Later, the re-increase of the temperatures allowed an astonishing renaissance of human activities, in sciences, arts, architecture, and other expressions of mankind's intellectual performances. In the 20th century, two main periods of warming have occurred in Europe. A second period of warming started in 1976–2000, recorded as the warmest one, and it is still in action. The evidence was an increase of approximately 1.2°C over the past 100 years, which is twice the average global rate. We are now in a warming period, and the trend is probably accelerated by the emissions due to human activities. Warning consequences are higher nighttime temperature, with limited differences between day and night, and fewer frost days in winter, associated with milder temperatures throughout the winter period, longer dry periods, and peaks of temperature. However, if we consider only the data referred to the tendencies of global planet changes, we can extrapolate a limited, even misleading, impression. In particular, considering Europe, we must focus on opposite effects: temperature increases were most marked in both Central Europe (Italy, Corsica, and the Balearic Islands) and Eastern Europe (western Bulgaria, northern Greece, Albania, Macedonia, Bosnia, Montenegro, and Croatia). In contrast, central Iberia and the region around the border between Morocco and Algeria have cooled. Simplifying the tendency obtained from models, Europe is warming and North Africa is cooling. This has resulted in changes in precipitation dividing Europe into two parts: the number of wet days have increased in North Europe and decreased in South Europe, increasing the existing tendency to desertification of several regions. North Europe, including the UK, northern Iberia, and Scandinavia, is becoming wetter, whereas southern Iberia, France, Germany, and Italy are becoming drier. The main problem in all this is people's attitudes, virtually all oriented to the defensive. However, we must consider seriously the necessity of adapting our strategies to climate change, and even consider the possibilities of obtain working places and inputs for the economy. I suggest abolishing the word "conservation," so beloved by my ecologist colleagues, usually the preview of a useless and unsustainable position of a rearguard. Vector-borne pathogens are particularly sensitive to climate, a fact that has led to widespread and continued speculations that anthropogenic climate change will increase the incidence and intensity of their transmission. Other climatic abiotic and biotic factors can also affect disease distribution. Diffusion can be very rapid and effective. Adult insects are usually not strong fliers, but they can be passively dispersed by the wind, possibly up to several kilometers in a single night, especially over the sea. Thus, natural barriers cannot be considered an efficient control of the diffusion. Otherwise, these insects can travel utilizing ancient transportations, like other animals (street ruminants), or new unexpected ones, e.g., inside old tires as happened for *A. albopictus*.

There is an urgent need for ecologically sound, equitable, and ethical pest management, based on control agents that are pest-specific, nontoxic to humans and other biota, biodegradable, less prone to pest resistance and resurgence, and relatively less expensive. The last aspect is fundamental for large-scale use in emerging countries.

The general feeling is that we are facing something exceptional and never before seen, but history tells us another tale.

It happened before, and it will happen again

History books are full of facts, but very few contain information about reasons for certain events, in particular when they are not referring to physical-chemical implications. Everything seems to be the consequence of human decision or of the arising of important personages influencing the destiny of communities. Nevertheless, we can reverse the paradigm and consider human activities as the result of the effects of natural forces.

Many history books begin with the story of two empires@ one in Egypt and the other in Mesopotamia, the land between the Tigris and Euphrates rivers. The Akkadic Empire began in the cradle of our civilization, wherein an alphanumeric system and geometry were already known, as well as advanced social organizations. In 4500 BCE the great towns of Akkad and Babylonia lost their independence to King Sargon, who was able to conquer other large territories. This led to the foundation of the first "universal" empire, meaning a single political-territorial entity, united despite the great geographic, ethnic, and cultural diversities. The northern part of the empire was relatively rich, profiting from fertile and productive fields, whereas the southern regions were semi-arid. Starting from the cornerstone in Mesopotamia, the empire extended to what are now Syria, Turkey, Iran, Iraq, and Saudi Arabia, conquering most of the civilized lands on the planet. However, after 150 years the Akkadic Empire collapsed (Weiss et al., 1993; Weiss, 2017). The reason for its disintegration is usually assigned to undetermined invaders from the north, pressing people to the south. However, something in this seems difficult to accept. The empire was already well-established and the populations in the northern regions should have had low interest to change in favor of the southern territories. Even in the case of an invasion from other populations, considering the empire's wealth, assimilation by the invaders appears more probable. Scientists started to speculate on the case, searching for evidence of the real sequence of events. This time, the research was focused on traces of very old time. It was a difficult and specialized search. Since the old reports were contradictory, researchers looked for indisputable traces, like those trapped in the rocks (Zanchetta et al., 2016; Kornei, 2019), consisting in evidences able to resist during 4000 years to the atmospheric aggressions and environmental changes, including the human activities. Although we have encountered so far towns, fertile fields, and arid deserts, the scene was taken by a deep cavern, wherein the scientific team, whose leader was the paleoclimate expert Stacey Carolin (Carolin et al., 2019), University of Oxford, found the trace they wanted (published in PNAS in 2019). Inside the cavern of Gol-e-Zard (*c.*5000 m distant from Teheran and protected by a great mountain from the northern winds), the team was able to obtain a sort of archive of the climatic changes in that area. They obtained information, measuring the quantities of magnesium present in the stratification of the stalagmites produced by the deposition of minerals coming from the water percolating from the upper soils. This is a reliable method to measure the quantity of rain and powder present on the surface of an exposed area over a determined period, but these data must be cross-referenced with the chronologic system of dating, based on uranium/thorium concentrations. In such a way, the team obtained precise information about the quantity of powder and sands present at a determined time in the zone, as well as the quantity of rain. The results evidenced the occurrence of two long periods of extreme drought; the first one started in 4510 BCE and lasted for 110 years, and the second started in 4260 BCE and lasted for 290 years. These data were confirmed by other independent researches, such as those on corals. The idea is that powder and sand were transported by the wind to be deposited in such quantities as to make any cultivation impossible, and the fertile fields of the north thus became unproductive. In such a way, the previous equilibria between north and south were completely thrown out of balance, causing a great migration of people in search of food. Once the scientific determination was obtained, it was possible to compare this with historical information. The beginning of the second period coincides with the start of the collapse of the empire. In a fragment of a document of that period, named "The Damnation of Akkad," we can read: "the great cultivable fields were not able to produce wheat any more, the flooded fields did not produce fish, the orchards did not produce syrup o wine, no rain from the dense clouds." This may have been the first human mass migration registered in history as caused by climate changes. The rest of the story is always the same. The exodus from north to south generated conflicts with the local populations, and the consequence was the construction of a wall 180 km long between the Tigris and Euphrates, which was named the "Wall of Amorrei."

The Trump Wall is planned to run for *c.*3000 km, coast to coast. Already a third of the boundary is divided by various kinds of barrier, for pedestrians and/or vehicles. The wall is going to cross several types of habitats, such as mountains, rivers, deserts, and plains. Now, we can try to leave aside any political considerations (fortunately they are beyond the aims of this book) and the environmental effects (there is always an ecological impact) involving more than 200 plants and animal species (according to the online 2016 Information for Planning and Conservation) and the production of 7.4 million tonnes realized in the atmosphere, the consequences of genetic breeding, and many other arguments. We can consider everything as a necessary collateral effect, and let us evaluate only whether the wall will be able to act as an effective antimigration barrier, and thus achieve its objective. In the light of what is going on in other countries, stopping terrestrial human migration is not easy, but is arguably largely possible. However, there are then the coasts and the seas. In particular, California's coast is highly populated and extended, with plenty of transportations possible through water. A confirmation of this trend comes from the situation in Europe, with a continuous migratory flux from Africa. Nobody, so far, apart the official declaration of hostility against migrants, was able to stop it and avoid the loss of lives during the desperate trip. How many boats or other kind of ships full of migrants were prevented from reaching Italian harbours? As far as we know, the number is zero. There are many ethical and juridical considerations, but once again the physical concern is the main one. From the point of view of insect-borne diseases, the Trump Wall can have few effects, since insects can fly and eggs can easily be transported by the wind and by any type of vehicles—unless the wall will be able to stop all molecules.

The role of secondary metabolites

Four olive cultivars, Cellina di Nardò, Ogliarola di Lecce, Frantoio, and Leccino, resulted positive in response to the Xylella infection, accounting for most cultivars present in Salento. However, the response to the disease was not equal, based on the qPCR assays. Cvs. Cellina di Nardò and Ogliarola di Lecce showed higher disease resistance than Frantoio and Leccino. The HPLC-ESI-TOF-MS analysis showed a correlation with modification of phenolic content, focusing on two key metabolites: hydroxytyrosol and quinic acid. As evidenced in a parallel study on grape cultivars, these substances act as a reservoir in the pathway production of a quantity of phenolic compounds, whose importance in antioxidative and defensive mechanisms is well-known. In particular, constitutive levels of hydroxytyrosol glucoside turned out to be strongly reduced in infected plants (more than 94%), although the effects were greater in cvs. Frantoia and Leccino. Regarding quinic acid, the situation was different: levels did not indicate a significant change in cultivars, but in the infected aforementioned cultivars the contents of quinic acid increased four- and five-fold. The results refer to the leaves, which are most affected by the disease. It is conceivable that these intermediate metabolites, in the case of an attack, are converted into other derivatives. Hydroxytyrosol glucoside is a precursor of oleuropein, which can be considered the most important secondary metabolite in olive production for its antiROS properties. Quinic acid is fundamental in the production of caffeoylquinic compounds, also considered important for the same reasons. These studies suggest a significant response of the infected plants, based mainly on the secondary metabolism, as expected, and could explain the different sensitivities of species and varieties. This molecular response by the plant could be improved and offer the key to the tree's capacity to react to the disease.

Another consideration concerns information about the possible manifestation of OQDS in other parts of the planet, as well as differences in evidence of the disease. California is the sole producer of olives in the USA with ~17,800 ha planted and a production value estimated at US$130 million per year (Overall and Rebek, 2017). Increasing occurrences of dieback and leaf scorching symptoms in California olive trees suggested the possible occurrence of OQDS in California (Purcell and Saunders, 1999). A total 198 samples of olive trees showing branch dieback and leaf scorch symptoms were collected and investigated, but in only 33 samples (~17%) was it possible to detect the presence of the bacterium. These data were confirmed by DNA analysis and laboratory tests on grapevines and almonds, showing that the Xylella strain recovered from olives can be assigned to a subgroup known to cause almond leaf scorch but not Pierce's disease. Furthermore, bacterial cells obtained from laboratory cultures were inoculated into healthy olive plants. After 1 year, the analyses of tested olive trees did not show any symptom of disease development and the infection tended to be self-limiting. All the experiments indicated that Xylella is unlikely to be the causal agent of olive leaf scorch/branch dieback, which must thus be attributed to other agents or causes. Nonetheless, it is conceivable that infected olive trees may serve as a reservoir for the bacterium and insect vectors, contributing to the epidemiology of diseases caused by Xylella.

The research on California olive trees also demonstrated by vector transmission assays and insect trapping that the glassy-winged sharpshooter, *Homalodisca vitripennis* (Hemiptera: Cicadellidae), and green sharpshooter, *Draeculacephala minerva*, were active in olive orchards, contributing to transmission of strains of both subspecies *multiplex* and *fastidiosa* to olives, though at low efficiency. Another approach was inspired by the consideration that olive trees may serve as an alternative (albeit suboptimal) host of *X. fastidiosa* in the case of absence or low availability of the preferred hosts.

Conclusion

In 2013, I was among a team of experts participating in a restricted first meeting to evaluate the Xylella problem. My opinions, not accepted by the other participants, were as follows: (a) the situation must be interpreted as a severe epidemic; (b) the attack cannot be solved only by the ordinary methods, but needs an extraordinary effort to produce new solutions; and (c) considering the relevance of the situation, a program of information and general mobilization of farmers and the population should immediately start. Regarding the first consideration, everyone can see directly what happened by looking at the photos and reports (Figs. 5.25 and 5.26). As for the second, we will see in the near future, but so far the signs are not positive. About the third one, we lost 2 years to silence and when the alert exploded, the result was that the population was against eradication, with the usual pathetic solution of reimbursing farmers for the damage. Nobody will reimburse all of mankind for the environmental damage. Meanwhile, smart merchants, like vultures, are ready to profit of the disaster, selling cultivars of olive trees, presented as untachable by the Xylella.

The moral of the Xylella affair is clear. The good news is that we have powerful methods to detect the presence and diffusion of the pathogen's strains. The bad news is that the chances of preventing or predicting an invasion are still practically zero, as evident in the current situation (Fig. 5.27). Therefore, we are still able to act when there are signs of emergence. The terrible news is that pathogens are very fast indeed compared to our capacity to respond. Most insect-borne diseases are decreasing, but this is probably dependent on mankind's health, and the other fronts are in a different situation. The research front is moving in accordance with requirements and producing valuable proposals, but it needs financial support and, more importantly, the chance to realize concrete results.

Fig. 5.27 The current look of several parts of Apulia after the Xylella attack.

References

Almeida, R.P.P., Blua, M.J., Lopes, J.R.S., Purcell, A., 2005. A vector transmission of *Xylella fastidiosa*: applying fundamental knowledge to generate disease management strategies. Ann. Entomol. Soc. Am. 98, 775–786.

Appleby, A.B., 1980. The disappearance of plague: a continuing puzzle. Econ. History Rev. 33 (2), 161–173.

Aziz, A.T., et al., 2015. Insecticide susceptibility in larval populations of the West Nile vector Culex pipiens L. (Diptera: Culicidae) in Saudi Arabia. Asian Pac. J. Trop. Med. 6 (5), 390–395.

Baliraine, F.N., Rosenthal, P.J., D'Alessandro U., 2011. Quinine, an old anti-malarial drug in a modern world: role in the treatment of malaria. Malar. J. 10, 144–154 (2011).

Barker, G.M., Pottinger, R.P., 1986. Diel activity of the adult argentine stem weevil. New Zeal. J. Zool. 13 (2), 199–201.

Barker, G.M., Pottinger, R.P., Addison, P.J., 1989. Population dynamics of the argentine stem weevil (*Listronotus bonariensis*) in pastures of Waikato, New Zealand, Agriculture. Ecosyst. Environ. 26 (2), 79–115.

Bendict, M.Q., Levine, S.R., Hawley, W.A., Lounibos, L.P. 2007. Spread of the tiger: global risk of invasion by the mosquito Aedes albopictus. Vector Borne Zoonotic Dis. 7, 76–85 (2007).

Benedictow, O.J., 2004. The Black Death 1346–1353: The Complete History. Boydell Press, NY.

Benelli, G., 2016. Spread of Zika virus: the key role of mosquito vector control. Asian Pac. J. Trop. Med. 6 (6), 468–471.

Benelli, G., 2018. Tracking the vector *Xyllella fastidiosa*: geo-statistical analysis of long-term observation host plants influencing the distribution of *Phylenus spumarius* nymphs. Environ. Sci. Pollut. Res. https://doi.org/10.1007/s11356-018-3870-5.

Benelli, G., et al., 2018. Bluetongue outbreaks: Looking for effective control strategies against Culicoides vectors. Res. Vet. Sci. 115, 263–270.

Birney, E., et al., 2007. Identification and analysis of functional elements in 1 per cent of the human genome by the ENCODE pilot project. Nature 447 (7146), 799–816.

Bodino, N., et al., 2019. Phenology, seasonal abundance and stage-structure of spittlebug (Hemiptera: Aphrophoridae) populations in olive groves in Italy. Sci. Rep. 9, 1972.

Bowne, J.G., 1971. Bluetongue disease. Adv. Vet. Sci. Comp. Med. 15, 1–46.

Calistri, P., Caporale, V., 2003. Bluetongue in Italy: a brief description of the epidemiological situation and the control measures applied. Bull.-Off. Int. Epizoot., 15–17.

Caminade, C., et al., 2012. Suitability of European climate for the Asian tiger mosquito Aedes aldopictus: recent trends and future scenario. J. R. Soc. Interface 9, 2708–2717.

Carlucci, A., Lops, F., Marchi, G., 2013. Has Xylella fastidiosa "chosen" olive trees to establish in the Mediterranean basin? Phytopathol. Mediterr. 52 (3), 541–544.

Carolin, et al., 2019. Precise timing of abrupt increase in dust activity in the Middle East coincident with 4.2 ka social change. PNAS 116 (1), 67–72.

Carter, R., Mendis, K.N., 2002. Evolutionary and historical aspects of the burden of malaria. Clin. Microbiol. Rev. 15 (4), 564–594.

Chatterjee, S., Almeida, R.P., Lindow, S., 2008. Living in two worlds: the plant and insect lifestyles of Xylella fastidiosa. Annu. Rev. Phytopathol. 46, 243–271.

Chen, Q., Schlichtherle, M., Waklgren, M., 2000. Molecular aspect of severe malaria. Clin. Microbiol. Rev. 13, 439–450.

Costa, H.S., Blua, M.S., Bethke, J.A., Redak, R.A., 2000. OmmendiaTransmission of Xylella fastidiosa to oleander by the Glassywinged sharpshooter. Homalodisca coagulata. Am. Soc. Horticult. Sci. 35 (7), 1265–1267.

Davis, M.J., Purcell, A.H., Thomson, S.V., 1978. Pierce's disease of grapevines: Isolation of the causal bacterium. Science 199, 75–77.

Dongiovanni, G., et al., 2018. Evaluation of efficacy of different insecticides against Philaenus spumarius L. In: Vector of Xylella fastidiosa in Olive Orchards in Southern Italy, 2015–17. https://doi.org/10.1093/amt/tsy034.

Dongiovanni, C., et al., 2019. Plant selection and population trend of Spittlebug Immatures (Hemiptera: Aphrophoridae) in olive groves of the Apulia region in Italy. J. Econ. Entomol. 112 (1), 67–74.

Donin, A.S., et al., 2018. Takeaway meal consumption and risk markers for coronary heart disease, type 2 diabetes and obesity in children aged 9–10 years: A cross-sectional study. Arch. Dis. Child. 103 (5), 431–436.

EPPO/OEPP, 2004. Diagnostic protocol, Xylella fastidiosa. Bulletin OEPP/EPPO Bulletin. 34, 187–192.

EPPO/OEPP, 2016. *Xylella fastidiosa*. EPPO data sheets on quarantine organisms. No. 166. EPPO Reporting Service 500/02, 505/13, and 1998/9.

Fox, D., 1982. Methemoglobinemia: a puzzling case study. Ann. Emerg. Med. 11 (4), 21421.

Foxi, F., Delrio, G., 2010. Larval habitats and seasonal abundance of Culicoides biting midges found in association with sheep in northern Sardinia, Italy. Med. Vet. Entomol. 24, 199–209.

Foxi, F., Delrio, G., 2013. Efficacy of neem cake for the control of *Culicoides* biting midges larvae. Pharmacol. Archiv. 3, 110–114.

Gensini, G.F., Conti, A.A., Lippi, D., 2007. The contributions of Paul Ehrlich to infectious disease. J. Infect. 54 (3), 221–224.

Goldson, S., Proffitt, J.R., Baird, D.B., 2001. Unexpected presence of larvae of Argentine stem weevil, Listronotus bonariensis (Kuschel) (Coleoptera: Curculionidae). New Zealand Pasture Thatch Aust. J. Entomol. 40 (2), 158–162.

Govindachari, T.R., et al., 1992. Chemical and biological investigations on *Azadirachta indica* (neem tree). Curr. Sci. 63, 117–122.

Hernandez-Martinez, R., Costa, H., Dumenyo, C.K., Cooksey, D.A., 2006. Differentiation of strains of *Xylella fastidiosa* infecting grape, almonds and oleander using a multiprimer PCR assay. Plant Dis. 90, 1382–1388.

Hernandez-Martinez, R., de la Cerda, K.A., Costa, H.S., Cooksey, D.A, Wong, F., 2007. Phylogenetic relationships of Xylella fastidiosa strains isolated from landscape ornamentals in Southern California. Phytopathology 97, 857–64.

Hong-Juan, P., et al., 2012. A local outbreak of dengue caused by an imported case in Dongguan China. BMC Public Health 12, 83.

Howell, P.G., Verwoerd, D.W., 1971. Bluetongue virus. Virol. Monogr. 9, 35–74.

Jackson, A.P., 2015. The evolution of parasite genomes and the origins of parasitism. Parasitology 142 (Suppl 1), S1–S5.

Janse, J., Obradovic, A., 2010. Xilella fastidiosa: Its biology, diagnosis, control and risks. J Plant Patho 92, 35–48.

Kant, A.K., Whitley, M.I., Graubard, B.I., 2015. Away from home meals: associations with biomarkers of chronic disease and dietary intake in American adults, NHANES 2005–2010. Int. J. Obes. (Lond) 39 (5), 820–827.

Karthika, P., et al., 2016. DNA barcoding and evolutionary lineage of 15 insect pests of horticultural crops in South India. Karbala Int. J. Mod. Sci. 2 (3), 156–168.

Kelly, O.J., Gilman, J.C., Kim, Y., Ilich, J.Z., 2016. Micronutrient intake in the etiology, prevention and treatment of Osteosarcopenic obesity. Curr. Aging Sci. 9 (4), 260–278.

Kornei, K., 2019. The Akkadian Empire—Felled by Dust? Eos 100. https://doi.org/10.1029/2019EO115307.

Krugner, R., Sisterson, M.S., Chen, J., Stenger, D.C., Johnson, M.W., 2014. Evaluation of olive as a host of *Xylella fastidiosa* and associated sharpshooter vectors. Plant Dis. 98 (9), 1186–1193.

Latini, A., et al., 2019. Tacking the vector of Xylella fastidiosa: Geo-statistical analysis of long-term field observations on host plants influencing the distribution of Phylaenus spumarius nymphs. Environ. Sci. Pollut. Res. https://doi.org/10.1007/s11356-018-3870-5.

Lindsay, R.L., Hough, B., Blandina, P., Haas, H.L., 2012. Chapter 16—Histamine. Basic Neurochemistry. In: Principles of Molecular, Cellular, and Medical Neurobiology. eighth ed. American Society for Neurochemistry. Lippincott-Raven, Philadelphia, USA, pp. 323–341.

Mann, M.E., et al., 2009. Global signatures and dynamical origins of the little ice age and medieval climate anomaly. Science 326, 1256–1260.

Marshall, G.A.K., 1937. New Curculionidae (Col.) in New Zealand. Proc. R. Soc. N. Z. 67, 316–340.

McCormick, M., 2003. Rats, Communications, and Plague: Toward an Ecological History. J. Interdiscip. History 34 (1), 1–25.

Melior, P.S., Wittiman, E.J., 2002. Bluetongue virus in the Mediterranean basin, 1998-2001. Vet. J. 164, 20–37.

Morente, M., Cornara, D., Moreno, A., Fereres, A., 2018. Continuous indoor rearing of Phylaenus spumarius, the main European vector of Xylella fastidiosa. J. Appl. Entomol. 142, 901–904.

Murugan, K., et al., 2016. DNA barcoding and molecular evolution of mosquito vectors of medical and veterinary importance. Parasitol. Res. 115 (1), 107–121.

Nicoletti, M., Murugan, K., Benelli, G., 2016. Emerging Insect-Borne Diseases of Agricultural, Medical and Veterinary Importance, Insecticides Resistance, Stanislav Trdan. Rejika IntechOpen, Croatia, pp. 219–241 (Chapter 11).
O'Meara, G.F., Evans Jr., L.F., Gettnan, A.D., Cuda, J.P., 1995. Spread of Aedes albopictus and decline of Ae. Aegypti (Diptera: Culicidae) in Florida. J. Med. Entomol. 32, 544–562.
Otto, T.D., et al., 2014. A comprehensive evaluation of rodent malaria parasite genomes and gene expression. BMC Biol. 12, 86.
Otto, T.D., et al., 2018. Genomes of an entire *Plasmodium* subgenus reveal paths to virulent human malaria. Nat. Microbiol. 3, 687–697.
Overall, L.M., Rebek, E.J., 2017. Insect vectors and current management strategies for diseases caused by *Xylella fastidiosa* in the southern United States. J. Integr. Pest Manage. 8 (1), 12.
Pappas, G., Kiriaze, I.J., Falagas, M.E., 2008. Insights into infectious disease in the era of Hippocrates. Int. J. Infect. Dis. 12 (4), 347–350.
Pioz, M., et al., 2012. Why did bluetongue spread the way it did? Environmental factors influencing the velocity of bluetongue virus serotype 8 epizootic wave in France. PLoS One. 7, e43360.
Poinar, G., 2005. Plasmodium dominicana n. sp. (Plasmodiidae: Haemospororida) from tertiary Dominican amber. Syst. Parasitol. 61 (1), 47–52.
Purcell, A.H., 1997. *Xylella fastidiosa*, a regional problem or global threat? J. Plant Pathol. 79, 99–105.
Purcell, A.H., Saunders, S.R., 1999. Fate of Pierce's disease strains of *Xylella fastidiosa* in common riparian plants in California. Plant Dis. 83, 825–830.
Purse, B.V., et al., 2005. Climate change and the recent emergence of bluetongue in Europe. Nat. Rev. Microbiol. 3, 171–181.
Reiter, R., 1998. Global warning and vector disease in temperate regions and at high latitudes. Lancet 351, 839–849.
Reiter, R., 2001. Climate change and mosquito-borne disease. Environ. Health Perspect. 109 (Suppl. 1), 141–161.
Riethmiller, S., 2005. From Atoxyl to Salvarsan: searching for the magic bullet. Chemotherapy 51 (5), 234–242.
Rodriguez, A., et al., 2019. Herbivore accumulation on invasive plants increases the distribution of range of generalist herbivorous insects and supports proliferation of non-nature pests. Biol. Invasions, DOI: doi/org https://doi.org/10.1007/s10530-01913-1.
Roiz, D., 2001. Climatic factors driving invasion of the tiger mosquito (Aedes albopictus) into new areas of Trentino, northern Italy. PLoS One. 6, e14800.
Rolshausen, G., 2019. Contemporary evolution of reproduction isolation and Phenotipic divergence in Sympatry along a migratory divide. Curr. Biol. 19 (24), 2027–2041.
Romi, R., 1995. History and updating of the spread of Aedes albopictus in Italy. Parassitologia 37, 99–103.
Romi, R., 2010. Arthropod borne diseases in Italy: From a neglected matter to an emerging health problem. Ann. Ist. Super. Sanita 46 (4), 436–443.
Rosen, W., 2007. Justinean's Flea. Empire and the Birth of Europe. Pimlico, London (UK).
Saegerman, S., Berkvens, D., Mellor, P.S., 2008. Bluetongue epidemiology in the European Union. Emerg. Infect. Dis. 14, 539–544.
Sallares, R., Bouwman, A., Anderung, C., 2004. The spread of malaria to southern Europe in antiquity: new approaches to old problems. Med. Hist. 48 (3), 311–328.
Steinbiss, S., et al., 2016. Companion: a web server for annotation and analysis of parasite genomes. Nucleic Acids Res. 44 (W1) W29–34.
Teri, T.A., et al., 2009. Finding the missing heritability of complex diseases. Nature 461, 747–753.

Trost, C. (Ed.), 1982. The blue people of troublesome creek. Science 82, 34–39.
Tu, Y., 2011. The discovery of artemisinin (qinghaosu) and gifts from Chinese medicine. Nat. Med. 17, 1217–1220.
Turner, W.F., Pollard, H.N., 1955. Additional leafhopper vectors of phony peach. J. Econ. Entomol. 48, 771–772.
Turner, W.F., Pollard, H.N., 1959. 1959. Insect transmission of phony peach disease. United States Department of Agriculture. TEC Bull. 1193, 1–27.
Verwoerd, D.W., Els, H.J., De Villiers, E.-M., Huismans, H., 1972. Structure of the bluetongue virus capsid. J. Virol. 10 (4), 783–794.
Weiss, H., 2017. In: Höflmayer, F. (Ed.), Seventeen Kings Who Lived in Tents. The Late Third Millennium in the Ancient near East: Chronology, C14, and Climate Change. Univ Chicago, Chicago.
Weiss, H., et al., 1993. The genesis and collapse of third millennium north Mesopotamian civilization. Science 261, 995–1004.
WHO 2019. World Malaria report 2019.
Winau, F., Westphal, O., Winau, R., 2004. Paul Ehrlich-in search of the magic bullet. Microbes Infect. 6 (8), 786–789.
Wong, R., et al., 2005. Update of *Xylella fastidiosa* in landscape. Plant Host. (Proc. Pierce's Res. Sympos, December 5–7, San Diego. CA).
Xiao, J.-P., He, J.-F., et al., 2016. Characterization of large outbreak of dengue fever in Guangdong Province, China. Infect. Dis. Poverty 5, 44–87.
Zanchetta, Z., et al., 2016. The so-called "4.2 event" in the Central Mediterranean and its climatic teleconnections. Alp Mediterr Quat 29, 5–17.

CHAPTER SIX

Novel solutions to insect-borne diseases in action

Image courtesy of Shutterstock

Plant metabolites as a natural resource

Lead compounds for antibacterial chemotherapy, as for all chemotherapy, are obtained from two sources: de novo chemical synthesis and natural products from living organisms. For antibacterials, natural products have historically been by far the more important, with only three clinically used classes having a purely synthetic heritage. The remaining classes of antibacterials all have their roots in natural products obtained from microbial sources. One view is that the production of these compounds may have evolved over millennia to enable competitive fitness of their microbial producers at the expense of less competitive organisms; their efficacy in antagonizing bacteria is, therefore, clear. It is thus unsurprising that the isolation and evaluation of

Insect-Borne Diseases in the 21st Century
https://doi.org/10.1016/B978-0-12-818706-7.00006-1

these bioactive compounds has proved a very fruitful line of investigation for medicinal chemistry.

Most of the antibacterials, introduced during the golden age of these drugs, were discovered as a result of screening natural products obtained from microbial fermentations. However, the current development of novel antibacterials continues to represent an unattractive investment for big pharma. Chemical space is basically infinite, comprising all possible molecules, which has been estimated to exceed 10^{60} compounds even when only small (less than 500 Da) carbon-based compounds are considered. A large component of biologically relevant chemical space is occupied by natural products. Natural products can be regarded as prevalidated by nature. They have a unique and vast chemical diversity and have been optimized for optimal interactions with biological macromolecules through evolutionary selection. Virtually all biosynthesized compounds have biological activity.

In this part of the chapter, we shall look at the natural products world and their potential in control of insect-borne diseases. This is one of the key arguments of the book. To achieve a shift from synthetic to bioorganic insecticides, it is necessary to have a clear idea of what natural products are and what they can do. First, we should try to define natural products. I had occasion to participate in a scientific chat about this argument. Responses included: "They are antioxidant compounds," "They are plant constituents with medicinal properties," "They are produced by the secondary metabolism," etc. These definitions are more or less what you can find on the internet or in books, where you often have old definitions and consolidated information. These sentences are too superficial and elementary and can act as intellectual poison for people in search of their own opinion. Let us consider some other points of view.

First, the denomination natural products is misleading. Everything is natural, perhaps excluding something generated by our imaginations. We would say organic or bio-products, meaning produced by living organisms, versus synthetic products created in a test tube or out of any metabolism. An organic product is the chemical reply to a selective pressure and the result of a cellular chemical mechanism that is different between an organism and organism. Therefore, organic products are the consequence of evolutionary pathways and the effect of the environment on the homeostatic needs of each individual as part of its species. We can consider natural products as the results of the specific metabolism of a taxon, such as a distinctive characteristic like number of legs or type of fruit, but in this case we are talking strictly of chemistry.

You are CO_2 and to CO_2 you shall return

Before going into detail, let us start from some very general considerations, as the cornice of the picture. The universe works by changes, or more accurately turnovers, of energy and matter. This is its only *raison d'être*, magisterially unscripted many times ago in the mot "*panta rei*" ("everything flows"). From superblack holes to the timid first movements of a newborn bacterium, the rules (or, if you like, equations) dominating all of matter since time began must be the same ones, everywhere. As far as we know, the changes are governed and regulated by the second law of thermodynamics. Thus, differences between inorganic and organic matter are less relevant than usually reported. Living organisms are specialized in swift and efficient turnovers, necessary to maximize the profit of the energy around.

Biomachines need special energy to work. In particular, they are collectors of selected negative entropy from the environment. The collection is obtained by the selection, introduction, and transformation of the available energy around (for us, air and food). Any organism tends to optimize its job by a personal molecular mechanism, called metabolism. During its homeostatic transit, which we call life, the possibilities of an organism to personalize its metabolism are quite restricted. Cell metabolisms, although different according the organism, are derived from a unique molecular model. Practically they are descendant from the original one, based on the primary metabolites, like proteins, carbohydrates, nucleic acids, and lipids. This model emerged a long time ago during the life sunrise, and later limited variations occurred as the result of adaptation to the different environmental situations. In practical terms, everything was already written in the nucleic acid of LUCA. This limitation is useful, since it allows the exchange of materials between organisms. Our reality is the consequence of a series of evolution steps, generally losing and rarely adding previous characters. Evolution is often the amplification of some aspects of the metabolism resulting in difference in forms. The consequence is that most changes, even in a living organism, do not need selection but are the consequence of the aforementioned forces. Thus, autoassemblement, the dominant molecular process in cell metabolism, of a nucleic acid is the result of the same ruling forces acting in the formation of a snow crystal.

Therefore, we can summarize these first steps to reconsider natural products. The need to change is the only common tendency of matter in the universe. The dicotomic separation between inorganic and organic matter is mainly an invention, or more accurately a forcing act, which is a consequence of our ignorance of the forces governing the universe and a pretension of hierarchy. Living organisms are the result of complex molecular assembling, devoted to optimize the turnover of energy and matter to their own advantage. To obtain this result, it is necessary to develop rapid, efficient, bold, smart biomachines, able to be in synthesis with environmental changes through their internal molecular mechanism. In fact, the increase of the capacity to react to environmental changes was a key of the passage between the inorganic and the organic matter. The advent of mankind exalted this ability, with extreme attempts to adapt the environment to human necessities without any limit.

Therefore, we must consider a common basis of metabolism, directly derived from the first original cell, and subsequent modifications of the initial molecules produced by the primary metabolism into a series of molecules in the opposite direction, synthetized to furnish to any organism the necessary individuality and capacity to be in accordance with environmental changes. For obvious metabolic reasons, to synthetize the personalized molecules it is necessary to utilize the products of the primary metabolism. Therefore, the further formed molecules can be considered part of a secondary (derived) metabolism. No hierarchy, no level of importance, no total separation inside the cell: primary means only first produced. Primary metabolites can be considered an integral part of the whole universe of living organisms; secondary metabolites are additional and personalized. Their presence is usually not essential but it can represent a fundamental improvement in the fight for selection and competition. The consequence is that living organisms are mixtures of primary and secondary products, organized and personalized. It is important to note that viruses (not considered as organisms but acting similarly in many aspects) are made only by primary molecules, i.e., nucleic acids, lipids, and proteins.

The variation, meaning adaptation, is obtained mainly by the sacrifice of the natural products of the secondary metabolism. These are small molecules in contrast to those of the primary metabolism, which are usually macromolecules, but their biosynthesis is more complicated. Macromolecules (polymers) are the results of a repetitive act of junction between units (monomers) of the same chemical class: a pool of a few units generating myriad possible combinations. A secondary metabolite must be synthetized ad hoc, following a logical series of chemical steps. Once its role is terminated, its synthesis disappears along with its importance.

Once synthetized by the orders due to a response to the environmental pressure, the secondary metabolite is therefore ready to go to its final destination, consisting of the expected reaction, and later to come back again and again in the flux of matter, the same flux that generated the molecule. At the end of its pathway, the secondary metabolite is destroyed at the point that it completes its functionality. Whereas enzymes are regenerated, secondary metabolites are destined for a short life, but while they are useful they will be produced, because no sacrifice of any molecule or organism is pointless. Once its temporary experience is terminated, the first law of the universe will claim its possession to give rise to another molecule, another life, another possibility to be different, until the negative entropy that generates the universe will be consumed.

The importance of natural products

Previous considerations must be revised in the light of cellular needs. The activities of natural products cannot be generalized or linked to our will by considering their utility to us, but they are related to taxonomy and evolution, which means the place and the role of the organism in its habitat and its previous ancestral history over the centuries. The distinction between primary and secondary metabolism is a didactic invention to introduce order in the apparent chaos of metabolism and biosynthesis, in times when the knowledge of the chemistry of living organisms was still very deficient. The reality is that there is only one metabolism, with no target on a natural product or limit in its utilization or hierarchy of importance. Steroids are considered typical secondary products; however, they are fundamental for cell membrane stability as well as chemical mediators for interactions with herbivorous insects. Therefore, the two levels of metabolism are necessary to understand and classify the metabolic situation, but the reality is that there is only a unique metabolism and any sequences or series of steps are the results of mandatory activity of the genome.

In Table 6.1 (Mora et al., 2011), we can see that the major producers of natural products are small and unable to move at all or for long distances. The situation of microorganisms is misleading since the limited number of species is the result of our inability to distinguish between the various taxa. Everyone who has had occasion to work with microorganisms knows perfectly their superior capacities to change and adapt rapidly and radically. This exception confirms that interpretation. It is also noteworthy that these numbers are not totally informative, since the taxonomic distance should also be considered. Several species in insects and plants are so similar that only experts of the taxon are able to identify the differences. However, the imbalance is evident with the prevalence of invertebrates species, with a total

Table 6.1 The correlation between secondary metabolism production and the biodiversity in selected taxa (values from different sources).

Taxon	Classified species	% of the total still probably to be classified	Production of secondary metabolism
Microorganisms	5700	24	Abundant
Invertebrates (most of which insects)	1250000	24	Abundant
Plants	322700	33	Abundant
Amphibians, reptiles	18700	5–10	Low
Fish	33600	17–1	Low
Birds	9000–10000	6–1	Low
Mammals	4956	5–10	Low

number of invertebrate species that could be 5 million, 10 million, or even 30 million, compared to just 60,000 vertebrates. This means that the medieval hierarchy about the success of models in animals should be totally revised, including the prevalence of humans, who however are doing all that they can to be a great part of the planetary biomass.

Usually natural products are considered the secondary metabolites produced by plant species. These compounds are easier to extract, separate, and study, and are produced in great quantities in many plants and provided with physiological activity. In fact, the first inputs of interest were for the alimentary and medicinal properties of these molecules. Phytochemistry allowed the knowledge of more than 1000 compounds, but the molecules produced by mankind in similarity or as derivatives to implement the original activity number more than 400,000.

The rationality of natural products

Now we can focus on the biosynthesis of plant natural products, a central argument in phytochemistry. The final general picture resembles that of a chemical clepsydra. By photosynthesis, plants and other chlorophyllian organisms transform CO_2 and water into sugars to trap solar energy, and they are already converted into amino acids, nucleic acids, and fatty compounds. These are utilized as they are or segmented to give rise to the pool of intermediate metabolites. Interestingly, this pool of intermediate metabolites is composed of a few molecular pieces, which are joined together and converted in the precursors of different classes of natural products. The boom of a successful structural model, consisting in the repetition of the initial precursor in a series of similar compounds, is absolutely similar to that acting on the natural selection of living organisms. The model is repeated obsessively with little difference between the derived metabolites of the same type, generating an explosion of molecules (or organisms) very similar if not practically identical. However, exactly the same phenomenon is also present in the inorganic matter, confirming once more the possibility of evolution without selection. It is therefore necessary to know the biosynthetic origin to recognize and valuate a natural product. Also, in this case, there are some tricks. In chemistry, molecular structures of natural products tell us their story, how they are born, and why they are in nature. Like looking at figures of a book, we can interpret nature's tale by the structures of natural products. Intermediate metabolites are utilized as molecular units to construct the secondary metabolites. The mechanism is similar to that of Lego pieces. In the plant metabolism, there is a simple and universal unit, consisting in the acetyl coenzyme A. This intermediate metabolite is a piece of two C, resembling the little brick of the Lego, and it can easily assembly forming the skeleton of any natural product. In addition, there other intermediate metabolites, let's say the special units, like the decarboxylated amino acids, leading to the alkaloids, and the deaminated amino acids, leading to phenols and other aromatic metabolites. Usually, the main part of the secondary metabolite is made from the junction of C2 units derived from AcCoA, but other units can be added to obtain a series of precursors giving rise to a cascade of derivatives.

Students of chemistry first know the separation in two initial courses: inorganic and organic chemistry. The first one is considered general chemistry, concerning all the atoms in the Periodic Table, whereas the second is focused on molecules where carbon atoms are prevalent. Simplifying, inorganic chemistry is the chemistry of all the elements, and organic chemistry is the chemistry of some compounds containing C, which are produced by living organisms. However, let us consider two simple derivatives of C: CO_2 is considered an inorganic molecule and CH_4 an organic one, but we have plenty of carbon dioxide, and methane is toxic. Natural products, as present in living organisms, are characterized by a molecular skeleton of carbon atoms, which are mainly saturated by hydrogens. In other words, to recognize a natural product, you can examine the structure and check the presence of a hydrocarbon skeleton. This rule is so strict that organic chemistry has evolved a special approach useful to simplify the writing of the structures of the organic molecules, by a sort of design. A chemical structure can be considered as an ideogram representation, useful to recognize or evidence the functional groups and the class to assign the metabolite on the basis of the skeleton type. Of course, the design must be interpreted and needs some experience, but once adopted it was universally accepted and used. However, this writing system is imperfect, since, adopting the sheet of paper, the molecule is constricted to two dimensions, losing all the conformational states that make it act as living.

The enormous numbers (imagine if we add the constituents of the creatures that lived in the past) of known natural products are considered typical of life, but most of their characters are common to all matter. If it is true that there is not a leaf equal to another leaf living or lived and each zebra possesses its own unique skin pattern, each snowflake is unique, the general characters can still be assimilated to a common model of reference. The common hexagonal scheme of all snowflakes do not prevent the formation of thousands of dendritic forms. The hexagonal (trigonal) system is the constant of the forms of the crystals of calcite, but variability allows the existence of more than 2000 combinations of crystallographic structures. The hexagonal cycle is present in most cyclic structures of natural products and it is the icon of the particular chemical tautomerism typical of benzene and the plethora of aromatic compounds. The ratio of the obsessive multiplication of models, either in chemical structures as in organisms, can be explained on the basis of the axiom, that we reported in the start of this chapter.

However, it is possible to argue that if differences can distinguish one organism from another, the same is not true in the case of a molecule. Once the structure is determined, a molecule is a molecule, even if generated by the metabolism or by a synthetic route, and therefore living rules cannot be applied to the molecular world. This is true only in part. We must consider, in particular for natural products, the conformational forms, which are different ways for the same molecule to react. An organic molecule can change its 3D structure easily to adapt to environmental needs, just like we open and close our hands. Furthermore, once inside the metabolism, the molecule, even if inorganic, is integrated in the organic network and it is forced to collaborate with an integrated system. The cell environment is very different from the test tube of the chemist. In practice, the molecule is simply part of an advanced complex integrated dynamic system, limiting its freedom. Besides the effects of the different physical forces working as tensors in the assembling of the matter, as we can know and distinguish a giraffe from a wizard, in the same way we can study and classify natural compounds on the basis of their structures and assign their places in phytochemistry and their role in metabolism. About the role, we must remember that natural products are chemical mediators inside the environment, meaning that they must act on the receptor of the target organisms (Pagare et al., 2015). Therefore, the structures of natural products are derived from their activity—the same argument we use in the consideration of a pharmacological drug. Here, the importance of natural products inspires all the molecules tailored to affect living organisms. In particular, among the detected activities in plant species, defense against herbivorous activity is highly reported. Plants cannot counteract by movements like animals and therefore an arsenal of chemical weapons is essential in the fight for survival. The consequence is that in many cases plant natural compounds are toxic or repellent to phytophagous, like the insects. Natural products with such properties should be extracted and found in the complex and confused reservoir of secondary metabolites, and their presence in the plant is not sufficient. It is necessary to realize the mechanisms of action and their potentiality to be used as market products. The pathway to explore the utilization of a natural product should comprehend the discovery, the isolation, the selection of the best source and the possibility of production in high quantity, the environmental impact, and the cost of production.

However, the preliminary approach in the utilization of this type of compounds is the choice about a single active constituent or a total/enriched extract. In the second case, the synergic action of many constituents could be important to obtain an equilibrium of different forces as in nature, whereas in the first case it is easier to obtain the determination of the doses and the mechanism of action, as well the target. The single molecule approach allows the study of pharmacologic action, as well as the determination of the necessary doses, whereas in the use of an extract, whose composition is variable and more or less known, the same approach is difficult to apply and the results variable. In recent years, there has been conflict between these two approaches, also in consideration of the massive introduction into the market of new plant products, generally known as nutraceuticals, with the apparent intent to act in health promotion, favored by increasing demand from consumers, clients, and aficionados. However, this current debate will appear several times in many sections during the next chapter.

References

Mora, C., et al., 2011. How many species are there on earth and in the ocean? PLoS Biol. 9(8), e1001127.

Pagare, S., Bhatia, M., Tripathi, N., Bansal, Y.K., 2015. Secondary metabolites of plants and their role: overview. Curr. Trends Biotechnol. Pharm. 9 (3), 293–304.

CHAPTER SEVEN

New solutions using natural products

Image courtesy of Shutterstock

Dr. Bruce Reid in his *Principles of Botany* wrote: "Genes determine what an organism can do, the environment determines which of the things that an organism can do the organism will do. A given species may be constructed differently in different environments." Therefore, genes = can, environment = will. Natural products are the bridge between the environmental wellspring of experience and the yearning for changing the world in one's own favor. These products are the result of the combined effects of the possibilities and limits of the genome and the environmental inputs. Each organism, as well as creating its best and adequate form, synthetizes its chemical arsenal, necessary to realize and consolidate its individuality inside the habitat. This arsenal is mainly dedicated to the interactions with abiotic

Insect-Borne Diseases in the 21st Century
https://doi.org/10.1016/B978-0-12-818706-7.00007-3

and biotic targets, which are important to the individual homeostasis and survival. One form of evidence in accordance with this interpretation is the strong correlation between natural products' producers and biodiversity, therefore indicating a matter of adaptation.

However, it is possible to argue that if differences can characterize each organism from another, the some is not true in case of a molecule. Once determined the structure, a molecule is a molecule, even if generated by the metabolism or by a synthetic route, and therefore the living rules cannot be applied to the molecular world. This is true only in part. We must consider, in particular for natural products, the conformational forms, which are different ways of the same molecule to react. An organic molecule can change easily its 3D structure adapting to the environmental needs, like we open and close our hands. Furthermore, once inside the metabolism, the molecule, even if inorganic, is integrated in the organic network and it is forced to collaborate to an integrated living system. The cell environment is far different from the test tube of the chemist. In practice, the molecule is simply part of an advanced complex integrated dynamic system, limiting its freedom. Beside the consideration that physical forces work in the same way in the assembling matter, as we can know and distinguish a giraffe from a wizard, we can study natural compounds and distinguish them on basis of the structures and assign their place and role. About the role, we must remember that natural products are chemical mediators inside the environment, meaning that they must act on the receptor of the target organisms. Therefore, the structures of natural products are derived from their activity, the same argument we use in the consideration of a pharmacological drug. Here, the importance of natural products inspiring all the molecules tailored to affect living organisms. In particular, among the detected activities in plant species, the defensive against herbivorous is highly reported. Plants cannot counteract by movements like animals and therefore an arsenal of chemical weapons is essential in their fight for survival. Therefore, it is highly possible to find natural products toxic or repellent to phytophagous, like insects, as well as active constituents against pathogens. Natural products with such properties should be extracted and found in the complex and confused reservoir of secondary metabolites, but the presence into the plant is not sufficient. It is necessary to realize the mechanism of action and the potentiality to be used as marketed products. Therefore, the pathway should comprehend the discovery, the isolation, the sources, the bioactivities, and the possibility to be obtained in high quantity, the method of utilization, the environmental impact and the cost of production.

Essential oils as natural insecticides

The natural products have been already reported for antibiotic activity. In particular, essential oils and phenols are stated in many papers as being responsible of antibacterial and insecticide properties (Ghosh et al., 2012; Gibbons, 2008). However, the activity seems to be too general. Most essential oils are composed of the same main constituents, the difference being mainly quantitative for each compound or hidden deeply inside the plethora of secondary constituents. Furthermore, excluding the phenols typical of essential oils, the quantity of known phenols is very large and complicated, with thousands of structures reported and different activities connected. Therefore, special research must be performed, including the possibility to test new substances never reported.

Among secondary natural products with insecticide activity, a special place must be assigned to essential oils. In this regard, several papers report the utilization of essential oils as the active ingredient against pests. The antimicrobial activity of essential oils is also well-reported and evident in nature. Several plants accumulate essential oils in the inner parts of roots and rhizomes, in order to avoid the devastating attack of micropathogens, wherein the plant often accumulates precious reserve substances. In other cases, the aerial parts are focused on defensive or cooperative actions. Some birds defend their clutch by surrounding the nest with aromatic plants. The human utilization of essential oils of different kinds against insects has a long story. Herodotus reported the use in ancient Egypt of mosquito nets and towers impregnated with fish odor to avoid mosquito bites.

An essential oil is a complex chemical mixture of substances volatile at ordinary temperature (Figs. 7.1 and 7.2), and therefore the constituents must have low molecular weight. In other words, they are micromolecules, with average molecular weight of 120–160 uma and hydrocarbon prevalence. Essential oils can be extracted from the raw materials by utilizing their volatile properties, such as in the steam distillation method. The antimicrobial activity of essential oils is usually a consequence of the content of phenols, but other properties must be considered. On the basis of the structures of the active constituents, there are two types of essential oils. The first one, mainly present in less advanced Angiosperm dicotyledons, like Magnolidae, contains mainly root and fruit drugs rich in simple phenolic phenylpropanoids, which are mainly utilized by the plant in protection of pathogens. In Rosidae and Sympetalae, the terpenoids progressively

Fig. 7.1 A typical current apparatus for the production of essential oils by steam distillation.

become predominant. The new volatile constituents, in addition to the protective and toxic effects, afford a positive attraction on pollinatory agents, evidencing the plant position and allowing a memory of the selected species. In this way, the scenario of the interaction between animals and plants changes from defensive to collaborative. The new plants to be selected and appreciated can enrich the offer to the collaborative animals

Fig. 7.2 A historical apparatus for the production of essential oils. Also shown is the typical mask once used for centuries in Europe in the case of plague.

with fruits of inebriant flavors, colored flowers and other nice experiences. Therefore, utilized essential oils are mainly complex mixtures of volatile plant secondary metabolism and consist mainly of monoterpenes and sesquiterpenes, which means lipid secondary metabolites, and, to a lesser extent, of aromatic compounds. The choice of essential oil depends firstly on the taxonomy of the selected plant and the effects depend on the nature of the constituents of the essential oil. The conclusion, based also on direct experiments, is the presence of a general antibiotic and insecticide activity; however, another real need is a selective toxicity in favor of the useful organisms. Therefore, some kind of activity is expected for an essential oil, but there is a necessity to maximize its effectiveness. They are exploited in several fields, such as perfumery, food, pharmaceutics, and cosmetics, but essential oils have also long-standing uses in the treatment of infectious diseases and parasitosis in humans and animals.

Essential oils, currently more than 300 of which are known, are highly variable in their complex composition. Usually, at least a mixture of more than main 30 different constituents of low molecular weight is present. Among the single species, the qualitative composition of the essential oil is respected, although a quantitative variability is common between population according to the environmental pressures. However, some terpenes can be easily found, like the hydrocarbons (myrcene, pinene, terpinene, limonene, cymene, α- and β-phellandrene) and the oxygenated ones, like the alcohols (geraniol, linalool, menthol, terpineol, borneol), the aldehydes (citral, citronellal), ketones (menthone, pulegone, carvone), bicyclic monoterpene ketones (thujone, verbenone, fenchone), acids (citronellic acid, cinnamic acid), oxides (1,8-cineole) and esters (linalyl acetate), but the aromatic phenols (carvacrol, thymol, safrole, eugenol) are also common. A few essential oils may also contain sulfur-containing constituents, methyl anthranilate, coumarins, and special sesquiterpenes such as zingiberene, curcumin, farnesol, sesquiphellandrene, turmerone, nerolidol, etc. Often these components are at low concentrations (less than 1% each), but the opposite is also possible, with major compounds that can represent up to 70% of the total volume of oil, as much as 90% like eucalyptol in *Eucalyptus* or limonene in *Citrus* or pinenes in turpentine of *Pinus*. Therefore, the antiparasitic activity of an essential oil can vary according to differences in its chemical composition, but it is usually present. Nowadays, there is an increasing interest in the utilization of essential oils against endoparasites and ectoparasites of animals and humans, in particular when they are resistant to conventional drugs. However, the use of essential oils is in general restricted for the high cost and considering that usually they are not adequately specified for the considered target (Bagavan and Rahuman, 2010; Shaalan et al., 2005; Tikar et al., 2018). Again, also in the case of essential oils, the insurgence of the insecticide resistance must be considered (Brown, 1986), although usually less common in these cases. Therefore, in consideration of their general but also weaker effectiveness of essential oils in comparison with synthetic insecticides, their utilization requires the insecticidal properties of essential oils to be investigated in different approaches of selection of the studied plants and their uses. Some examples of researches, in which I had occasion to participate, involving essential oils in insect-borne diseases are reported here. The leading idea was to utilize the essential oil properties in an innovative way, such as in mixture or selected types.

In 2017 (Benelli et al., 2017a,b), the activities of five essential oils were investigated. The essential oils were obtained from different plants: *Pinus nigra* var. *italica* (Pinaceae), *Hyssopus officinalis* subsp. *aristatus* (Lamiaceae), *Satureja montana* subsp. *montana* (Lamiaceae), *Aloysia citriodora* (Verbenaceae), and *Pelargonium graveolens* (Geraniaceae) against *Culex quinquefasciatus* (Diptera: Culicidae), which is a vector of lymphatic filariasis and of dangerous arboviral diseases, such as West Nile and St. Louis encephalitis. The research was original in its focus on the potential synergistic and antagonistic effects, testing them in binary mixtures on *C. quinquefasciatus* larvae. Mixtures of essential oils are very easy to obtain, since the constituents are perfectly soluble in the final solution and the selected oils were cheap and easy to find on the market. In such a way, knowing the composition, it is possible to combine constituents, enhancing the range and the quality of activity. The pool of the investigated species was highly varied, but this was considered a positive factor. First, the chemical composition of each essential oil was investigated by GC-MS analysis, which is the best analytic method in such mixtures of volatile compounds. Therefore, it was also necessary to test the activity of each essential oil and later to try the best combination on the basis of its effectiveness. The highest effectiveness was obtained by *S. montana* subsp. *montana* essential oil ($LC_{50}=25.6\,\mu L\,L^{-1}$), followed by *P. nigra* var. *italica* ($LC_{50}=49.8\,\mu L\,L^{-1}$), and *A. citriodora* ($LC_{50}=65.6\,\mu L\,L^{-1}$). It was possible to obtain an enhancement of the larvicidal activity by preparing simple binary mixtures of essential oils (ratio 1:1), such as *S. montana* + *A. citriodora*, which showed higher larvicidal toxicity ($LC_{50}=18.3\,\mu L\,L^{-1}$). On the other hand, testing *S. montana* + *P. nigra* (1:1), an antagonistic effect was detected, leading to an LC_{50} ($72.5\,\mu L\,L^{-1}$) higher than the LC_{50} values calculated for the two oils tested separately. Therefore, these results indicate the extreme need for innovation and imagination in natural products research, against many papers repeating the same procedure that change only the plant used.

Another work (Pavela et al., 2016a,b) was based on geographic distribution and the traditional use. Six medicinal and aromatic plants—*Azadirachta indica* (see later in this chapter), *Aframomum melegueta*, *Aframomum daniellii*, *Clausena anisata*, *Dichrostachys cinerea*, and *Echinops giganteus*—have been traditionally used in Cameroon to treat several disorders, including infections and parasitic diseases. The aim was to evaluate the activity of the essential oils of these plants against *Trypanosma brucei* TC221 and determine their selectivity with Balb/3T3 (mouse embryonic fibroblast cell line) cells as a reference. Essential oils from *A. indica*, *A. daniellii*, and *E. giganteus* proved to be the most active ones, with half maximal inhibitory concentration (IC_{50}) values of 15.21, 7.65, and 10.50 μg/mL, respectively. These essential oils were characterized by different chemical compounds, including monoterpenes and sesquiterpene hydrocarbons and oxygenated sesquiterpenes. Some of their main components were assayed as well on *T. brucei* TC221, and their effects were linked to those of essential oils. In this way, the research partially confirmed the ethnopharmacological indications, validating their traditional use and confirming the utility of popular information in the search for useful plants.

The synergic action of binary mixtures of similar constituents of essential oils against larvae of the filariasis vector *Culex quinquefasciatus* was also the inspiration behind research (Benelli et al., 2017a,b) on four Apiaceae species: *Trachyspermum ammi*, *Smyrnium olusatrum*, *Pimpinella anisum*, and *Helosciadium nodiflorum*. Initially, all the essential oils proved to be highly toxic to the larvae, but short-term exposure to both binary mixtures strongly reduced emergence rates, fertility, and natality of the *C. quinquefasciatus* that survived after the treatment at the larval stage. In addition, larvicidal acute toxicity of essential oils main constituents, i.e., germacrone, isofuranodiene, and (*E*)-anethole, were carried out, with LC_{50} being $18.6\,mg\,L^{-1}$, $33.7\,mg\,L^{-1}$, and $24.8\,\mu L\,L^{-1}$. The results demonstrated the promise of these essential oils and their constituents to develop cheap and effective mosquito larvicides.

In another paper (Pavela et al., 2016b) published in the same year, the vector target was the same but the selection of the plant totally different, as endemic to Madagascar. The reason is that in some parts of the world, there are interesting examples of endemic flora whose species could contain different essential oils and therefore different activity. For this reason, pharmaceutical companies often explore remote parts of Amazonia or isolated zones in search of new active compounds. There were examples of exploitation of rare African *Rauwolfia* species to obtain their indole alkaloids. Working with endemic species is important considering that in many cases, populations are in limited numbers and at risk of extinction, and we need to identify their molecular treasure before they disappear.

This was also the motivation for my trips to several parts of the world, focusing in particular on deserts and islands, in search of special plants. Madagascar's fauna and flora are diverse and unique. When the unique Gondwana continent braked up in several pieces, India started to move to Asia living Africa. However, a consistent block remained near to Africa, becoming a great island, now known as Madagascar. This happened more than 100 million years ago. The isolation of Madagascar gave rise to a particular case of biodiversity. This is the story of the beginning of Madagascar, as far as we know. Here, it is important to report that the island is characterized by at least seven very different habitats, each with different endemisms. The potentiality of *Cinnamosma madagascariensis*, an endemic species widely present in the forests of Madagascar, was reported to us thanks to the exceptional collaboration with Professor Philippe Rasoanaivo, who had a deep knowledge of the flora of the island and their economic importance. This plant has important traditional uses ranging from management of dementia, epilepsy, and headache to malaria (Rakotosaona et al., 2015). Few data have been reported about the chemical composition of its essential oils, and no studies have been published on its bioactivity against mosquitoes. Once again, we first investigated the chemical composition of essential oils extracted from stem bark and leaves of the plant, and later their larvicidal potential against the filariasis vector *Culex quinquefasciatus*. The reason was that when you have little information, you must consider that different parts of a plant can contain very different essential oils. In fact, GC-MS analysis revealed differences between the volatile profiles of leaves and bark oils. In the former, linalool (30.1%), limonene (12.0%), myrcene (8.9%), and α-pinene (8.4%) were the major constituents, while in the latter one, β-pinene (33.3%), α-pinene (19.3%), and limonene (12.0%) were the most representative compounds. Acute toxicity experiments conducted on larvae of the filariasis vector *C. quinquefasciatus* led to an LC_{50} of 61.6 and 80.1 $\mu L L^{-1}$ for the bark and leaf essential oils, respectively. Overall, *Cinnamosma madagascariensis* bark and leaf essential oils against filariasis vectors proved to be promising, since they are effective at moderate doses.

The insecticidal activity of the essential oil of another Malagasy plant was also studied (Benelli et al., 2020). *Hazomalania voyronii* is popularly known as hazomalana and its use to repel mosquitoes and resist against insect attacks has been handed down from generation to generation in Madagascar. The property of the essential oils obtained from the stem wood, fresh and dry bark of *H. voyronii* were able to repel important mosquito vectors (*Aedes aegypti* and *Culex quinquefasciatus*). Furthermore, the toxicity of the aforementioned essential oils was investigated by WHO on three insect species of agricultural and public health importance (*Cx. quinquefasciatus*, *Musca domestica*, and *Spodoptera littoralis*), respectively, as well as the adequate topical application methods and compared with the commercial repellent *N,N*-diethyl-*m*-toluamide (DEET). Repellence assay revealed almost complete protection (>80%) from both mosquito species for 30 min when pure fresh bark essential oil was applied on the volunteers' arms, while DEET 10% repelled more than 80% of the mosquitoes up to 120 min from application. The research validated the traditional use of the bark essential oil to repel insects, although an extended-release formulation based on *H. voyronii* essential oils is needed to increase the repellent effect over time. Furthermore, it evidenced the wide spectrum of insecticidal plants potentially useful in the fabrication of green repellents and insecticides useful to control mosquito vectors and agricultural pests, avoiding the utilization of synthetic products.

Another interesting study (Benelli et al., 2017b) was dedicated to *Helichrysum faradifani* (Asteraceae), which is a perennial endemic shrub growing in rocky and sandy places of Madagascar. The ethnopharmacological about Malagasy traditional medicine reports that this plant is used as a wound-healing agent, disinfectant, and for the treatment of syphilis, diarrhea, cough, and headache. The chemical composition of the essential oil distilled from the aerial parts of *H. faradifani*, and analyzed by GC-MS, evidenced that monoterpene hydrocarbons (51.6%) were the major fraction of the essential oil, with bicyclic α-fenchene (35.6%) being the predominant component. Sesquiterpene hydrocarbons (34.0%) were the second major group characterizing the oil, with γ-curcumene (17.7%) being the most abundant component. Its insecticidal activity was evaluated against second, third, and fourth instar larvae of the lymphatic filariasis vector *Culex quinquefasciatus* by acute toxicity assays. The most sensitive were second instar ($LC_{50}=85.7\,\mu L\,L^{-1}$) larvae. For the third and fourth instar larvae, the estimated LC_{50} were 156.8 and $134.1\,\mu L\,L^{-1}$, respectively.

Finally, a different approach to volatile substances was performed, considering that smoke is often traditionally used against mosquitos (Ansari and Razdan, 1996). Therefore, the larvicidal, pupicidal, and smoke toxicity of *Senna occidentalis* and *Ocimum basilicum* leaf extracts against the malaria vector *Anopheles stephensi* were evaluated (Murugan et al., 2015). In larvicidal and pupicidal experiments, *S. occidentalis* LC_{50} ranged from 31.05 (I instar larvae) to 75.15 ppm (pupae), and *O. basilicum* LC_{50} ranged from 29.69 (I instar larvae) to 69 ppm (pupae). Smoke toxicity experiments conducted against adults showed that *S. occidentalis* and *O. basilicum* coils evoked mortality rates comparable to the pyrethrin-based positive control (38%, 52%, and 42%, respectively). Furthermore, the antiplasmodial activity of these plant extracts in antiplasmodial assays was evaluated against chloroquine (CQ)-resistant (CQ-r) and CQ-sensitive (CQ-s) strains of *Plasmodium falciparum*. The *S. occidentalis* 50% inhibitory concentrations (IC_{50}) were $48.80\,\mu g\,mL^{-1}$ (CQ-s) and $54.28\,\mu g\,mL^{-1}$ (CQ-r), while those for *O. basilicum* IC_{50} were $68.14\,\mu g\,mL^{-1}$ (CQ-s) and $67.27\,\mu g\,mL^{-1}$ (CQ-r). The high potentiality of the reported data must be considered. These smokes, as the essential oils, can be quite easily obtained in good quantity and low cost and therefore locally produced and directly utilized. This is important for countries with limited economic resources.

The natural range

The distribution of individuals in accordance with the Boltzmann curve is the result of the current chemical-physical environmental pressure, concentrating the organisms of the species in the most adapted form. However, sooner or later situations are destined to change, and some of the individuals confined in the wings of the curve are ready to profit of the change and enter in the center, as soon as the conditions will be favorable to them. Another consequence of this typical statistical distribution is the careful preservation of individual types inside the population. In practice, on the genetic point of view, the best species or the favored community do not exist in absolute, and any declaration or pseudo-scientific argumentation about the primacy of a race, also human, must be considered as a guilty stretch. As confirmation, this is also in accordance with the distribution of the constituents of matter at a subatomic level. The final consideration is that the chemical composition of a plant is limited, being the expression of the genome of the species, but it may change at any time in response to internal and external stimuli.

Let us use these concepts to evaluate insecticides used in insect-borne diseases. The interest in the use of biocidal products of natural origin began in the 1930s and grew until the 1950s, when it was obscured by the arrival of synthetic insecticides on the scene. For a long time, the pesticides scenario was dominated by synthetic products, until several factors caused a decline in their utilization. However, in the last 20 years, interest in natural products has reappeared intensely, especially for the control of noxious insects at larval stage. This situation has matured, as is well known, following the indiscriminate (and not always necessary) use of excessive amounts of pesticides which, once released into the environment, are difficult to eliminate, as evidenced by the paradigmatic case of DDT (see Chapter 1). At the same time, incidence of insect resistance has increased, resulting in partial product inactivity and/or increasingly massive dosage requirements. All this led to a need for the formulation of a new generation of pesticides, and to focus research and production efforts on natural products. In 1997, the World Health Assembly reported in Resolution 50.13, Section 2.4, the need to develop bio-insecticides. Slowly but inexorably, the pesticides market registered the rise of biopesticides from natural products.

The change in favor of natural products is the result of two concomitant facts: the evidence of the environmental damage due to massive utilization of synthetic products, and a new and growing sensibility in favor of respect for habitats, asking for more compatible solutions. In the current search for the production of a new generation of pesticides, useful for mankind's battle against superbugs and other threats, to face challenges to food supply and health, plant sources play a relevant role. The current prevalence of natural products evidences that a consistent number of biocides and antibiotics have been obtained from substances produced by living organisms, which are part of the great book of Mother Nature, whose lessons are still useful. That probably means that attention in chemistry is finally moving from the free synthetic approach to natural products, already selected during the long story of molecular evolution.

In an article that appeared in ACS' *Journal of Natural Products*, Charles L. Cantrell and colleagues pointed out the impact of natural products—substances produced by living plants, animals, and other organisms—on the production of pesticides. The article reports the percentages for registered insecticides obtained from 277 new active ingredients in the period 1997–2010 (Cantrel et al., 2012). The paper's aim was focused on the impact of natural product and natural product-based pesticides on the U.S. market, obtained on the basis of NAI registrations of new active ingredient registrations with the U.S. Environmental Protection Agency (EPA). The ingredients are categorized into four categories: biological (B), natural product (NP), synthetic (S), and synthetic natural derived (SND). In particular, NPs are considered substances produced by living plants, animals, and other organisms. The report evidences that NPs, SNDs, and Bs all have origins in natural product research. NPs accounted for 35.7%, Ss for 30.7%, Bs for 27.4%, and SNDs for 6.1%, arising from the combination of conventional pesticides and biopesticides. In the registered conventional pesticides, the category of biopesticides alone registered an evident majority of NPS (with 54.8%), followed by Bs (44.6%), SNDs (0.6%), and Ss (0%). In contrast, on the conventional pesticides alone, the category S clearly dominated with 78%, followed by SND with 14.7%, NP 6.4%, and B 0.9%. The review indicates that in the same period, more natural products were registered as NAIs for conventional pesticides and biopesticides than any other type of ingredient. The authors report that when biological ingredients and natural products recreated in laboratories are included, more than 69% of all NAIs registered in that time frame have natural origins. More than two out of every three new insecticides approved in the last years are directly derived from natural substances produced in plants or animals or have significant roots in them. It is noteworthy that these numbers are very similar to those obtained if we compare with a similar projection concerning registered medical drugs in a similar period, and published in the same scientific journal.

It is also noteworthy that a similar trend can be observed in the case of the registration of medical drugs in the period 1981–2014, as previously reported in the same scientific journal by David J. Newman and Gordon M. Cragg (Newman and Cragg, 2016): 36% of registered drugs directly or indirectly derived from NP of secondary metabolism, 16% from B, 11% from SND, and only 31% from S. Again, the details reveal differences in the sectors, with NP and B dominating in anticancer and antibiotics, whereas the opposite concerns antiinflammatories, with S clearly dominating. These data, obtained on a total number of 1562 new approved drugs, are the results of several reviews that confirmed these percentages. In particular, the authors stress the role of microbes in the production of new drugs derived from natural products: "We wish to draw the attention of readers to the rapidly evolving recognition that a significant number of natural product drugs/leads are actually produced by microbes and/or microbial interactions with the "host from whence it was isolated," and therefore "it is considered that this area of natural product research should be expanded significantly."

In other words, the future of pharmaceutical drugs could be related to natural products obtained by natural synthesis. Once conceivable that the shift from synthetic pesticides to biopesticides seems to be incontrovertible, as fueled by the resistance phenomenon and the general tendency for "natural," the key argument is the choice of the raw material for the best bioinsecticide. As evident from the above reviews, bioinsecticides could be extracted from a living organism, like a plant, or produced by a living organism, like a bacterial strain by hemisynthesis, or obtained by synthesis in accordance with the structure of the active natural product. Here, the debate is open between those affirming that "a molecule is a molecule" and those in favor of "original" natural products. Anyway, in the case of an extract, the complexity of ingredients cannot be reproduced or performed by synthesis.

The main characters of a natural "ideal" insecticide should be: biodegradability, environmental care, sustainability (obtained from renewable materials) and selectively (harmful to beneficial insects). It should also satisfy some conditions to be economic appealing and relevant, like be easy to produce, low cost, derived from raw materials that available and abundant in the country where the insecticide should be utilized. The last conditions are important to avoid accumulation by multinational agencies, as is happening for coffee and cacao. Finally, but not in order of importance, the ideal natural insecticide must be able to compete in the market with the insecticides currently in use. The research for the ideal bioinsecticide is open, starting from the plant to be used.

The neem's world

Scientific classification of neem

Reign: Plantae
Division: Magnoliophyta
Class: Magnoliopsida
Clade: Angiosperm Eudicotyledons
Superorder: Rosidae
Order: Sapindales
Family: Meliaceae
Genus: *Azadirachta*
Species: *A. indica* A. Juss, 1830
Synonymous: *Azadirachta indica* var. *minor* Valeton; *Azadirachta indica* var. *siamensis* Valeton; *Azadirachta indica* subsp. *vartakii* Kothari, Londhe & N.P. Singh; *Melia azadirachta* L.; *Melia indica* (A. Juss.) Brandis

My interest in the neem tree started several years ago from curiosity. I asked myself about the most important medicinal plant worldwide. Consulting the literature, it was evident that this role should be assigned to neem. At that time, the tree was practically unknown in the Occident, since its distribution and utilization were confined to the Indian subcontinent. Therefore, I wrote an article about neem to explain the importance of this plant, and some years later, I was contacted by an Italian industry using neem oil because of the curiosity and interest raised by that article.

The references about this plant and its use appear since time immemorial, as reported in the Ayurveda and Unani systems of traditional medicines, and even in the earliest Sanskrit writings referring to the medical uses of fruits, seeds, oil, leaves, roots, and bark (Gupta, 2001). Neem can be found all over the Indian subcontinent since the time of the Sanskrit-speaking Aryans. So important was the species for the Aryans, they even mentioned and included it in their sacred *Ayurveda*, which is the sacred book of Indian medicine. The species was later dispersed throughout the Old Tropics, including Indonesia, either naturally or brought back by the ancient Austronesian sailors after visiting and trading in India at least around 2000 years ago. Through the centuries (Kumar and Navaratnam, 2000), the medical importance of neem never waned in the Indian subcontinent and it is now considered the "Village Pharmacy" for its importance in the ordinary life of Indians, who use this plant to treat several illnesses (Nix, 2007; Girish and Bhat, 2008). Many other news and references increased my interest.

The marvelous tree, the problem-solving tree, the divine tree, India's tree of life, Nature's drugstore, the pharmacy tree, the panacea for all diseases—these are just some of the terms used to evidence the respect for this plant and its importance (Ruskin, 1992; Brahmachari, 2004; Puri, 1999; National Research Council, 1992). Neem's relevance and its beneficial properties has been reported by the WHO/UNEP (1989), which considered neem as an effective source of environmentally powerful natural pesticide and one of the most promising trees of the 21st century for its great potential in pest management, environmental protection, and medicine (Nicoletti and Murugan, 2013; Koul and Wahab, 2007). Furthermore, the U.S. National Academy of Sciences dedicated a report to neem, significantly titled "Neem—A Tree for Solving Global Problems" (NAS, 1992). The importance of neem has increased exponentially in recent years. Considering the enormous quantity of results and scientific data concerning the validation of medicinal and biological properties, the international scientific community included neem on the list of the top 10 plants to investigate and use for the sustainable development of the planet and the health of mankind (Tewari, 1992; Foster and Moser, 2000). However, in the Occident, insecticidal activity is the most common application for neem oil and its derived products.

The plant, besides neem, is also known as nimba, nimtree, margosa, and Indian lilac. In botany, neem is *Azadirachta indica* A. Juss and belongs to the Meliaceae family. Meliaceae are Angiosperm Rosidae, closely related to Simaroubaceae and Rutaceae (Schumutterer, 2002). The family includes 51 genera and c.600 species. These are woody plants, like trees, shrubs, and shrublets, pantropically present, with a few temperate representatives in China and South Africa. It is estimated that c.39 million years ago, two subfamilies diverged, Cedreloideae with 14 genera and Melioideae with 37 genera, including *Aglaia*, dominating with c.120 species, and *Azadirachta* with only two species. Meliaceae are commonly known as the mahogany family, being known mainly for some important timber species, such as the true mahogany (*Swietenia mahagony*). However, losses due to overexploitation and genetic erosion, as well as toxic effects on workers, limited the use of this true mahogany, nowadays widely replaced by Spanish mahogany (*S. macrophylla*).

Azadirachta indica is an evergreen tree that grows up to a height of 15–20 m, but in favorable conditions it can to a height of about 20–35 m, with a trunk diameter up to 2.5 m (Fig. 7.3). The leaves of neem, composed of 9–19 leaflets, meaning 4–8 leaflets with a single terminal, are abundant, suspended by a strong and long petiole which lacks stipules, crowded near the ends of the branches. The leaflets are toothed, deeply serrated, their margins irregularly serrated, sharply pointed, and curved like a scythe. Young leaves are pale, tender green, and tinted with rust, but during the favorable season the tree profits from the fresh, green color and shining surface of the leaves, giving a delicate and charming appearance. White small flowers are abundant, very fragrant, bisexual, or staminate in male exemplars.

Fig. 7.3 The neem tree.

They are arranged in clusters at the axils of the leaves. They present five separated petals arranged in the form of a star. They appear in spring, and open in the afternoon giving out a delicate smell, which increases during the night. The fruit is a smooth, yellow-green, small, round drupe with a sweet-flavored pulp. During the monsoon, when the flowers have fallen and the tree is in full foliage, the curved, toothed leaves, massed round the branches, have a distinctive appearance, which is easy to recognize. From March to May, the flowers, with five whitish petals, appear in great numbers on long, drooping stems. Flowers are used to produce a bitter honey. The fleshy fruits are purplish-black, single-seeded drupes, which turn yellowish when ripe. Elliptical in shape, they have a sweet-tasting juice loved by birds and bees. However, after the rains, the fruits change, giving off a strong unpleasant smell. In autumn, fruits fall down in great quantities if not harvested.

The tree is believed to be native at least of North-East India and Burma, or Indonesia, but now widely distributed in the Indian subcontinent, and it grows naturally throughout the dry regions of the country. It is usually planted along roads and avenues in towns and villages, because it grows fast and easily, and has an irregularly rounded crown with a canopy of leaves, making it a useful shade tree (Fig. 7.4). The central regions of India are considered the patria of neem. If you go to Coimbatore, in Tamil Nadu, you can find neem trees everywhere, in towns and in the countryside. It is mainly used for shade, lining streets or in most people's back yards. In India, it grows throughout the states of Uttar Pradesh, Bihar, West Bengal, Orissa, Delhi, Maharashtra, Gujarat, and Andhra Pradesh, but the original natural distribution is obscured by widespread cultivation. Cultivation is easy, since neem usually grown from seed but can be propagated also from cuttings or root suckers, and it is a fast-growing species. Potential utilizations of neem concern human, animal, and environmental health. The last one is a recent but very important acquisition. Neem is cultivated for two main reasons: environmental care and the production of seed neem oil.

Environmentally beneficial neem

The tree's tolerance and adaptation to hot and dry climates has made it one of the most commonly planted species in arid and semi-arid areas (Tiwari et al., 2014). The survival capacity of neem is mainly due to its highly expanded root system. Neem trees are extremely useful to counteract desertification and furnish the only source of wood in arid and nutrient-

Fig. 7.4 Neem trees are often used to arborize streets and squares in towns.

deficient zones. The plant does not need particular care and grows rapidly up to 30 m tall. Neem trees attain maturity in 6–7 years in areas where the sunlight is intense, weather is warm, and good well-drained soil is found. The tree is stable in windy zones and can live for 200 years or more. To survive in arid climates, neem depends on a wide strong root system with a deep tap root and extensive lateral roots, which are ideal for soil conservation.

Furthermore, its planetary presence can contribute to positive carbon sequestration to minimize climate changes, considering that adult neem trees can retain ± 2.2 g of CO_2 per m^2 and per hour, which means 40–50 tons of CO_2 per hectare. Therefore, neem trees can be considered to be air purifiers as well as air fresheners. For all these reasons, neem is widely cultivated in warm countries, and its areal distribution is expanding rapidly by massive cultivations in sub-tropical regions of America (Caribbean Cuba, Central and Southern America), Asia (Nepal, Pakistan, Bangladesh, Sri Lanka, Myanmar, Thailand, Malaysia, Indonesia, Iran, China, Turkey, Indonesia),

Table 7.1 Current data on neem presence in many countries.

Country	**Number of neem trees (million)**
India	75–80
China	20–25
Australia	1.2
Myanmar	1.1
Dominican Republic	0.75
Thailand	0.75
Indonesia	0.6
Haiti	0.6
Brazil	0.5–0.7
Rest of the world	3.5–4.3
Total	*104–115*

Africa (Kenya, Cameroon), and Australia. The cultivation of neem trees is in particular increasing in the drought-prone areas, like in South Arabia (Arafat Valley) and in the UAE. In 1978, northern Nigeria, thanks to the governmental project Arid Zone Afforestation (AZAP), saw 700,000 neem trees being planted. In Europe, some cultivations are reported only in Southern Spain and Portugal (Sara and Folorunso, 2002).

The presumed current global neem trees presence and production are reported in Table 7.1. The data were obtained by cross-referencing several sources and are only indicative, considering that a real census was never completed in many countries (in particular in China) and many plantations are in progress. India is still by far the homeland of neem, but the scenario has changed rapidly in the last years and will continue to do so in the future, as can be deduced from the data in Table 7.1.

Visiting Oman, I noted the presence of planted neem trees in areas where acacia was the only other woody plant (usually the only plant) able to survive. The neem tree is resistant to drought and it grows in many different types of soil, but it thrives best in well-drained deep and sandy soils. Normally, it flourishes in areas wherein the neem is a life-giving tree, especially for dry coastal, southern districts. Its capacity to survive in arid zones improved cultivation in sub-arid to sub-humid conditions, with annual rainfalls between 400 and 1200 mm. It can tolerate high to very high temperatures, but it does not tolerate temperatures below 4°C, making cultivation in temperate climate very difficult. However, the future worldwide distribution of neem is not predictable. Indians are convinced that it will be difficult to obtain similar ideal conditions to their country and others in the Orient; however, the story of the cultivation of Cinchona by Dutch botanists in Indonesia tells us that the results of the cultivation can be successful and the results even superior.

In the complete Linnaean binomial name of neem we found A. Juss, which is the mark of the author of this species, designating the scientist who first published the name and a complete and scientifically reliable description of the species. A. Juss refers to Antoine Laurent de Jussieu (1748–1836), a great French botanist, who was the first to publish a complete and valid natural classification of flowering plants, named *Genera plantarum*, published in 1789, the same year as the French Revolution, surpassing the sexual system presented by Linneus. Antoine Laurent Jussieu was the member of a family of a plant enthusiasts; his uncle was the botanist Bernard de Jussieu, whose transferred knowledge and unpublished work were the starting point of the book of Antoine Laurent, and his son Andrien-Henri also became a botanist. The merit of his work was the use of multiple characters to define taxa. In this approach, he achieved a significant improvement over the "artificial" system of Linneus, mainly based on the number of the reproductive characters, i.e., stamens and pistils. Many people know the work and name of Linnaeus, but the impact of Jussieu's work was fundamental in taxonomy, founding the principles that served as the foundation of plant classification in a natural system. Many present-day plant families are still attributed to him, as the species is to Linnaeus.

That's what who can find in ordinary sources of information. The consequent idea is that Jussieu, though living in France at a historical revolutionary time, was totally dedicated to botany, but his work was also revolutionary, through a strange pathway. Deeper information about his life is a source of important lessons. His uncle, Bernard de Jussieu, invited the young Antoine to Paris, where he was trained in medicine for 4 years. However, his uncle, via his position as a demonstrator at the Jardin du Roi (Royal Garden), had other plans for Antoine. He guided his nephew's studies and prepared him for a lecturer's position at the Garden, which was soon to become vacant. At just 22-years old, Antoine was transferred to that position and his botanical training was limited because the subject then was viewed only as an accessory to his medical course. The inexperienced Jussieu had to study botanical topics by night, since he had to teach by day. Using the plants in the garden to teach plant morphology, which were arranged according to the current artificial system of Joseph Pitton de Tournefort, Jussieu started to realize the inadequacy of that system. This progressively changed his interest into a true passion for botany, and classification began until, as part of an application for a place at the Academy of

Sciences, he produced a treatment of the Ranunculaceae, starting a complete revision of the plant taxonomy system. Continuing in his study, it became apparent to Jussieu that the artificial system of Tournefort was inadequate, and from 1774 he began arranging the plants in the Royal Garden in his own way and finally transferring the knowledge in the construction of his own system of plant classification. Despite the initial success obtained by Linnaeus' sexual system, it was clear to Jussieu that the Swedish naturalist had used a counting method, whereas it was necessary to grade the characters, considering some of them more important than others, depending on how variable they are within a species. This was a necessary lesson to consider the variability of living organisms correctly. However, he continued practicing medicine, chiefly devoting himself to the health of very poor people. In 1790, he was put in charge of the hospitals and charities of Paris. In the final years of his life, Jussieu, by then almost blind as well as deaf, dedicated the last part of his extraordinary life to meditation and prayer.

Also neem can be in danger	
Kingdom	Animalia
Phylum	Arthropoda
Class	Insecta
Order	Hemiptera
Family	Diaspididae
Tribe	Aspidiotini
Genus	*Aonidiella*
Species	*A. orientalis*

The neem tree has few diseases and enemies (Boa, 1995; Schmutterer, 1998). In general, it is considered a very resistant and healthy plant (Nicoletti et al., 2017). However, it is possible that after its spread around the world, something is changing. We must move our focus from India to the new settlements, in Africa. In Northern Nigeria, neem now is planted in towns and villages, as a highly evaluated source of shade and firewood, as well in the establishment of shelterbelts. In Nigeria, like in other parts of Africa, the small twigs of neem are used to clean and whiten teeth, in consideration of its antibacterial properties. In particular, reports of pests and damages comes from East Africa, like gall mites (*Phyllocoptes* sp.) on older plants, but the most potentially dangerous pest is *Aonidiella orientalis*, known as oriental scale (Ofek et al., 1998; Lale, 1998; Elder et al., 1998). This insect was widespread in western Kenya but is not currently harmful. Likely to be native to Asia, it has been introduced to many regions via shipments of plants and then began its slow spread. Some ports check for this and other scales in plant shipments. In Africa, neem has been widely planted in the Sahel region. Oriental scale first appeared in the Sahel during the mid-1970s and caused widespread damage to neem trees planted there. Infestations were first detected in Nigeria in 1987 along its border with Cameroon and by the mid-1990s, widespread damage had been reported to neem trees throughout northeastern Nigeria.

The oriental scale is a flattened, circular or oblong insect, about 1.6 mm in diameter. It varies in color from yellow to light reddish brown. It frequently forms large colonies and sucks the sap of small stems and branches, which is phloematic sap, rich in sugars and other organic substances. Infestations often spread to the foliage, fruits, and even seeds. Feeding damage causes the foliage to die, giving infested trees a burnt appearance. This is followed by progressive dieback of branches and eventual tree mortality. The heaviest infestations appear to be on large trees located either in marketplaces or around human settlements. The female attaches to the surface of a plant and causes the disease by the larvae, which roam the plant, feeding on sap by inserting their stylets, sucking sap, and weakening the plant progressively. The physical damage includes discoloration and deformation of leaves. Flowers and fruits fail to develop. It is noteworthy that the effects of the pest attacks are very similar to those already reported for olive trees, including exsiccation of leaves and wood. The lesson is that, in this outbreak, we have an explosive negative mixture of alien species and man's activity on the habitat. The enormous spread of neem trees in recent years is an unusual phenomenon, whose consequences should be better monitored.

Chemistry of neem

Before going into detail about the constituents, we must consider the typologies of the forms of the marketed products containing natural products. We are referring to insecticides, but the same considerations are simply applicable to nutraceuticals, phytomedicines, cosmetics, and even food. It is possible to find the plant drug, meaning a part of the plant utilized. The plant drug is usually utilized exsiccated, or as a derived product, like an extract, resin or oil, which can be obtained as such, or be enriched in one or more constituents, which are considered responsible for the activity. In this sequence, the original starting point, which makes the plant useful, has been betrayed in favor of increased efficacy. However, following this treatment, there are products registered as food supplements containing 90% of a pure substance and usually the origin is not at all natural (although this is not declared on the label). In such cases, the product is more similar to a medicinal drug than an extract and it should be considered and used in medicinal form. The distinction between such marketed products is not evident and not reported to the unaware consumer, who may well prefer the product due to its apparently "natural" origin.

Knowledge of the chemistry is therefore fundamental and the basis of any decision about the appropriate use of a plant. However, we cannot know exactly what is inside a plant. All our methods of investigation are limited and may be misleading, although papers and books are full of information about the compounds contained in plants or their derived products. This is the consequence of plants' extreme chemical complexity. In a single leaf of hemp, more than 400 constituents have been detected, considering the secondary metabolites alone, and hemp can contain high levels of THC or cannabinoids can be practically absent. In such cases, the morphology does not give any help and only a reliable chemical analysis can indicate what kind of hemp we are handling.

Since its beginning, phytochemistry has sought knowledge of all natural products. In about one hundred of years of activity, innumerable analyses were made and an enormous quantity of data collected in a very useful data base, but recent advancement in analytical devices and novel interpretation ask for a revision of the result of this job (Kaushik et al., 2014; Forin et al., 2011). We must remember that the molecular world is not detectable by our senses and therefore we have secondhand and probably only partial information. Sometimes, this information is considered sufficient to assign the compound/activity relationship, but only until other analyses confirm the presence of other molecular candidates.

The seeds of neem contain at least 100 identified biologically active compounds (Govindachari, 1992). Among them, major constituents are nortriterpenes, named limonoids, i.e., azadirachtin, nimbin, nimbidin, nimbolides, and many others (Ragasa et al., 1997). However, each year other new limonoids are discovered. More constituents mean more possible activities.

Preparations from the leaves or oils of the seeds are used as general antiseptics (Mossini et al., 2009). Due to neem's antibacterial properties, it is effective in fighting most epidermal dysfunction, such as acne, psoriasis, and eczema. Ancient ayurvedic practitioners believed high sugar levels in the body caused skin disease. Neem's bitter quality was said to counteract this sweetness. During the last Desert Locust Plague in Africa, it was noticed that these insect, *Schistocerca gregaria* (Forskal), ate almost any vegetal around, leaving a bare landscape when they fly away, except for neem trees, probably because of the antifeedant effect of the very bitter leaves, due to the presence of limonoids. Traditionally, Indians bathed in neem leaves steeped in hot water. Since there have been no reports of topical application of neem causing adverse side effects, this is a common procedure to cure skin ailments or allergic reactions. Neem also may provide antiviral treatment for smallpox, chicken pox, and warts, especially when applied directly to the skin. Its twigs are commonly used to clean and disinfect teeth. The brushing of teeth with neem to prevent gum diseases and for teeth whitening is very common, and not only in India. There are also various kinds of natural toothpaste on the market that contain neem extracts. It is also possible to prepare a homemade toothpaste to achieve shiny, cleaner teeth. Neem powder is made by grinding dried neem leaves, which is traditionally used by mixing one teaspoon of neem powder with one teaspoon of baking soda and enough water to make a useful paste. These preparations can help to avoid the plaque and tartar that build up in gums, which are the root causes of bad breath. In addition to cosmetic uses, neem's antimicrobial activity to maintain dental health is also worth noting (Chava et al., 2012). A preparation can be obtained by boiling neem leaves in water until they reduce to a quarter of the original volume. Finally, gargling with this concoction contributes to good breath and whiter teeth, as it kills the bacteria inside the mouth. The teas of neem leaves are utilized in Indonesia and Oman for their digestive properties (Sujarwo et al., 2016).

Neem's effectiveness is due in part to its ability to inhibit pathogens from multiplying and spreading (Benelli et al., 2016). Neem produces pain-relieving, antiinflammatory, and fever-reducing compounds that can aid in the healing of cuts, burns, sprains, earaches, and headaches, as well as fevers (Chopra et al., 1952). Several studies of neem extracts in suppressing malaria have been conducted, all supporting its use in treatment. Neem has broad applications to human and animal health, as well as organic farming (Bhowmik et al., 2010). It is reported as a powerful antiviral and antibacterial, with peculiarities that set it apart from other herbs in that class of broad antimicrobials (Sandanasamy et al., 2013). Neem oil is also commonly added to a variety of creams and salves. It is effective against a broad spectrum of skin diseases including eczema, psoriasis, dry skin, wrinkles, rashes, and dandruff. These are just a few examples of the possible utilization of neem, and the potentiality of neem is considered by everybody, including the ONU and other institutions, to be very high, but so far little has been done to develop appropriate products from it to help mankind. In particular, considering insect-borne diseases, in vivo activity of neem seed oil (NSO) against malaria *Plasmodium* has also been reported (Dahiya et al., 2016; Trapanelli et al., 2016).

Today's exploding growth in human population is seriously depleting the world's natural reserves and economic resources. Unless the runaway human population growth rate is slowed down, there will be little hope for raising everyone out of poverty in the developing world. Besides educational constraints, the nonavailability of inexpensive methods of contraception, which do not cause trauma or impose on the esthetic, cultural, and religious sensitivities of people, limit the success of birth regulation programs. However, recent findings indicate that some neem derivatives may serve as affordable and widely available contraceptives. A recent controlled study in the Indian army proved the efficacy of neem as a contraceptive. In 2020, the report of the Washington-based International Food Policy Research Institute predicted a world even more unequal than the present, with food surpluses in the industrialized world and with chronic instability and food shortages in the global south, particularly in African countries.

The state of the art of neem oil

The main product of neem seed oil (NSO) is the fixed oil obtained by expressing the seeds, still enclosed in the kernels. Therefore, the fleshy pulp is removed or dried, to obtain the inner part. NSO can be obtained by different extraction methods. Most NSO is produced in India by small-scale producers at ordinary temperatures using very simple machinery, which is utilized in other periods of the years for other oily extraction, like arachnids or soya seeds (Figs. 7.5–7.9). However, modern apparatus is also used in India and many other countries are now producing and refining NSOs. Therefore, considering also the possible different geographical origin of the raw material, combined pre- and postharvesting factors can result in great differences in constituents present in marketed NSOs, as already reported.

Therefore, despite the common definition for all the oils obtained from kernels of neem, it is necessary to consider that there is no single NSO, but many NSOs differing in shape, color, viscosity, chemical constitution, and

Fig. 7.5 A farm in India producing neem oil.

Fig. 7.6 Simple apparatus for expression of neem kernels.

activity. Medicinal and cosmetic utilizations are relevant and continuously increasing. Cold pressed neem oil is commercially known as margose oil and considered as pressed directly from seeds. There are hundreds of marketed products worldwide based on margose oil. Neem, or margose, oil has a brown color, a bitter taste, and a garlic/sulfuric smell. The oil is usually obtained by simple pressing, but extraction by organic solvents, like hexane, is also used, though in such cases traces of the residue of solvent are always present in the final product. The type of insecticide is commonly

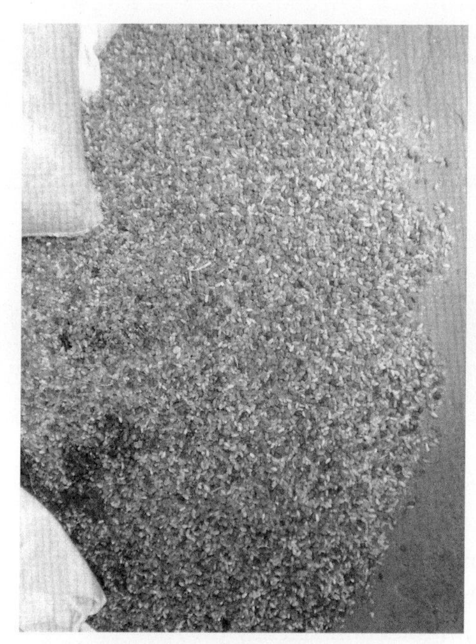

Fig. 7.7 Kernels of neem, the raw material for neem oil.

Fig. 7.8 Production of neem oil.

registered as biocide, insect repellent, and antifeedant, intended for use on outdoor and greenhouse agricultural food and ornamental crops as a repellent and insect growth regulator. The products are considered to have no risk to human health because of their low toxicity via all routes of exposure.

Fig. 7.9 Neem cake resulting from the extraction process.

There is no reason to believe that any nontarget organisms, including honeybees and other beneficial insects, would be adversely affected, as tested by the Environmental Protection Agency (EPA).

Industrial applications of neem oil

The Environmental Protection Agency (EPA) is probably the most important agency in charge of environmental care and health. It is an agency of the United States federal government whose mission is to protect human and environmental health. The EPA is also in charge of the regulations of carbon emissions from power plants, automobiles, and other contributors to climate change. The EPA became popular in Europe because of a civil enforcement case against Volkswagen and other car manufacturers, subject to reservations set forth in each of the partial settlements. In 2017, the U.S. Department of Justice resolved a criminal case against Volkswagen AG with a plea agreement for the offenses of conspiracy, obstruction of justice, and entry of goods by false statement, and the U.S. Customs and Border Protection resolved civil fraud claims with Volkswagen arising from the illegal importation of affected vehicles.

The EPA was established in December 1970 by an executive order of President Richard Nixon, with headquarters in Washington, D.C., in response to widespread public environmental concerns that gained momentum in the 1950s and 1960s. EPA is a giant public agency with a budget of 9 billion US $, and it is born as reaction to the public movement in favor of the environment due to the Carson's book (see Chapter 1). It is considered reliable because independent. The EPA is responsible for creating standards and laws to protect and preserve the natural environment and improve the health of humans by researching the effects of and mandating limits on the use of pollutants. The EPA's aims include the regulation of the manufacturing, processing, distribution, and use of chemicals and other pollutants. In addition, the EPA is charged with determining safe tolerance levels for chemicals and other pollutants in food, animal feed, and water.

The best presentation of neem insecticide properties and the rationale of its utilization is the report of the EPA about the registration of cold pressed neem oil, concerning a product named plasma neem. The report has a significant subtitle: "Reasons Why Neem Oil is an Effective Way to Control Insect Hoppers" (EPA (US Environmental Protection Agency), 2012).

The target of the product cold pressed neem oil is insect hoppers, because of their special liking for cash crops. The consequence is a crisis because of insect hoppers infesting these cash crops by chewing and sucking the leaves, as reported by many farmers. The reasons to use neem are focused on the effects of chemical fertilizers, which "provide a remedy but tend to kill beneficial insects as well." The report identifies three reasons why neem oil should be considered an effective way to control insect hoppers which feed on cash crops.

Reason 1: Selectivity. "Neem oil is usually sprayed on the leaves and stalks of a cash crop. So it is aimed at only the insect hoppers who spoil the cash crop by chewing the leaves or biting off bits of the stalk. The insect hopper which attacks the plant by consuming it will end up consuming the neem oil as well and hence die as an aftermath. Beneficial insects, who replenish the soil in a natural fashion, do not consume the plant and hence do not get affected by neem oil."

Reason 2: Eco-friendly efficacy. The content of azadirachtin in neem oil is considered very good at controlling and eliminating insect hoppers, avoiding the negative effects on the productivity of cash crops. Furthermore, the neem oil exerts no damage to the soil or disruption of the chemical composition of a fertile soil. The product affects biological functioning of the insect hopper, and therefore the insect hopper forgets to feed or breed after devouring a cash crop with neem oil sprayed on it. The conclusion about this point is clear: "This leads to the complete elimination of the insect hopper and the infestation cycle without any other adverse chemical side effects." The action is mainly larvicide, but also deterrent and adult insecticide in most species.

Reason 3: Cost-effectiveness. The utilization of neem oil for preserving cash crops from insect hoppers is considered very cost-effective. In comparison to chemical fertilizers, neem oil is expensive, but other positive effects must be evaluated. There is another consideration about the cost/benefit effect: eliminating negative insects and worms by neem oil, farmers need not spend money on buying supplements for the enrichment of soil. The heavy investment to preserve soil quality is totally avoided when a farmer uses neem oil. In conclusion: "Overall, it can be said that neem oil as a source to control insect hoppers in cash crops is extremely beneficial for farmers."

The final consideration of the EPA is the advice not to use the content of the report as an endorsement of NSO. In fact, it is important to remember that the report is based on scientific and experimental data. The EPA is not working in favor of the industry or the market, but its judgments must be considered objective although its aim is environmental improvement. NSO, obtained by the cold expression method, is the only natural product-derived insecticide, whose registration was approved by EPA. The EPA's authority is so far internationally recognized.

Therefore, in the report there are three main key items of information: the evaluation of the selective insecticide activity; the preference for synthetic products in consideration of the eco-friendly properties; and the indication of azadirachtin as necessary for the activity. Let us consider the report as a guide and evaluate each of these points. Later, we will add other considerations.

It is noteworthy that all these indications can be found already reported and present in the information about the ethnobotany and traditional medicine of neem. In the cultivation of rice, when the plantlets are underwater, the addition of extracts of neem to the usual fertilizer, not only decreases the mosquitoes number, but also has beneficial effects on the production. The utilization of neem to increase soil fertility and treat medicinally plants and livestock is recommended in the *Upavanavinos*, an ancient Sanskrit book on agriculture. The dried leaves and the oil are used as preservatives against insects and microorganisms for the postharvesting conservation of foods. Conservation is guaranteed for more than a year. Panels derived by the extraction of the oil are used as food additives for livestock, and the animals are washed with this diluted oil to prevent the attacks of parasites or harmful insects.

In the last decade, the focus was on the potentiality of a new analytical technique, HPTLC (high performance thin layer chromatography)

Fig. 7.10 Part of the HPTLC devices.

(Fig. 7.10). HPTLC is the last evolution of planar chromatography and allows the evidence of most of the constituents of an extract in an identifying track, named fingerprint, wherein identification of constituents can be obtained by direct comparison in the same plate with the correct standards, utilizing the Rf value and the reaction with adequate derivatization (Nicoletti et al., 2012a,b). Improvements in separation and visualization of the spots are obtained by reduction of the size of the particles of the silica gel, constituting the fixed phase. The mobile phase can be selected on the ample repertory of solvents and their mixtures, as well as the several methods of derivatization and detection. The final effect, in comparison with ordinary planar chromatography (TLC), is like a myopic person wearing glasses. The advantages in comparison to the TLC are in the total control of the environmental conditions and the automatization of the procedures. Each step of the analysis is performed entirely by a series of devices, and the operator is only asked to produce the program of actions by the software. Plates can be visualized and derived in several ways, obtaining multiple information (Figs. 7.11 and 7.12). They can be easily preserved and stored as digitalized images inside the computer, and immediately sent everywhere or compared with a data bank. However, the most important feature of HPTLC is that the results of the analyses are very clear, thanks to careful preparation work that enabled optimal chromatographic conditions, as demonstrated by the quality of the images. In other words, our idea was to "see the molecules" (Nicoletti, 2013) and obtain simple and clear evidence of the

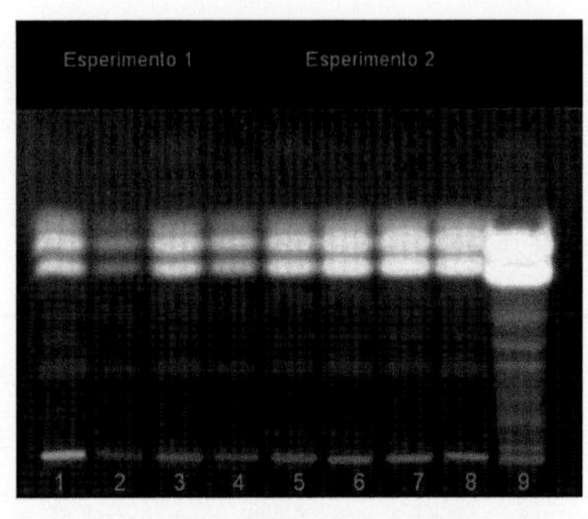

Fig. 7.11 A HPTLC analysis of NSOs, demonstrating the different constitution of marketed oil neems.

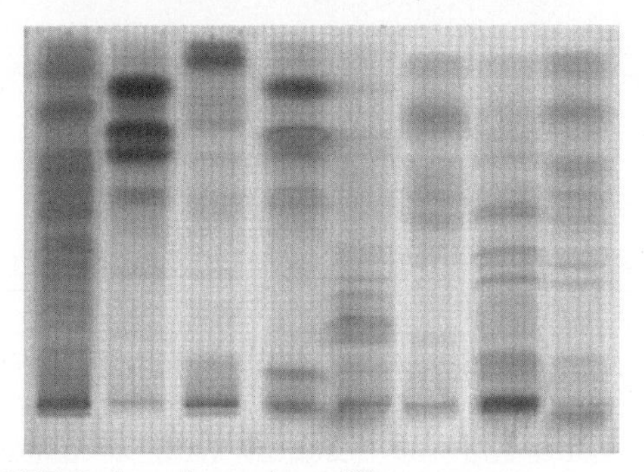

Fig. 7.12 Another HPTLC plate shown in a different way.

metabolic production. Molecules are too small to feel their presence with our senses or directly by any sort of device, but it is possible to evidence their chemical properties in the plate, recording the Rf value, the fluorescence, and the color after derivatization. An HPTLC or NMR graphic needs an expert for correct interpretation, such as any specialized analysis, like a cardiogram (Nicoletti and Toniolo, 2015). In HPTLC, without any knowledge of chemistry, the presence or absence of a determined spot is evident (Figs. 7.11 and 7.12).

HPTLC was selected to obtain a metabolomic approach, meaning the study of many constituents as possible, focusing on secondary metabolites (Toniolo et al., 2013). Metabolomic is one of the -omic sciences generated by the dissection of the dogma of biology, based on the sequence DNA➔RNA➔proteins.

It is necessary to propose some comments about the dogma of biology and why its crisis generated a series of other points of view. First of all, the use of the word "dogma" should be avoided in biology, since the matter is more complicated than a simple sequence, as actually happened. The central dogma of biology was first proposed in 1958 by Francis Crick, as a consequence of his discovery of the structure of DNA. The dogma describes process by which the instructions contained in DNA are converted into a functional product. Another definition is: "the coded genetic information hard-wired into DNA is transcribed into individual transportable cassettes, composed of messenger RNA (mRNA); each mRNA cassette contains the program for synthesis of a particular protein (or small number of proteins)" (sources: definition from Chapter 1: The Dynamic Cell, of *Molecular Cell Biology*). The flow of genetic information within a cell follows the sequence: DNA codes for RNA via the process of transcription (occurring within the nucleus), RNA codes for protein via the process of translation (occurring at the ribosomes), and proteins are responsible for the synthesis of the other metabolites (proteins are spread everywhere). Cell data are organized within the database of DNA and reversed in the metabolic flux, through RNA. Although clearly deficient, the central dogma of biology dominated genetics for decades, but through ongoing research, many exceptions were discovered. For example, most DNA is silent, since it does not encode proteins. Retroviruses, which are relevant for our arguments, present the possibility that RNA transcribes into DNA through the use of a special enzyme called reverse transcriptase, and other cases of deviance can be reported. However, the biggest revolution consists in the direction of the arrows. It is necessary that information could follow also the reverse pathway, allowing an appropriate response by the genome potentiality. Therefore, at least the dogma must be rewritten with two-way arrows.

In principle, a metabolomics study should be the determination of the pathway of cell production from the genome through transcription, but the term "metabolome" is now used to evidence the whole pool of metabolites, in particular for natural products, whereas transcriptomics is related to proteins. Transcriptomics involves serious difficulty to obtain reliable results. A protein seems perfectly comfortable inside the cytoplasm, but outside, irreversible denaturation causes definitive degradation and consequent difficulty in understanding the protein's functionality. In contrast, small molecules are more stable in any environment and their molecular structures at least can be determined by phytochemical analysis (Nicoletti and Toniolo, 2012; Toniolo et al., 2014). However, in the metabolome we have hundreds of thousands of different constituents to be studied, and the classic approach to study the molecules one by one is impracticable, and other methods must be utilized. The lesson is that the role of any metabolite cannot be discarded a priori, and also a secondary influence in the evaluation of the property of an extract can be important to definite and obtain the final reaction. Once again, the "magic bullet" paradigm is under discussion, but the total utilization of plant extracts must also be considered an unsatisfactory solution.

The aim of our approach was to adapt the method to other subjects outside the pharmaceutical applications. Therefore, our first studies focused on the determination of adulterants in nutraceuticals and other pharmaceutical products, like the "green viagras." Later, we adapted the method and the devices to use the metabolome as a source of information about what is going on in a complex system in which living organisms are acting. Therefore, we are able to study the effect of environmental factors, like ozone, on the quality of wine (Valletta et al., 2015). However, probably the most impressive application was the study of the environmental effects of the *Costa Concordia* disaster (Toniolo et al., 2018).

On the night of January 13, 2012, the *Costa Concordia*, a giant yacht with approximately 1500 cabins, 3229 passengers, and 1023 crew, was wrecked off the rocks of the Italian coast a few hundred meters from the port of Giglio, a little island on the Tyrrhenian coast in Tuscany. Like an injured helpless mastodon, the cruise ship inclined dangerously, until the inclination stopped with most of its starboard side under water. Because of the inclination and the amount of people, the overnight evacuation of the *Costa Concordia* was a challenging process, and 32 people died. The cost of removing the ship was US$799 million. For scientists like us, interested in environmental damage, it was a unique occasion. For 1 year, 9 months, and 4 days, the enormous hull of the boat altered the underlying marine habitat, interrupting the normal flow of sunlight over a surface of more than 10,000 m^2. The seagrass *Posidonia oceanica* was chosen as the target organism of the impact evaluation, since, like in other parts of the Mediterranean Sea, it forms large underwater meadows. Using HPTLC analysis, it was possible to determine the health of each collected plant and make a map of the metabolic damage, which accorded with the shadowed area. However, albeit the negative conditions, the rhizomes turned out to be mostly still alive and able to reproduce the meadow again. Therefore, the final task of our research was simply to wait until nature carried out its work. However, there is a further chapter of this story, written after our study. To remove the ship, a platform was transported from north Europe. The problem was that the bottom of the platform was full of mytilus. When the platform was exposed to the hot Mediterranean sea temperature, the mussels died, releasing their bodies, covering down the sea background and causing a further source of damage. A clear example of human stupidity and superficiality. Anyway, devoted to our task, we are now repeating our analyses to understand what happened and what is still going on, relying on the quality and reliability of our indisputable results.

The study of neem oil was based on the experiences obtained by improving the HPTLC devices via the metabolomics approach. The central idea was to collect as much information as possible about the constituents of the neem products, without any preference for any kind of metabolite, considering any product and any extract like a unique molecular system.

In the HPTLC analyses on NSOs, the objective was to achieve the total chemical characterization of the used oil, and then the derived products, by means of the production of a chromatographic reference profile of the metabolites' production. This objective is not easy to achieve due to the complexity and variability of neem oil. Neem products are subjected to great variation in composition, due to preharvesting factors, like environmental situation, genomic differences, influences of the habitat, and others, including postharvesting situations, like harvesting and stocking, treatment of the raw material, separation of different parts, extraction methods, production of the final product, and others. In fact, analyzing different marketed neem oil from several productions and countries, we decided that it was not possible to refer to a single neem oil, but to neem oils in the plural, due the great differences in composition. Therefore, we decided to obtain and adopt a reliable reference metabolomic HPTLC profile for the neem extracts or products to be utilized in our biological experiments in vitro and in the field.

In fact, one of the typical problems in activity tests is the differences in raw material giving rise necessarily to different results in activity and utilization. Another important aspect of our metabolomic study was that the complexity of the neem profile was even greater than expected. This result is the consequence of the generalist approach. In other molecular chromatographic or spectroscopic analyses, like HPLC, the result is *sub judice* on the detector's settlement. Therefore, if the molecule does not possess the adapt chromophore, the molecule, even if it is the main component, is invisible to the detector. In HPTLC, there are universal derivatization methods, like H_2SO_4, to reveal the organic substances, but it is possible when necessary to adopt a particular agent. In this way, it was possible to exclude the presence of a relevant percentage of azadirachtins and the occurrence of other constituents relevant for the activity. This is another recurring lesson for those studying natural products. Although a plant has been the object of several phytochemical studies, new constituents can be obtained. An example is the discovery of gossypol in cotton oil.

Insecticidal activity of neem

Insecticidal activity is reported in a hundred or so published papers, concerning a wide range of species of arthropods, as confirmations of many traditional uses (Schumutterer, 1995; Amirthalingam, 1998; Jones et al., 1989; Van der Nat et al., 1991; Biswas et al., 2002). Leaves are used in houses to repel and keep away insects. When half of a sample of soya leaves are sprayed with NSO and offered as food to the Japanese coleopteran (*Popillia japonica*), the insects feed only on the nontreated parts of the leaves. In Nicaragua, farmers spray their cultures with an aqueous extract obtained by leaving the seeds for 12 h in water.

In general, NSO-based products have proven to be very effective against a huge range of pests of medical and veterinary importance, mainly including mosquitoes. The insecticidal properties of neem and its many formulations are based on experimented antifeedant, fecundity suppression, ovicidal and larvicidal activities, including growth regulation and repellence against a great number (around 600) of different insects, also at very low dosages, whereas useful insects were shown to be unaffected (Nicoletti et al., 2014; Isman, 1997; Sharma et al., 1993; Schumetter and Singh, 1995; Forin et al., 2011). The deterrent activity was also important, which can be easily determined as reported in Fig. 7.13. Other studies, like the molting and the growing of the selection under investigation, need special devices, as those reported in Fig. 7.14.

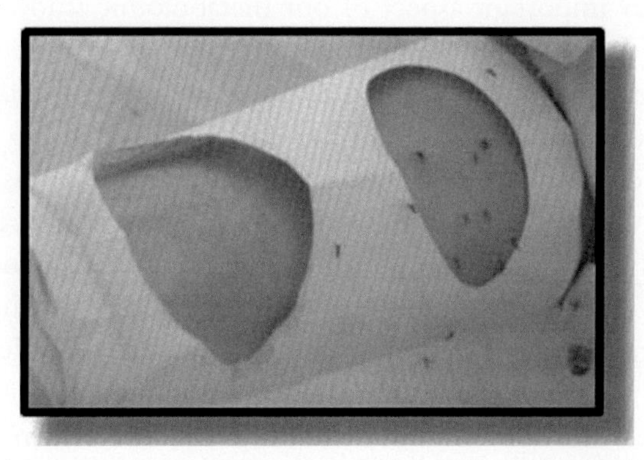

Fig. 7.13 The deterrent activity of oil neem can be easily and directly determined.

Fig. 7.14 Other activities of neem oil need special chambers to follow the growing and molting of the insects.

In particular, a concentrated extract of neem seeds, named MiteStop, developed by the university spin-off company Alpha-Biocare (Düsseldorf, Germany), proved to be very effective against a huge range of pests of medical and veterinary importance, including *Ixodes* and *Rhipicephalus* ticks, house dust mites, cockroaches (*Blatta*, *Blattella*, and *Gomphadorhina*), raptor bugs (*Triatoma*), cat fleas, bed bugs, biting and bloodsucking lice, poultry mites, and beetle larvae parasitizing the plumage of poultry. Neem leaves can also be used to protect stored woolen and silk clothes from insects.

Concerning mosquitoes, emulsified formulations of *A. indica* oil showed excellent larvicidal potential against different mosquito genera, including *Aedes*, *Anopheles*, and *Culex*, also under field conditions. Insect growth regulatory activity of neem-borne molecules alter or block the metamorphoses of larvae (Toniolo et al., 2014). Neem weakens the cuticle defense system of the young instars, causing easy penetration of pathogenic organisms, or interferes with the molting mechanism. Concerning biological control, an increase of the control of *Aedes* populations was observed after the combined application of predatory copepods and neem-based larvicidal products, since repeated application of NSO does not affect populations of predatory copepods. However, relevant limitations are related to the relatively high cost of refined products and the low persistence on treated surfaces exposed to sunlight. In the soil, the half-life of azadiracthins, meaning the time necessary to degradate the compounds, is from 48 min to 4 days, depending on the environmental conditions, like moisture, high temperature, and sun. The breakdown is faster on plant leaves, due to the exposure and the surface.

In the attempt to assign the active constituents of NSO, we must consider that the chemistry of neem is very complicated in terms of numbers and types of constituents. Despite the great quantity of dedicated research, chemical research is far from complete. Hundreds of compounds have been isolated and identified from various parts of neem, with seeds being the most investigated for their commercial value. The seeds may contain approximately 45% of a brownish yellow oil, mainly constituted by several fatty acids, i.e., oleic acid *cis*-9-ottadecenoic (50-60%), palmitic esadecanoic acid (13%–15%), stearic acid ottadecanoic (14%–19%), linoleic acid *cis, cis*-9,12-ottadecadienoic (8%–16%), and arachidic acid (1%–3%), although several other compositions have been reported. After a certain time, fatty constituents tend to separate and appear as white amorphous material. The main characteristics of the oil are its unpleasant strong alliaceous odor and acrid taste, attributed to sulfurous constituents. The shape and consistence can be very different according to the extraction method and the source. In fact, the composition of neem oil is highly variable, depending on preharvesting factors, like the cultivar, the geographic and environmental origin of the raw material, collection seasons, and postharvesting, like the extraction method, preservation, and conservation. Extraction can be executed with different apparatus, temperatures, pressures, and methods, affecting the yield as well as the content. As later reported, these aspects have been deeply considered and HPTLC analyses can be utilized to ensure the chemical composition of the neem oil utilized in the biological experiments.

Among the c.300 compounds characterized from the neem seeds, more than one-third of them are nortriterpenoids, which are triterpenoids lacking some carbon atoms (Kaushik et al., 2014). Nortriterpenoids are chemotaxomically well located in a few related families of Rosidae Angiosperm Dicotyledons, i.e., Rutaceae, Simarubaceae, Cucurbitaceae, and Meliaceae, within the Rutales order. Generally, in the plants of the Rutales order, the partial loss of the lateral chain is followed by a complicated rearranging of the remaining part, giving rise to different polycyclic molecular skeletons, full of oxygenated functional groups, partially acylated. Syntheses of complex natural compounds are costly and therefore they are usually used only for special activities. It is necessary to consider this point, which is in favor of the use of the plants as source of these compounds, since the synthesis can reproduce the chirality of nortriterpenoids only with extreme difficulty and cost.

Nortriterpenes are a very interesting part of the plant's chemical ability, that we call biosynthesis, to produce active complex molecules.

Nortriterpenes present very complicated structures and high numbers of active parts. We must remember that in natural products, activity is based on the presence of functional groups, made by heteroatoms, which means mainly O and/or N. If nitrogen is present, you have alkaloids, otherwise the range is higher, comprising phenols, alcohols, ketones, and others. However, the introduction of an oxygen inside a derivative usually is obtained to increase the activity, but also introduces instability in the molecule. We must remember that a natural product started from CO_2 and is likely to become CO_2 again at the end of its life. This is the necessary turnover of atoms and energy in organic matter. The first process accumulates energy and it is based on reductive reactions (endothermic reactions), whereas the second one is based on oxidation and produces energy (exothermic reactions). In other words, life is based on subtraction of negative entropy from the habitat, and at the end of its life, the organism releases this energy to the system. During its life, the molecule is expected to carry out its role inside the biosystem, which is the reason for its synthesis inside the plant. To understand the role, the nature of the target is essential. In insect-borne diseases, the natural product should interfere in the life of herbivorous insects or dangerous pathogens. In the case of neem, the activity is mainly larvicidal, blocking the metamorphosis to the next pupal stage. The larvae are unable to develop and change their state. To obtain this result, a lot of chemical and finalized activity are necessary, in this case consisting of hormonal interference in the insect metamorphosis process. In other words, the molecule must be able to mimic the internal complex chemical apparatus that allows drastic changes in the forms of the insect, until it stops the process.

Let us consider in particular the class of nortriterpenes. We have already had occasion to meet monoterpenes in the essential oil constituents. Owing to their biosynthetic origin derived from progressive accumulation of isoprene units, each made by five carbon atoms, terpenes are classified according to the increasing molecular weight in monoterpenes (C10), diterpenes (C20), triterpenes (C30), and tetraterpenes (C40). Squalene, the unique precursor of all triterpenes, is a linear unsaturated hydrocarbon, but its derivatives, for stability reason, are all cyclized with hexagonal and pentagonal cycles in the final structure. Among triterpenes, the most famous class is certainly the steroidal one. Steroids are present in any organisms, where they carry out several fundamental roles. Without steroids, starting from the cholesterol stabilization of cell membranes to the influence on metabolism, no cell, and therefore no organism, could survive. However, each organism synthetizes its own steroids. In fact, animals possess simpler steroids usually with few oxygenated functional groups, bacteria produces steroidal triterpenes mainly dedicated to the stability of the cell envelope, and plants biosynthetize quite complex structures, named phytosterols. Generally, phytosterols possess more cycles respect to the ordinary structure of the steroid model and an increase of the number of functional groups (Roy and Saraf, 2006). Animal steroids are based on a simple, easy-to-remember sequence of four fused cycles: three hexagonal and the last pentagonal. The basic structure of a steroid is quite easy to write and remember by students, including the stereochemistry, whereas the structures of limonoids are very complex and not so easy to remember. The problem is that in the basic structure of a steroid the sequence of the cycles is linear, whereas in limonoids there are complicated re-arrangements, causing a circular total structure. The insect's molt and metamorphosis are triggered and directed by hormones, usually consisting of steroids, such as prohormones (pheromones or juvenile hormones) and ecdysones. The term "ecdysone" was introduced by the German biochemist Peter Karlson (1918–2001) in 1956 in "Chemische Untersuchungen über die Metamorphosehormone der Insekten" (Karlson, 1956). The etymology of this word is interesting, from the Greek *ekdusis* "shedding," or more precisely ἐκ(ς) ("external" or "from inner to out") + δύω ("dress oneself") + *-si(s)* ("action") + *-ona* ("hormone"). However, ecdysones have typical steroidal structures, whereas limonoids are the result of a complex chemical rearrangement.

Thus, to interfere in insect life, special triterpenes are necessary. Several plants, like those in Rutales and Sapindales and related families, are specialized in the synthesis of such molecules, probably made to defend the plant from phytophagous insects. To obtain this result, great chemical ability is necessary. First, part of the typical hydrocarburic lateral chain, typical of most steroids, is lost ("nor" in chemistry means exactly this passage obtained by cutting C–C bonds), and the remaining part is both oxygenated and compressed in a complicated polycyclic structure, which is quite stable in the cell environment, but easily degradable in contact with atmospheric oxygen and sunlight. In this way, the nortriterpenes can be produced in the plant and transferred to the insect with mortal effects through a subtle toxic effect. The idea is to interfere with the growth regulators, interrupting the balance of hormones, named juvenile, in particular interrupting the transition process from the larval instar stages to pupae and adults by juvenile hormone analogues. Therefore, on learning this lesson from nature, we can find solutions and inspirations.

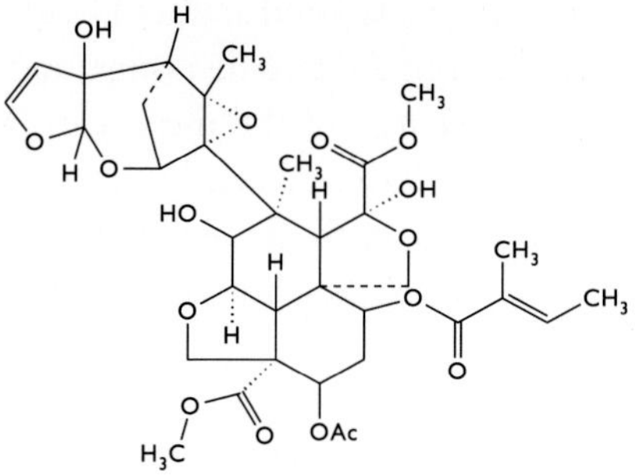

Azadirachtin A

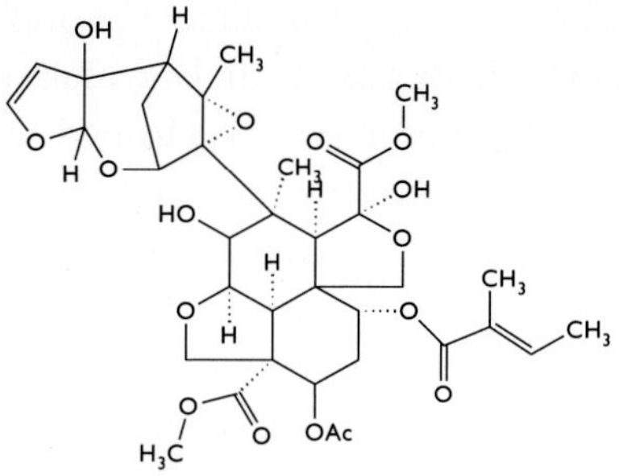

Azadirachtin B

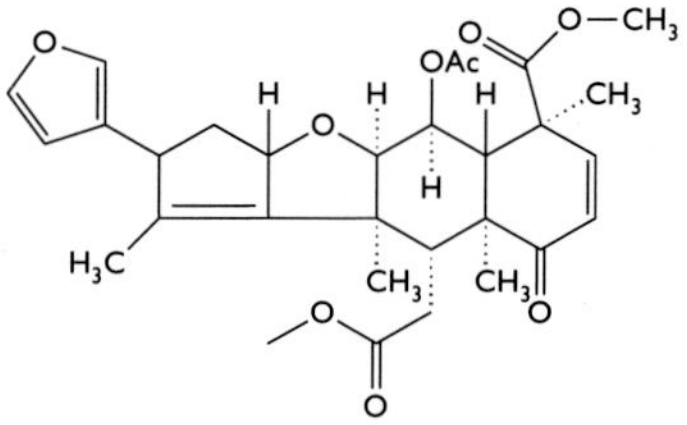

Nimbin

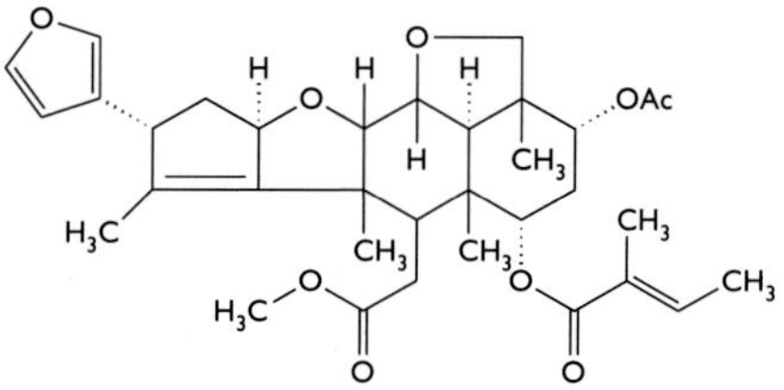

Salannin

Fig. 7.15 Structures of major limonoids of NSO.

On the basis of these considerations and the structural diversity (Fig. 7.15), nortriterpenes can be divided into two main groups: limonoids (C26), with partial loss of the lateral chain (Manners, 2007), and quassinoids (C20 and C19), with total loss of the lateral chain (Vieira and Braz-Filho, 2006). In ancient times, plants containing these kinds of compounds were mainly famous for the bitterness of their drugs, utilized in the production of tonics, digestifs, and medicines. Limonoids are part of our ordinary experience with some fruits and are crucial for the dissemination process. When we eat a citrus fruit, such as a lemon or an orange, we taste the agreeable flavor of the juice, but the seeds are discarded because they contain the nortriterpene limonin, which is very bitter and unpleasant. By throwing away the seed, we contribute to the reproduction of the plant. Several other properties of limonoids have also been reported, including antioxidant, antimicrobial, and antitumoral activities, the insecticidal of neem being so far the most important.

Limonoids are considered the most active ingredient of insecticide neem products. They are classified into nine basic structures, with three main skeleton types: (a) azadiracthins, highly polioxygenated and acylated, with a saturated first ring, a tetrahydrofurane ring between the two first rings, and a final dihydrofurane ring chained with the other part of the molecule; (b) nimbins, less oxygenated and acylated with a skeleton evidently similar to that of the steroids, the furane ring with only a link with the remaining part of the molecule; and (c) a third type similar to the azadirachtins one, but the polycyclic part containing the dihydrofurane ring is less complicated, giving rise to a more linear general skeleton. Such variability is necessary to sustain the large range of targets. In fact, these differences are just as important for the biological activity as for the decomposition. However, in terms of market considerations, azadirachtins, in particular azadirachtin A, are considered the reference constituents to evaluate the quality, and therefore the activity, of neem oil.

Azadirachtin A is a highly oxygenated C-secolimonoid, whose content in the seeds is highly variable (0.1%–1%), mainly depending on the producing zone and the seasonal trend. This compound acts as a biocidal on insects after ingestion or contact, with several effects: (a) interference on the growing processes, inhibiting the molting or blocking the hormone ecdysone synthesis; (b) antifeedant, with reduction of feeding; (c) negative effects on adult fecundity and egg fertility; and (d) diminution of the defense capacity of the cuticle, easing the penetration of pathogens. In particular, the larvicidal effect consists in the formation of the "permanent larvae," i.e., larvae are able to complete the molt as a consequence of destruction of the cuticle or of hormonal perturbation of the metamorphoses. This study consists in the careful observation of the larvae transformations and in the daily count of the consequence of azadirachtins and related compounds on the molting of phytophagous with buccal apparatus, either biting-sucking and chewing, comprised in all systematic categories: orthoptera including grasshoppers, locusts and crickets, etheroptera, homoptera, aphides, cicadellidae, hymenopterous, thysanoptera, aleurodidae, dipera, beetles, and others, including acarus and nematodes.

Neem's oil formulations usually show a range of different azadirachtins amounts, ranging from 1000 to 4000 mg/kg, meaning that products can be obtained either by using directly poor neem oil or a dilution process of neem extracts containing different quantity of azadirachtins, up to 5%. In addition to neem oil, azadirachtins are also marketed, in particular azadirachtin A. The amount of production of this substance amounts to about 64 tons, with 80% coming from India, and China as the second producer. Other data about the activities of NSO and its products can be found in the references at the end of this chapter.

Larvicidal activity of neem

Our first experiments clearly demonstrated strong larvicidal activity of NSO and neem cake on Asian tiger mosquito (Nicoletti et al., 2012a,b). However, our HPLC and HPTLC analyses showed a low content in azadirachtins in the NSO and in the methanol extract (Mariani et al., 2013). The result was interesting, since it is well-known that insecticidal activity is strictly related to the chemical composition, but in contrast to most reports evidenced a relation between the activity and the presence of these limonoids. This consideration prompted research to identify a relationship between composition and activity in the case of NSOs marketed by different producers. First, the HPTLC analysis indicated great differences in the fingerprints of the analyzed oils, with special reference to limonoids (Nicoletti, 2011; Toniolo et al., 2014). A second analytical step consisted of a fractionation of three selected neem oils in three fractions of increasing polarities (i.e., ethyl acetate fraction (EA), butanol fraction (BU), and water (WE)). The initial neem oil and the obtained fractions were evaluated for larvicidal toxicity and field oviposition deterrence against the Asian tiger mosquito, *Aedes albopictus*. The experiments showed good toxicity of the entire neem oil and EA fractions against *A. albopictus* fourth instar larvae (with LC50 values ranging from 142.28 to 209.73 ppm), while little toxicity was exerted by BU and WE fractions. The differences of activity were in accordance with the results of HPTLC analyses, since the NSOs more concentrated proved to be more active. These results were confirmed by deterrence of *A. albopictus* oviposition in the field (effective repellence values ranging from 98.55% to 70.10%), while no effectiveness of BU fractions was found. Concerning ovideterrent activity, no difference due to the production site was found. These experimental data evidenced the possible use of neem constituents against Culicidae in the field. The constituents must be found in the apolar fraction, but the HPTLC analysis showed a complex composition, wherein limonoids were not prevalent. Therefore, neem oil and EA fraction seem promising, since they are effective at lower doses, if compared to synthetic products currently marketed, and could be advantageous alternatives to develop newer and safer mosquito control tools, but other studies are necessary to obtain a better definition of the active constituents and tailor the neem products in accordance with the required utilization (Benelli et al., 2015c; Mariani et al., 2013).

Therefore, when we started our work on neem products, we found several incongruities between the reported studies in the literature and our results (Nicoletti et al., 2017; Mariani et al., 2013). In case of incongruence of the experimental data, two main interpretations are possible. The anomalous data could be the consequence of some error in the experimental procedure or the previous reported data must be reconsidered on the light of the new ones. In fact, many scientific important discoveries have been as consequences of unexpected results. There is a strong tendency in pharmacology to assign the activity of a plant drug to one constituent, or eventually a few of the same chemical class. This is mainly a consequence of the pharmacological tests, which are tailored on the magic bullet axiom and the difficulties in determining precisely the composition of an extract. However, in an extract, and consequently in the plant, there are hundreds of compounds, with effects on bioavailability, solubility, and synergic and antagonist activity. In opposition to the magic bullet, there is the approach of the phytocomplex, invoked by many researchers in phytochemistry and pharmacognosy.

An important part of research on neem was dedicated to increase its availability and properties, focusing in particular on stability and cost, toward the production of the ideal insecticide. The first aspect was assigned to the production of nanobioparticles containing neem extracts, which demonstrated clear larvicidal and deterrent activity on vectors, like *Ae. aldopictus*, also in field conditions (Chandramohan et al., 2016; Murugan et al., 2016).

Several factors must be considered in the case of a product based on natural substances. In theory, the plant could be available for everyone and therefore it cannot be patented. Therefore, so far natural products are available for everyone and thus have been explored very little. Natural products are the chemical part of the environmental interactions between living organisms and therefore they are natural candidates for the production of active drugs. The chemical production of a plant is strictly subjected to the environmental conditions that can highly influence this production. First, the exactly determined species must be used and determined in composition. Once the raw material is obtained, the process of transformation can significantly influence the composition of the product. The technological transformation is essential to the quality and efficacy of the product. Therefore, this second step is vital for the success of the product. The third step consists of the target being assigned to the product and the consequent marketing.

In future, natural products will be even more important in the production of new drugs and foods and feeds, able to face the challenges of a continuously changing market. Technology is key to this. The natural products market is expanding rapidly in previously unexplored areas, in particular as an alternative to products based on synthetic compounds. The prospects and possibilities in this situation are immense, but knowledge of nature and activity of natural products must be revised utilizing recent devices and research approaches. Importance and role of natural products will increase if the multidrug resistance continues, asking for new bacterial and insect possibilities of control.

The common composition of a botanical product is based on a single herb or on the combination of more species based on recipes and formulae mainly derived from the historical literature and empirical experiences. The long and accurate work of phytochemistry based on the sequence extraction-separation-identification, derived from the correspondence of one drug to one illness, generated a huge catalogue of identified natural substances that can be employed as useful standards to determine the composition of the botanical drug. The knowledge about composition must be as complete as possible; not a single constituent should be unused, and utility depends strictly on the utilization.

Natural products are derived mainly from plants as the result of coevolution between organisms and environment (Tehri and Singh, 2013). For this reason, they have been used for centuries in popular and traditional medicines, as well as often being employed as spices and insecticides. Unlike modern pharmacology and drug development, which are based on a single chemical entity, natural product preparations are multiingredient. A single herbal drug contains at least 100 compounds making a complex matrix, named a phytocomplex, in which the single active constituent is not considered solely responsible for the overall efficacy. The utilization of the phytocomplex is based on experimental basis, since many data afford the validity of this approach, although further confirmations could be obtained using modern pharmacological devices. In other words, the same botanical raw material can be used directly, or extracted in different ways or used as a source of selected substances, or modified according to the product and target.

In 2010, a mixed team of experts from MIT (Massachusetts Institute of Technology) and the Broad Institute of Harvard University, both in the USA, reported an interesting and innovative study for a scientific evaluation of the effectiveness of natural products. The argument is strictly inherent to the endless debate about the role of natural products and their efficacy, causing a fighting contrast, but useless and boring, between supporters of "natural" versus defenders of synthetic drugs. The key aim of the study was to understand what is going on between the two main levels of the metabolism (primary and secondary), on the basis of the consideration of the functional connection between genes and gene products, as well as between genes and targets. An innovative feature of the study is that the researchers decided to commit the argument to neutral judgment, submitting the elaboration of collected data to the computational work of artificial intelligence. The work was based on the comparison of cumulative connectivity distribution of small molecules, natural or synthetic, grouped according to connectivity associated with the target. Assuming that proteins form biological networks and that metabolism and health depend on these networks, we should be able to assign a role to the molecules considered as possible medicines. The result showed that natural products target the proteins with a high number of protein-protein functional interactions (higher network connectivity), whereas the synthetic ones act on a limited protein network. The conclusions of the study, based on a computational approach, were evident: "We observe that approved drug targets that are not also natural product targets exhibit a connection distribution much closer to the case for human disease genes that natural product targets, which remain the most highly connected targets." This sentence indicates a positive and useful consideration about the role and activity of natural products. Natural products tend to target more essential and general protein networks to an organism than other groups of small-molecule targets, like those more related to specific disease genes. Therefore, the dichotomy between natural and synthetic active constituents must be considered mainly as a consequence of a cultural heritage, unable to assign a complementary or differently appropriate role to the two classes of molecules. The results of the study are coherent with the nature of natural products, whose production is the consequence of environmental interactions, including defense against predators or pathogens. This kind of defense cannot be specific, and therefore natural products act on more highly connected network of proteins, interrupting or limiting the activity of the essential proteins in environmental competitors or invaders. They may be tailored for a positive or negative influence in physiologic activities and basic metabolism of an ample range of organic targets. These arguments are in favor of the potential use of natural products as insecticides.

In any case, there are several difficulties in assigning the activity to single constituents, causing several cases of wrong or misleading assignments of all the activity to single substances in the case of a plant extract, like in valerian (*Valeriana officinalis*), whose extracts are largely marketed and utilized for their mild sedative effects. With the discovery of valepotriates, the effects were assigned to these constituents, but after the evidence that extracts with low content in instable valepotriates also exerted similar action, the essential oil was considered additionally responsible for the effect. Another case consists of a current debate about hemp. Besides cannabinoids, its essential oil and other constituents are now considered important for the multiple activities of hemp. In other words, there are hundreds of marketed products of hemp and many related claimed activities, and this can be related to the complex cannabidioma and/or the different compositions of the products, although they are all derived from the same raw material. It is very important to stress that important new features can appear, also in the case of species highly studied in their chemical composition, as shown in the scientific literature. Recently, a new cannabinoid was isolated from *Cannabis sativa* (Citti, 2019). As is well-known, (−)-*Trans*-Δ^9-tetrahydrocannabinol (Δ^9-THC) is the main compound of hemp and it is considered the main one responsible for intoxicant activity. However, the chemical constitution of this species is subject to high differences in accordance with its varieties and cultivars. Cannabinoids possess a unique structure, derived by junction of a monoterpene and a polyketide unit. Most of them have a side alkyl chain, whose length influences the biological activity of this cannabinoid. In fact, analogues of Δ^9-THC with a longer side chain were synthetized and they have shown cannabimimetic properties far higher than Δ^9-THC itself (seven C against five). In this study, a new phytocannabinoid with the same structure of Δ^9-THC, but with a seven-term alkyl side chain, was isolated and identified, and its stereochemical configuration confirmed by a stereoselective synthesis. This new phytocannabinoid has been called (-)-*trans*-Δ^9-tetrahydrocannabiphorol (Δ^9-THCP). The binding activity of Δ^9-THCP against human CB_1 receptor in vitro ($K_i = 1.2\,nM$) proved to be similar to that of CP55940 ($K_i = 0.9\,nM$), a potent full CB_1 agonist. In the cannabinoid tetrad pharmacological test, Δ^9-THCP induced hypomotility, analgesia, catalepsy, and decreased rectal temperature, indicating a THC-like cannabimimetic activity. As confirmation, the corresponding cannabidiol (CBD) homolog with a seven-term side alkyl chain (CBDP) was also isolated and unambiguously identified by matching with its synthetic counterpart. The presence of this new phytocannabinoid could account for the pharmacological properties of some cannabis varieties that are difficult to explain by the presence of the sole Δ^9-THC and indicate the importance of the interaction between constituents of the so-called cannabidiome.

Therefore, we were not totally surprised when we found good larvicidal activity against *Aedes albopictus* also in NSOs with low content in azadirachtins (Mariani and Nicoletti, 2013). This was quite a novelty on the basis of the literature, but it is necessary to consider the importance of the metabolomics approach and the possibility with HPTLC to obtain several views of the same plates. Each view means a revelation of different compounds on the basis of their chemical structure and present functional groups. Using an appropriate revelation agent, it is possible to see compounds that are not visible with another derivatization. This approach is contrary to the tendency of current analytical chemistry to focus on a single class of compounds or even unique constituents, which obtain perfect and reliable but limited results. Another incongruence consisted of the presence of insecticide activity also in neem products after years of production, when limonoids should be highly degradated.

Neem's mechanism of action

The first experimental evidence we obtained on the activity of neem oil was the inability of the larvae of *Ae. albopictus* to complete the molt from larva to pupa. The larvae proved to be initially immature, their bodies imperfect, and finally before the third instar, most insects died and none was able to fly. The delicate mechanism of the development stage was jammed and the cruel destiny of the unfortunate insects assigned. Each organism has its weakness. Mosquitos, like any arthropod, possess a rigid exoskeleton, which offers efficient strong and secure protection, also against pesticides, which is one of the reasons for the success of these creatures.

The exoskeleton of insects is primarily made of proteins (sclerotin) and chitin (a polysaccharide), which are interwoven and linked together to form strong but flexible bundles. Interestingly, chitin is also the main constituent of the fungal cell wall. The ratio of the components of the exoskeleton varies from one body part to another on an insect. However, the exoskeleton is too rigid, and acts like a cover that encases the entire insect, and being a non-living formation, the exoskeleton does not change size and grow with the insect. The exoskeleton is too ridged to be recycled or modified, and it must be substituted, but it must also protect the insect until the new exoskeleton is ready. During the growth period, insects must shed the exoskeleton in order to assume a new form. As a result, it is necessary for the insect to shed its old exoskeleton to make way for a new, larger one through a process called molting. This is a hormone-controlled phenomenon. During the molting stages, the hormones are released to start and finalize each step of the metamorphosis, until the mature insect finally emerges. However, the chemical constitution of the exoskeleton is variable in each insect species and this is the reason for the selective toxic effects, such as those reported in the case of neem. Regarding the structures of insecticides acting as growth regulators, albeit in the case of ecdysones the relation with insect hormones is evident, in other cases the similarity is not so clear, as well as the real mechanism of action.

The stages between the subsequent molts are generally called instars. These correspond to altered body proportions, colors, patterns, and changes in the number of body segments or head width. For most insect species, an instar is the developmental stage of the larval forms, but an instar can be any developmental stage including pupa or imago. The larval stage is in particular a delicate stage of the insect metamorphosis.

However, we were totally aware that confirmation of the neem insecticide activity, albeit with a demonstrated chemical constitution, in a laboratory experiment was a weak starting point. The open questions were numerous: (a) how to obtain the same result in the field; (b) whether the larvicidal activity could be connected to other properties, in order to improve the use; (c) what the cost of neem oil would be, considering the large-scale spread of the insects; (d) how limited the stability of the active ingredients of neem would be; (e) what determination of chemical content of neem oil would be required, to be connected to the determination of the activity; and (f) what the ambit of utilization would be and the possible damage to the habitat.

Other advantages arising from the use of neem-based products are the rare induction of resistance, due to their multiple mode of action against pests, the low toxicity rates that have been detected against vertebrates, and finally the necessary environmental care.

The alternative to neem

There is a little confusion about the plant species named azedarach, and very similar denominations. The name azedarach was given by the famous Persian physician Avicenna (980–1037) to indicate some poisonous trees; however, *Azadirakhti* literally means "free book of India." In 1753, Linnaeus reported about *Melia azadarachta* in his *Species Plantarum* (1: 385 with habitat: India). In the same book (1: 384), we can find *Melia azedarach* (habitat: Syria) and *Melia azedarach* var. *sempervirens* (habitat: Zeylona). Actually there are two distinct species, *Azadirachta indica* A. Juss, attributed to neem (or nimba, meaning "who gives good health," as reported in the Sanskrit books) and *Melia azedaracht* Linneus, attributed to melia, a very similar tree. This is the typical taxonomic situation in botany and zoology. The differences between taxa are often very narrow and only specialists are able to find them. In any case, the problem of the significance of these differences is always a matter of debate. God bless taxonomists, because they are necessary to obtain order out chaos, but please do not spend your precious intellect on endless discussions with no final consistent result! In fact, the matter is complicated by synonymous, parental disputes, errors of any type, including wrong transliteration (i.e., gingko and ginko), disputable rules of the international codices, and more.

Neem and melia are very similar, but there are several tricks to distinguish between the two species. The first is commonly known also as Indian lilac and the second one as Persian lilac or simply melia. Neem has usually white flowers whereas melia presents an explosion of blue flowers; the fruits of the former have an elongated shape, whereas the latter's are totally rounded. If the trees do not have flowers or fruits, and you are not a botanist, you may be in trouble, but you can remember that neem cannot live in temperate climate regions, whereas melia can be easily cultivated in such places. Therefore, if you are in Europe or the USA, you can be 90% sure on the matter.

Melia azedarach is known by several common names, such as melia, chinaberry tree, Pride of India, bead-tree, Cape lilac, Syringa berrytree, Persian lilac, and others. It is usually a large tree growing up to 30 m tall, with leaves 2-pinnate, rarely 3-, with primary pinnae in two to six pairs, usually three to seven leaflets per pinna, narrowly ovate or subovate, serrate, acuminate, irregularly toothed, or crenate. Flowers are abundant and small, sweet-scented, in large axillary panicles. All parts of this tree are reported to have medicinal uses, but in particular, in terms of insecticide properties, seedlings are reported to present aphid attacks. A leaf used as a bookmark will deter insect pests. In Italy, the tree is known as the tree of rosary, since in the past, before the advent of plastic, its hard and round kernels were used to make the grains of a rosary.

Our research on *Melia azedarach*, as well as the references on this plant, evidences a significant difference in the chemical composition. Limonoids are present, but different from azadirachtins and other constituents make a marked relevant dissimilarity in composition. The initial conclusion was that melia probably cannot compete with neem as an insecticide, but other utilizations can be explored. However, once again a limit in the references is an irresistible task for a researcher in search of innovations.

In addition to the insecticidal properties, we were initially particularly interested in the antimicrobial activity. People often associate antimicrobial activity with infection and effects on their health, but microbes are everywhere and most damage affects cultivation of plants. Agricultural methods of reproduction of plants with economic value were totally transformed by the introduction of micropropagation and stem cell culture. Micropropagation allows the rapid cultivation of selected cultivars, saving time and resources. However, although the first steps of micropropagation were performed in aseptic conditions, the possibility of infection of calla, shoots, and seedlings is high. Avoiding the infection must be done via an appropriate and sensitive approach, avoiding damage to the delicate meristems—a typical job for natural products.

The antibacterial study (Marino et al., 2014) aimed to investigate the antibacterial activity of unripe fruits of *Melia azedarach* collected in different periods. The activity was tested on the shoots of a hybrid of *Prunus cerasifera* x *Prunus spinosa* and calla lily of *Zantedeschia aethiopica* against several bacterial species. The data reported evidenced a positive antibacterial activity and the absence of any negative effect on the growth of shoots surviving at the second subculture on a standard medium. HPTLC analysis showed the prevalence of polyphenols, such as chlorogenic and caffeic acids, which, on the basis of the literature, are consistent with the antimicrobial activity. This activity is important considering that many plant species of economic relevance are now obtained by micropropagation, and this cultivation in vitro is necessary to avoid any sort of contamination.

Further research is essentially the rational collection of most of the arguments previously considered, as evident in the title: "Green-synthesised nanoparticles from *Melia azedarach* seeds and the cyclopoid crustacean *Cyclops vernalis*: an eco-friendly route to control the malaria vector *Anopheles stephensi*?" (Anbu et al., 2017). In this research, once a single-step green-synthesis of silver nanoparticles (AgNP) using the seed extract of *M. azedarach* was obtained, we tested its mosquitocidal activity. In laboratory assays on *Anopheles stephensi*, Ag NP showed LC_{50} ranging from 2.897 (I instar larvae) to 14.548 ppm (pupae). In the field, the application of Ag NP ($10 \times LC_{50}$) led to complete elimination of larval populations after 72 h. Finally, we decided to test the nanoparticles on nontarget aquatic predators. The application of Ag NP in the aquatic environment did not show negative adverse effects on predatory efficiency of the mosquito natural enemy *Cyclops vernalis*. The reason for this additional research lies in the fact that numerous aquatic arthropods attack and devour preparasites. As we already know, the utilization of the insecticides, though with plant-derived active constituents, could be dangerous for the environmental equilibria. In particular, it could affect the natural biological control, based on the presence predators of the vector in the common habitat, remembering that all the insect stages, except the adult insect, need water. In such sites, there is fresh water everywhere, such as lakes, pools, and similar places, enabling life along the plant-covered banks of stagnant and slow-flowing bodies of water. In such places, mosquitos can proliferate as can any other predator, which in an aquatic environment is fundamental to limit the proliferation of the vector. In fact, after coupling, and the consequent blood feeding necessary to assume the proteins necessary for the eggs maturation, the female is looking for an appropriate place for the deposition of 100–500 eggs. A single *Anopheles*, like other insect, is able to produce a quantity of eggs and larvae enough to invade all the neighboring habitats, as in the classic case of a locust invasion. This is not possible only thanks to the natural enemies. The micro-aquatic environments are the scenario of a continuous fight for survival, where often two or more species of arthropods are involved, as predator or as prey. In our study, we selected the genus *Cyclops*, which is one of the most common of freshwater copepods, comprising more than 400 species. Copepods are very little crustaceans, commonly called water fleas. They have a single large eye, which may be either red or black, and therefore they are named for the Cyclops of Greek mythology. *Cyclops* prefers fresh water, and is less frequent in brackish water, where it feeds on small fragments of plant material, animals, or carrion. It swims with characteristic jerky movements and has the capacity to survive unsuitable conditions by forming a cloak of slime, with an average lifespan of about 3 months.

Several microscopic crustaceous, including copepod species, feed small and very small preys. In high-density, unstructured environments such as eutrophic lakes, predatory copepods commonly coexist with certain small-bodied prey, where encounters are frequent with ineptness on the part of the predator and counter-tactics by the prey. In particular, laboratory studies showed that copepods are effective predators on early-instar *Culex* larvae, involving an important role in suppressing mosquito populations, because of their feeding behavior and abundance. They are very efficient in this role, since the presence of alternative abundant food, like bacteria and protozoa, does not deter their attacks on their preferred prey. Copepods are capable of killing and eating at least four preparasites within 13 min and a predator density of 53 copepods/liter is expected to reduce the mosquito larvae by 50%, with the rate of predation inversely proportional to the water volume.

Neem cake: From by-product of an industrial process to multipurpose resource for a sustainable agriculture chain

During our research activity, we were highly interested in industrial plant-borne by-products, since they can offer new products to the market with lower cost and high usefulness. Our attention was immediately attracted by neem cake, a cheap by-product of NSO extraction, obtained as a residue after mechanical pressing of the neem seeds, considered of low economic value and utilized to enrich the soil of some mineral components, such as nitrogen.

The laboratory test indicated neem cake activity against *Aedes albopictus* and a number of *Culicidae* species (Nicoletti et al., 2010). In the case of biocidal treatments, it is important to demonstrate that insecticide activity is associated with antimicrobial activities in consideration of the high possibility of infection and the severe consequences for health, in particular in the case of the animals, both pets and livestock. The importance of insecticidal and antimicrobial activities for animal treatment has been evidenced in the experiments with NSO and neem cake further reported, including larvicide, deterrent, and repellent activities (Benelli et al., 2015a,b; Mariani and Nicoletti, 2013; Nicoletti et al., 2012a,b).

The complex range of different compounds in the neem seeds open the possibility to utilize the derived products to solve many current problems. The challenge now is to obtain marketed products tailored for different utilizations. The reported experiments evidenced these potentialities, which are only waiting a realization and a wider utilization.

Despite diseases, wars, and environmental disasters, the human population is growing. First of all, more people will need more food. This forecast shows in particular a massive increase in animal protein demand, needed to satisfy the growth in the human population, wherein billions of people require an increase of caloric input and better food. Therefore, attention is focused on the sources of feed protein and their suitability, quality, and safety for future supply. In addition, the quantitative production aspect is causing a series of problems. There will need to be a considerable increase in feed manufacture, requiring a thriving, successful, and modern feed industry, including a key aspect concerning the protection and preservation of the food produced and marketed. This aspect is strictly related to safety issues, which will remain paramount in the minds of consumers following recent food crises.

It is time to consider that the need of more food to feed due to increasing planet population perhaps cannot simply be solved by massive production, but reduction of food waste and conservation can increase food availability by 30%–40%. "Feed the Planet and Energy for Life" was the theme of World Expo 2015 in Milan, Italy. Among the activities occurring at the Expo, research and proposals concerning the utilization of neem products were presented in a call for projects in favor of sustainable progress and production of future foods. The Neem Project was selected as the best one due to its possible applications in the production of food and feed.

The Expo event projected feeding as the main challenge for humankind and showed the extreme urgency of elements of innovation in technology and science connected to the production and conservation of food. It was demonstrated how serious feed problems still plague several areas of the world today, and the possibility of new solutions was mentioned. The Neem Project was focused on the agricultural utilizations of neem cake concerning its advantages as soil fertilizer and as a natural ectoparasiticide for the treatment of sheep and goats.

Neem products were proposed as being able to affect the biotic composition of the soil. Neem cake must be preferred to neem oil for its cost and its form as a powder immediately available. Several experiments evidenced the improvements of the utilization of neem cake in agrarian ecosystems:

(1) availability of nutrients, in particular nitrogen and phosphorus, more consistent with the requirements of the crop;

(2) development of the microbial beneficial biomass of the soil, which increases in quantity and activity, but with selective influence against nematodes and other negative components. In agricultural practices, plants in addition to nutrients should count on a greater variety of useful microorganisms and on acquisition of nutrients themselves, through the activation of complex symbiotic systems. If you want to understand the state of health of a tree, you must look down, not up; and

(3) development of pest control system of insects and other arthropods of agricultural and livestock interest.

Neem cake, as an industrial by-product, is a heterogeneous material that maintains a high added value due in large part to its chemical composition, which confers its biological activity. Neem cake is widely available on the global market, considering the increasing presence of neem trees in the world and production of NSO. The exploitation of its characteristics in the food chain to improve consumer health, increase the productivity of agricultural products, and feed the planet is the logical consequence of the urgent need to develop new sustainable agricultural systems in a world where many highly polluting pesticides are no longer allowed to be used. However, more research, in particular in field conditions, is necessary to understand the real value of its microbiological, insecticide, fertilizer, and nematocide activity, involving collaborations between different experts in individual sectors—import companies, organic farms, and research institutions—in order to determine the manner and timing of land application of this valuable product of "waste," still underestimated.

Neem cake could lead to a revolutionary improvement in the fertilization of agricultural plants, adding to the characteristics of chemical fertilizers those of a soil improver. In agriculture, we could define neem cake as a prompt nutrient-release fertilizer, effective in allowing rapid absorption of nutrients and promoting development of the plant, with the capacity to increase the activity of the microbial biomass and organic matter, favoring the sequestration of carbon. The idea was to join the fertilizer, insecticide, and antimicrobial utilities of neem cake.

Exploitation of the use of neem cake as an insecticide came from this first test: some pots of impatiens (*Impatiens balsamina*) were fertilized with 3% by volume of neem cake, and 500 mosquito larvae were reared starting from eggs. The eggs were hatched in control and treated in pot saucers, but none of the newborn larvae survived in the water saucers of pots treated with neem cake, while in the water saucers of pots unfertilized with cake, the 500 control larvae completed their development in less than a week, becoming adult mosquitoes. Other major beneficiaries of the use of neem cake as insecticide are undoubtedly sheep farmers, who can use an organic product of natural origin and low cost that is simultaneously effective against the larvae of Culicoides and other pests, while respecting the natural biotic communities.

Direct beneficiaries of neem cake, as a fertilizer, are farmers seeking pest and nematode control, in particular for nematodes. Currently, some highly toxic products are still on the market by virtue of the absence of suitable alternatives. Particular attention must be focused on the changes on soil micro-composition, as evidenced in several field experiments.

In conclusion, we can report the following important advances in the use of neem cake as a functional fertilizer:

(a) energy saving flows from the use of a waste of an industrial chain;

(b) environmental sustainability, as documented by the analyses attesting the absence of heavy metals, aflatoxins and residues of pesticides;
(c) neem cake is an excellent alternative to methyl bromide (BM) (banned as being responsible for the "thinning" of the ozone layer of the atmosphere);
(d) neem cake is an excellent alternative to Temephos and other organophosphates used to treat water infested with disease-carrying insects including mosquitoes, midges, and blackfly larvae;
(e) neem cake is a great alternative to nematicides, like 1,3-dichloropropene; and
(f) neem cake in the field trials carried out in Sardinia had efficacy similar to azadirachtin biological products already established in organic farming, but were very expensive and not really effective. In addition, neem cake showed very low effect on "nontarget" insects that live in the same environments as Culicoides larvae.

Neem oil as an antibacterial natural product in pest control in livestock

Ectoparasites are organisms that inhabit the skin of another organism, causing significant infestations and pathologies. Many micropathogens can profit from the work of ectoparasites, either to colonize the skin injury and lesion, or be inserted in the host during the feeding. The vast majority of ectoparasites are arthropods, e.g., insects and arachnids. Again the triangle host-vector-etiological agent is reproduced.

Many ectoparasites are vectors of pathogens, which are typically transmitted while feeding on or from other hosts. Several ectoparasites (e.g., most lice) are host-specific, including livestock, pets, poultry, fish, and bees, but others parasitize a wide range of hosts, including humans. Typical effects of infection on the host are irritability, dermatitis, secondary infection (other parasites profit of the skin necrosis), fecal hemorrhages, blockage of orifices, inoculation of toxins, and exsanguination. As a consequence, the host's general health can be seriously affected with low weight gains, particularly important in livestock. Subdermally located parasitic larval stages of certain flies can be favored by the ectoparasited infection, causing a condition termed "myiasis." When insects (order Hemiptera) are involved, the infection mechanism is similar to that previously described for any insect-borne disease. The vector contains several hematophagous ectoparasites, including approximately 150 species of kissing ("cone-nose") bugs (Reduviidae, Triatominae) and bed bugs and bat bugs

(Cimicidae). These parasites make physical contact with the host principally when ingesting a blood meal. These kissing bugs usually prefer domestic animals, from which relatively large blood volumes may be imbibed; in such a way they can cause a great deal of damage and transmit important diseases.

Ectoparasites play a very detrimental role in terms of decreasing the productivity of livestock, such as sheep and goats. NSO was utilized in the field as an antibacterial in the case of ectoparasites' stings and bites resulting from goat wounds. Common external sheep and goat parasites include ticks, lice, and mites. They cause restlessness and irritation. Weight loss and reduction in milk production may occur as a result of nervousness and improper nutrition, because animals spend less time eating. Bites can damage sensitive areas of skin (teats, vagina, eyes, etc.). Some parasites feed on blood, causing anemia, especially in young animals. The bite and the sting of ectoparasites allow bacteria to proliferate in wounds from abrasions or lesions from scratching, and cause levels of tissue reaction of different entities, super-infection, and cervical lymphadenopathy.

Ectoparasites cause many problems in livestock production. They seriously damage sheep and goat skins, resulting in the rejection or downgrading of the skins. This causes huge economic loss, as this skin damage renders it unsuitable for the leather industry due to the decrease in quality. Lower production of meat is also a typical consequence. *Pseudomonas aeruginosa* wound infections are characterized by a change in the color of the skin around the wound area and the formation of lesions. The bacterium products and pigment cause yellow discoloration of wool and consequently reduced quality and market value.

NSO treatment in the field on as a natural ectoparasiticide for sheep and goats proved to be successful in preventing and curing the attacks of endoparasites (Fig. 7.16). The experiments were performed on selected livestock (Fig. 7.17) by a specialized team of CREA researchers (De Matteis et al., 2015). The effects on the parasites were evident (Figs. 7.18 and 7.19) and even after the first treatment with NSO, protection against ectoparasites was obtained. More important, the health of the treated livestock improved, as testified by the hematological profile of goats. *In vivo* and in vitro tests on blood cells from Siriana, Sanen, Cashmere, and Maltese goat (*Capra hircus*) breeds showed no significant difference ($P<.05$) between NSO treated and untreated goat hematological parameters at each sampling time considered. In addition, the NSO effect on goat PBMC cultured in RPMI medium was evaluated at 1:2 × 102 to 1:20 × 106 dilutions at 14, 21, and 40 h of exposure. The in vitro test revealed that the response of goat PBMC viability is

Fig. 7.16 The preparation neem oil solution for the experiment in field.

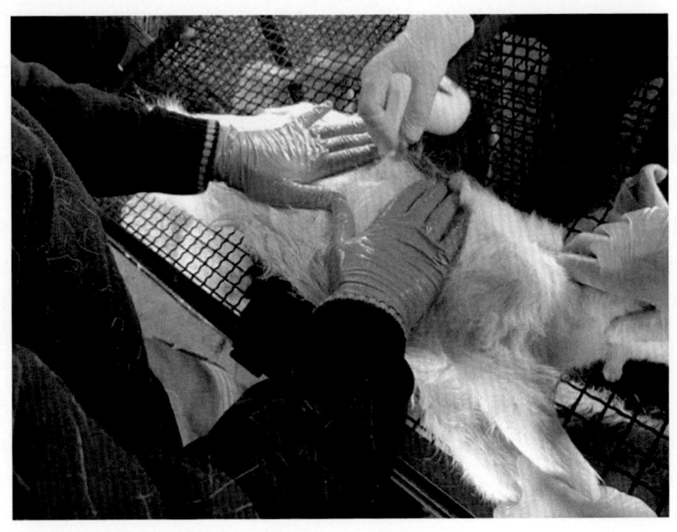

Fig. 7.17 The preparation of the ovine.

Fig. 7.18 Effect of NSO solution on the parasite (*Linognathus stenopsis*).

Fig. 7.19 Effect of NSO solution on another parasite (*Damalina caprae*).

dependent on concentration, incubation time, and NSO dose. In conclusion, the NSO should be considered useful, safe, and innovative for development of topical solutions for the care of wounds.

Among the most relevant typology of neem products, we focused on selectivity. The antibacterial activity of NSO was assayed (Del Serrone et al., 2015) against 48 isolates of *Escherichia coli*, considering that this bacterium can produce beneficial and pathologic populations. The molecular biology characterization showed that 14 isolates resulted in diarrheagenic *E. coli*. NSO showed biological activity against all isolates. However, there were significant differences between the antibacterial activities against pathogenic and nonpathogenic *E. coli*, as well as between NSO and ciprofloxacin activities. On the basis of the results obtained, NSO is able to counteract *E. coli* and also influence the virulence of *E. coli*-viable cells after treatment with NSO.

Saving food to save lives

The preservation of marketed food is an important aspect of the smart utilization of the produced food (Maruchecka et al., 2011). Furthermore, the consequent waste of unutilized food is a relevant problem in overcrowded towns (HLPE, 2014). A large quantity of food is lost or wasted throughout the supply chain, from initial agricultural production down to final household consumption (HLPE, 2014; Kader, 2005). The loss or waste for high perishable food, such as fresh fruit and vegetables, fish and livestock products, has been estimated at as much as half of all food grown before and after it reaches the consumer. Approximately one-third of all FFVs produced worldwide are lost during food supply chain production. Shelf life plays a central role in food spoilage. The impact of the enormous quantity of packaging is evident in any planet environments. Increase of the shelf life means reduction of cost and waste. Everything pivots around the material utilized for packaging, and new solutions are emerging (Otoni et al., 2016; Singh and Singh, 2005; Cooksey, 2005; Appendini and Hotchkiss, 2002), including passive packaging (Brockgreitens and Abbas, 2016; Ozdemir and Floros, 2004), active packaging (Coma, 2008), intelligent packaging (Lee et al., 2015; De Kruijf et al., 2002), and smart packaging (Dobrucka and Cierpiszewski, 2014). Although the results are not evident in our ordinary life, the galaxy of packaging is rapidly moving and increasing in research and proposals, based on new technologies and advanced techniques recently available, like nanotechnology and molecular biology. Efforts are focused on solving the food preservative problems, to extend the shelf life of perishable foods, by reducing the need for additives and preservatives.

"Smart packaging" is based on the production of functional methods to obtain the following goals: be tailored depending on the product being packaged, including several types of food, beverages, pharmaceuticals, household products, etc.; reduce food waste, increasing the shelf life; and maintain, and eventually enhance, product attributes (e.g., look, taste, flavor, aroma). The key words are protect, preserve, and present.

Several methods and approaches, such as oxygen scavenging and antimicrobial technologies associated to the production of modified films, have been considered (Del Serrone and Nicoletti, 2014). They are different solutions to serve the basic and fundamental properties of packaging. So far, the dominant packaging is the basic one, using low-cost material and involving no interaction with the food inside. This is passive packaging, wherein the traditional packaging systems are included, as the use of covering material characterized by some inherent insulating, protective, or ease-of-handling qualities. Usually, the ordinary packaging of food is mainly a used to method to attract and select the consumer, beside a preservation. The consequence is the enormous amount of waste, and the consequent damage to the environment. This situation is increasing due to the increasing numbers of consumers in emerging countries, where these consequences are not adequately considered. Packaging is considered active when it can interact in the same way and/or react to various stimuli, in order to keep the internal environment favorable for the maintenance of product quality. Several environmental, biotic, and abiotic factors must be considered, in order to respond to the degradation process successfully. The activity involved could be the presence of an oxygen scavenger (this can absorb high-energy oxygen inside a package and therefore increase the shelf life of a product) or an anti-ROS (a scavenger of radicals by oxygen or other origins), such as in the typical case of phenolic natural products. Smart packaging relies on the use of chemicals, electrical, electronic, or mechanical technology, or any combination of these. Technology is used to modify the packaging by adding constituents to change its features and properties (Kerry et al., 2006; Malhotra et al., 2015). Active and intelligent packaging is particularly dedicated to the preservation of fresh products, like vegetables, in accordance with increasing requirements for this kind of food (Nicoletti and Del Serrone, 2017; Nicoletti, 2014a,b). Intelligent packaging systems monitor the condition of packaged foods to give information regarding the quality of the packaged food during transport and storage (Aguilera et al., 2003). Probably the most innovative aspect of intelligent packaging is that it can be supported by the utilization of systems of detection in meat and meat products, obtained through the use of sensor technologies indicators (Thakur and Ragavan, 2013), including integrity, freshness, and time-temperature indicators (TTIs) and radio frequency identification (RFID). Therefore, active and smart packaging performs additional functions to the basic one by the introduction of innovations in the design of packaging, with the aim of increase the shelf life, but also to add conveniences for the user and usefulness for the consumer, to be introduced also in the supply chain. In this way, the product can respond not only to the need for a longer life, but also make the product more available, more useful, and more safe.

Since our invisible enemies are asked to play their role again, antibiotic activity is required. Packaging is mainly used to separate food from environmental conditions, utilizing simple material made of paper or plastic. However, it cannot prevent internal attacks by microorganisms, but can only limit or delay the effects. Therefore, additional treatments are required to limit their action, like the utilization of low temperatures, which involves additional costs and energy consumption and pollution. A new idea is to associate to the packaging some antimicrobial agent. Before and during packaging, storage, and shelf life, food is subjected to a continuous attack by microorganisms. These microorganisms are working to benefit themselves by demolishing progressively the molecular structure of the food, as soon and as completely as possible. Therefore, by preserving the food, we are working in a thermodynamically unfavorable situation. In term of shelf life, the food is in competition with its natural recycling, and, working to maintain as possible this limbo, we can utilize the food efficiently as it is possible.

The resistance phenomenon interests also zoonotic food- and water-borne pathogens becoming more resistant to antibiotics (Del Serrone et al., 2006). Resistant strains of pathogens have been isolated from food, causing an increasing incidence of food-borne diseases. Through the food, these microorganisms could be entering the human gastrointestinal tract on an almost daily basis. The antimicrobial activity of NSO and related products have already been reported (Palanappian and Holley, 2010; Baswa et al., 2001; SaiRam et al., 2002). A possible utilization of antibacterial activity of neem cake against meat spoilage bacteria was tested using a broth model meat system (Del Serrone and Nicoletti, 2013). The tests were positive, since the growth inhibition zone (mm) varied significantly ($P \geq .05$). With respect to ciprofloxacin activity, the antibiotic value ranged as follows: 11.33 ± 0.58 to 22.67 ± 0.58 mm and 23.41 ± 1.00 to 32.67 ± 2.89 mm, respectively. The percentage of bacterial growth reduction (GR%) also varied significantly ($P \geq .05$) in function of considered NCE concentrations (1:10–1:100,000), with the highest GR% for 10 μg NCE (79.75 ± 1.53 to 90.73 ± 1.53). The numbers of viable bacterial cells never significantly ($P \leq .05$) exceeded inocula concentrations used to contaminate the meat. All the results of the experiments showed that neem cake is able to counteract the main microorganisms responsible of meat spoilage, like strains of Gram-positive and Gram-negative, as well as facultative anaerobic bacteria. The antimicrobial activity of neem products was confirmed also for NSO against spoilage bacteria, such as *Carnobacterium maltaromaticum*, *Brochothrix thermosphacta*, *Escherichia coli*, *Pseudomonas fluorescens*, *Lactobacillus curvatus*, and *L. sakei*. After the second day after NSO, only *C. maltaromaticum*-viable bacterial cells were detected.

These data could be used to create new intelligent packaging. Utilizing a nanotechnology already employed for other materials, neem cake may be incorporated into the cavities of nanoparticles, maintaining its antiparasite activity. Once incorporated into the packaging material, the neem cake, also in minimum quantity, should be able to effect its preservative food action, acting against the demolishing microorganisms. The increase of the shelf life of meat should compensate for the additional cost of the packaging material, not considering the decrease of waste. It is possible that the first activity of LUCA was to find the energetic source for survival, and the second was to compete with the other LUCAs. The results are an endless transformation of forms and production of new molecules. The living organisms had a long time to organize their molecular weapons and the secondary metabolites are there, produced and organized to be considered and utilized in the right way. The neem tree is an example of nature's treasure. The advent of *Homo sapiens* (Lucy) changed in part the rules of the natural game, but natural products still remain a necessity for our life.

Malagasy plants as sources of chloroquine-potentiating agents

Recently, parenteral artesunate has replaced quinine and many other antimalarial products for the treatment of severe malaria. However, several reports have demonstrated the emergence of resistance to the efficacy of artemisinin-based combination therapy monotherapy, such as in western Cambodia and other regions in South-East Asia. To face the phenomenon, artemisinin-based combination therapies are now recommended by the WHO. The aim is to reduce the morbidity and mortality associated with malaria with artemisinin-based combination therapies, including chloroquine plus other drugs, like sulfadoxine-pyrimethamine. Meanwhile, with increasing resistance to chloroquine, quinine is reconsidered, being so far the only substance for which *Plasmodium* did not develop resistance. The consequences are that in Uganda quinine was prescribed for up to 90% of children younger than 5 years with uncomplicated malaria, and from 2009, 31 African countries recommended quinine as a second-line treatment for uncomplicated malaria, 38 as a first-line treatment for severe malaria, and 32 for treatment of malaria in the first trimester of pregnancy. Recent surveillance data from other sites are in accordance. However, quinine was substituted due to its limits and therefore in 2010, WHO (2010) guidelines recommended reinforcing quinine's activity by combining it with other antimalarial agents, like doxycycline, tetracycline, or clindamycin as a second-line treatment for uncomplicated malaria (to be used when the first-line drug fails or is not available) or quinine plus clindamycin for treatment of malaria in uncomplicated cases and in the first trimester of pregnancy. The development of effective cocktails is a current trend of medical treatment of several diseases, including forms of cancer. In addition, the combination of natural products and synthetic drugs is recommended.

Natural products can be utilized as resistance-modifiers or chemosensitizers, and may be able to restore chloroquine sensitivity in resistant strains of *Plasmodium*. The idea of 8 years of research from different research groups was that the antimalarial treatment combined with natural products could be based on lower doses of chloroquine, in order to minimize the resistance insurgence and to avoid the collateral effects in the case of prolonged use, necessary in areas where the disease is endemic. This approach came from an ethnopharmacological investigation by Professor Rasoanaivo (Rasoanaivo et al., 1992). Most people consider ethnopharmacology to be a collection of ancient utilizations of natural sources, and as knowledge that is going to disappear. On the contrary, in addition to traditional uses there are new ones emerging, even as consequences of the utilization of modern drugs. Considering that OMS reports that 80% of the planet population relies on traditional medicine, the utilization of medicinal plants is not limited to ancient times and past populations, but it changes according to needs and evolution of treatments. Ethnobotanical knowledge is still passed from one generation to another in the majority of populations living in rural areas, and in urban areas, where malaria has been revealed to be resistant or incurable by modern scientific medicines, people have turned to traditional treatments. It is therefore of paramount importance to preserve and transmit this ethnobotanical heritage. Therefore, this discipline must be regarded as a multidisciplinary science in movement, where botany, chemistry, and pharmacology play central roles for scientific evaluation and validation of popular uses. However, economic and social aspects must also be considered, in order to develop new drugs and treatments of both old and new diseases. Most antimalarial drugs currently in use belong to the classes of aminoquinolines (chloroquine, amodiaquine, primaquine), quimolinomethanol derivatives (quinine, mefloquine, halofantrine), diaminopyrimidines (pyrimethamine), sulfonamides (sulfadoxine, sulfadiazine), biguanides (proguanil and derivatives), antibiotics (tetracyclines, doxycyclin, clindamycin), sesquiterpenes (artemisinin, dihydroartemisinin, arteether, artemether, artesunate), and naphtoquinones (atovaquone). Among them, only quinine and artemisinins are natural products, but also a relevant part of the current antimalarial arsenal. The potentiality of natural products is very high. A review by Willcox and Bodeker (2004) on traditional herbal medicines for malaria in three continents reported 1277 plant species from 160 families. However, the clinical trials are largely lacking, since only eight clinically controlled trials have been reported, involving *P. falciparum* and *P. vivax*.

In the case of malaria, alkaloids are the first candidates responsible for the activity. There is a long tradition in popular medicine of plants containing these compounds to control fever. These plants also have a bitter taste, which is usually connected to the alkaloid presence, as already reported for the aforementioned quinine bark case.

Two important considerations attracted our attention, in view of the possibility to explore new strategies: the special endemic flora of Madagascar and the occurrence of information about a popular treatment of malaria as yet unreported. Madagascar is a land of endemism, consisting about 13,000 species of vascular plants, of which 80% are endemic, and eight families totally endemic.

Malaria is practically endemic in all Madagascar and therefore the population harbors a very rich and unique knowledge on antimalarial plants. After a resurgence of malaria in the early 1980s, as a consequence of *Plasmodium falciparum* resistance and due to the high costs of conventional drugs, local populations returned to the uses of herbal remedies. Two hundred and thirty-nine plant species, of which about 30% are endemic to Madagascar, have been reported as having antimalarial uses in Malagasy traditional medicine. Prof. Rasoanaivo discovered the use by some populations in Madagascar of decoctions of some local plants in association with low doses of chloroquine to complement chloroquine action against chronic malaria (Nicoletti et al., 2018). The lower use, one or two tablets of chloroquine (100–200 mg), is probably adopted to avoid collateral effects due to prolonged use of chloroquine, but such a dose could be considered inadequate to favor chloroquine resistance. Therefore, we have a mixture of recent learning and ancient knowledge, evidencing the reality of ethnopharmacology. However, popular uses of medicinal plants need scientific validation with advanced tools. Therefore, research started from the knowledge that some populations in Madagascar use decoctions of some local plants in association with low doses of chloroquine to complement chloroquine action against chronic malaria. In such a way, resistance insurgence and collateral effects are both lowered.

On the basis of the ethnobotanical work conducted by Rasoanaivo and his collaborators, 24 plants were selected and investigated for in vitro and in vivo antimalarial activity and a chloroquine-potentiating effect. In the case of validation of the activity, the determination of the active constituents followed. The results of these selections were that the alkaloids of Loganiaceae, Menispermiaceae, and Rutaceae were the most promising compounds showing significant effects, some of them potentiating the action of chloroquine.

From a phytochemical point of view, alkaloids are in pole position among natural products utilized in traditional medicine against malaria. Mono- and bis-indole alkaloids have been isolated from several plants that are traditionally used to treat malaria on different continents. The most active compounds are those that originate from plants belonging to the genera *Strychnos* (Loganiaceae) and *Alstonia* (Apocynaceae). A review covering the indole alkaloids that have high antiplasmodial activities in vitro and in vivo, and favorable selectivity indices ($SI = CC_{50}/IC_{50}$), was published by Frederich et al. (2003).

In the case of malaria, alkaloids are the first candidates as being potentially responsible for the activity. There is a long tradition in popular medicine of plants containing these compounds to control fever. These plants also have a bitter taste, which it is usually connected to the alkaloid presence, as already reported for the aforementioned quinine bark case.

Two important considerations attracted our attention, in view of a possibility to explore new strategies: the special endemic flora of Madagascar and the occurrence of information about a popular treatment of malaria so far never reported. Madagascar is land of endemism, consisting about 13,000 species of vascular plants, whose 80% are endemic, and even 8 families totally endemic.

Malaria is practically endemic in all Madagascar and therefore the population harbours a very rich and unique knowledge on antimalarial plants. After resurgence of malaria in the early 80's, as a consequence of *Plasmodium falciparum* resistance and due to high costs of conventional drugs, local populations back to the use of herbal remedies (Blanchard, 1901; Maggi et al., 2017). Two hundreds and thirty-nine plant species, of which about 30% are endemic to Madagascar, have been reported as having antimalarial uses in the Malagasy traditional medicine. Prof. Rasoanaivo discovered the use by some populations in Madagascar of decoctions of some local plants in association with low doses of chloroquine to complement CQ action against chronic malaria (Benelli et al., 2018). The lower use, one or two tablets of chloroquine (100-200 mg), is probably adopted to avoid the collateral effects due to a prolonged utilization of chloroquine, but such dose could be presumed inadequate to favour chloroquine resistance. Therefore, we have a mixture of recent acquirement and ancient knowledge, evidencing the actuality of ethnopharmacology. However, popular uses medicinal plants need a scientific validation with advanced tools.

Therefore, the researches started from the knowledge that some populations in Madagascar use decoctions of some local plants in association with low doses of chloroquine to complement chloroquine action against chronic malaria. In such way, resistance insurgence is lowered, as well as collateral effects. On the basis of the ethnobotanical work conducted by Rasoanaivo and his collaborators 24 plants were selected and therefore investigated for in vitro and in vivo antimalarial activity and a chloroquine-potentiating effect (Maggi et al., 2017). In case of validation of the activity, the determination of the active constituents followed. The results of this selection were that the alkaloids of Loganiaceae, Menispermiaceae and Rutaceae were the most promising compounds showing significant effects, some of them potentiating the action of chloroquine.

Mono- and bis-indole alkaloids are traditionally used to treat malaria in different continents (Ramanitrahasimbola et al., 2001, 2006). The most active compounds were mainly related to the genera *Strychnos* (Loganiaceae) and *Alstonia* (Apocynaceae). A review covering the indole alkaloids that have high antiplasmodial activities in vitro and in vivo, and favourable selectivity indices ($SI = CC_{50}/IC_{50}$) was published by Frederich et al. (2003).

Strychnos is a pantropic genus, with about 200 species, present in three continents: 75 in Africa, 73 in America, and 44 in Asia and Oceania (only *S. potatorum* is present in both Asia and Africa). Asiatic species are mainly small trees, whereas in the New World lianes are generally dominant. The most famous *Strychnos* species is the Asiatic *S. nux-vomica*, because of strychnine contained in the seeds with 12 other related alkaloids. Strychnine is also known and used for its bitter taste. South American species are characterized by different mono and bisindole alkaloids, important as constituents of some curare preparations of Indios Amazonia tribes (see Introduction). During the preparation of curare, the tribe curandero selects local plants and extracts the mixture by hot water. Finally, the extract is filtered on leaves and concentrated to obtain a paste, which is preserved into a container, like a calebassa or a tube, maiden by a cane, or a pottery. Active constituents in curare are bis-indole alkaloids from bark of *Strychnos* ssp. and bis-tetrahydroisoquinoline alkaloids from Menispermaceae.

The genus *Strychnos* is represented in Madagascar by 14 species, of which five are endemic to the island. Among them, *S. diplotrocha* Leeuwenberg and *S. myrtoides* Gilg & Bussse are used as antimalarial in the northeastern part of the country (Rasoanaivo et al., 2004). The phytochemical analysis allowed the separation and the structural determination of several indole alkaloids, some already known and others never reported, including mixtures of epimers, which is very unusual in the same plant (Rasoanaivo et al., 1991, 1996, 2001). The in vitro and in vivo chloroquine-potentiating effect of the crude extract of dried and powdered stem barks of *S. myrtoides* exerted chloroquine-potentiating effects on *P. falciparum* FCM29, but it was devoid of intrinsic antimalarial activity. The extract was also devoid of cytotoxic effects on HeLa and L 929 fibroblast cells. The two compounds exhibit a closely related structure but different basicity. Therefore, the latter parameter can be excluded from the factors affecting the chloroquine-potentiating effect.

These results were confirmed by other experiments, demonstrating that the crude extract of *S. myrtoides* showed higher chloroquine-enhancing activity than its major bioactive constituents. These data support the use of the plant as a phytomedicine to treat malaria, but minor components of the extract may act synergistically. Among the main isolated alkaloids, malagashanine was very interesting. Malagashanine is an unusual indole alkaloid of the *Strychnos* type. Its pentacyclic structure contains seven consecutive stereogenic centers and, most important, a transfusion between the C and D rings, against all the other similar natural alkaloids.

Therefore, malagashanine is the parent compound of a new type of indole alkaloids (Fig. 7.20) (Kong et al., 2016), named $N_bC(21)$-secocuran, isolated so far from the Malagasy *Strychnos* species, which are traditionally used as chloroquine adjuvants in the treatment of chronic malaria

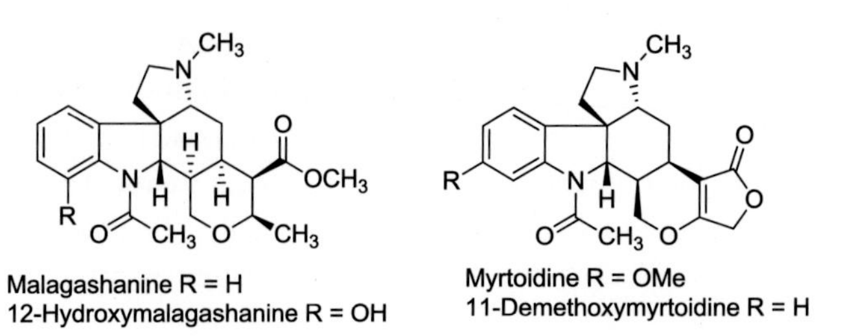

Fig. 7.20 Malagashanine and other related indole alkaloids from Malagasy *Strychnos.*

(Rasoanaivo et al., 1996a, 2001). Malagashanine showed only weak in vitro intrinsic antiplasmodial activity ($IC_{50} = 146.5 \pm 0.2\,\mu M$), but did display marked in vitro chloroquine-potentiating action against the FcM29 chloroquine-resistant strain of *Plasmodium falciparum*.

Another study allowed clarification of the mechanism of action of the major constituent, malagashanine, being able to prevent chloroquine efflux from the cell, and stimulates chloroquine uptake into drug-resistant *P. falciparum* strains. Malagashanine appears able to act more on plasma membrane than inside the parasite, allowing the toxicity of chloroquine against *Plasmodium*, even at sublethal doses. In the attempt to confirm the reversal of chloroquine resistance by the bark of *S. myrtoides*, a double-blind randomized controlled clinical trial of a standardized alkaloid extract titrated at 20% malagashanine took place in a government-run outpatient clinic in the town of Ankazobe (northwest central highlands of Madagascar), but the results of the treatment showed no significant efficacy, indicating a need for other confirmations. However, in conclusion, the approach, in accordance with recent tendencies on multidrug resistance control, based on mixtures of natural products and classic antimalarial drugs, with a relevant coincidence between the ethnobotanical reports and the scientific evidence, may offer interesting possible solutions for the treatment of malaria.

Many aspects about the mechanism of action of malagashanine as chloroquine adjuvant to reverse the resistance need further study. Malagashanine could increase drug accumulation by interacting with a dysregulated ion exchanger, avoiding the decrease inside the food vacuole, or acts by a mechanism related to drug binding to hematin (Perisco et al., 2017; Rafatro et al., 2000). In particular, in relation to the pH role in the blood red cell, it would be necessary to determine if malagashanine acts inside or outside the food vacuole, including the membrane periphery. The capacity of malagashanine to reverse CQ resistance may be related to the well-known properties of verapamil (Fig. 7.21) and related substances (Martin et al., 1987; Martiney

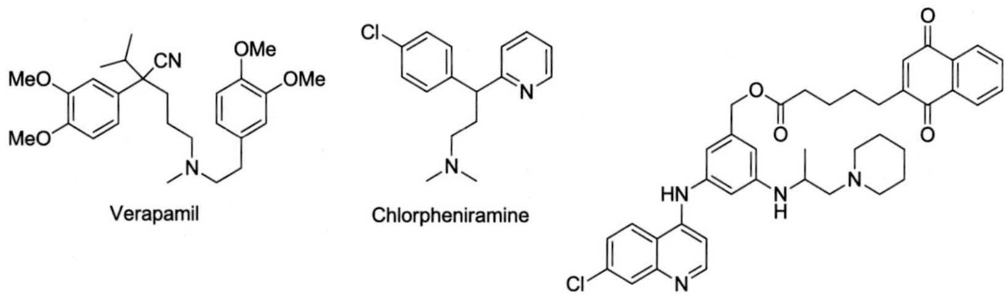

Fig. 7.21 Verapamil and other compounds studied for CQ-resistant reversal by membrane calcium channels blocking.

et al., 1995; Adovelande et al., 1998). Verapamil was the first calcium channel antagonist to be introduced into therapy in the early 1960s. It is a phenylalkylamine calcium channel blocker used in the treatment of high blood pressure, heart arrhythmias, and angina. In short-term incubations, verapamil was found to increase chloroquine accumulation in the lysosome of erythrocytes infected with both chloroquine-sensitive and -resistant organisms, but only to affect the chloroquine susceptibility of the latter. Verapamil works independently of the overall pH gradient concentrating CQ into a trophozoite's digestive vacuole. The activity is therefore related to the inhibition of membrane ion channels, interfering in the chloroquine transit within the parasite's cytoplasm. Other substances like chlorpheniramine and others are reported as candidates for CQ-resistant reversers. In any case, again the key role of natural products and ethnoparmacology information, such as for quinine (*Cinchona* sp.) and artemisinin (*Artemisia annua*), is fully confirmed.

Another attempt to explain the activity of Malagasy plants alkaloids explored the role of glutathione. L-Glutathione reduced (GSH) (Fig. 7.22) is a simple tripeptide, consisting of glutamic acid, cysteine, and glycine. It is considered one of the most powerful endogenous antioxidants, capable of preventing damage to cellular components caused by reactive forms of oxygen, radicals, and heavy metals, although its role in stress management and efficient defense against pathogens are still under study (Mangoyi et al., 2010). Besides its antioxidant defense and free radical scavenging, glutathione regenerates important antioxidants such as vitamins C and E. GSH exists in every cell of the human body, but it is also present in many other organisms, including fungi and bacteria.

There is a linkage between GSH and malaria. Some parasites are super-protected by GSH. They are endowed with powerful and host-independent mechanisms, which de novo synthesize or regenerate GSH and protect the parasites from oxidative damage and other outside attacks. GSH in particular protects the gametocytes against oxidative stress and inhibits the action of arginine, which produces NO and expels it from the food vacuole. At the trophozoite stage of *P. falciparum* in human erythrocytes, GSH takes

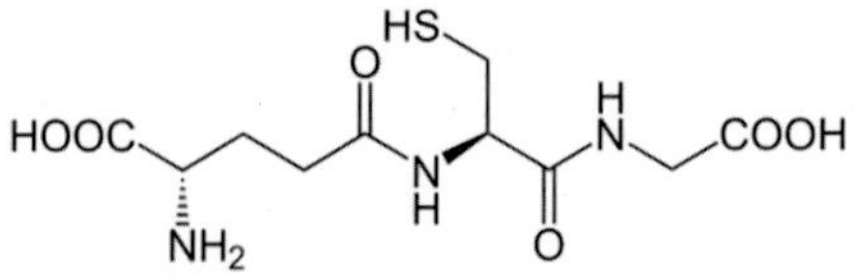

Fig. 7.22 Structure of glutathione.

part in detoxifying processes of heme, produced by hemoglobin digestion, by polymerizing some 30% of heme to insoluble hemozoin. Some authors suggest that the nonpolymerized heme, existing in the food vacuole, is subsequently degraded by GSH, increasing the role of this metabolite. Chloroquine could interact with GSH, competitively inhibiting the degradation of heme by GSH or allowing toxic heme to accumulate in membranes and damaging parasites. This argument merits some explanation. The pro-oxidant damage and inflammation process created by excessive heme, hemozoin, and fragments from rupture of the digestive vacuoles in blood vessels and plasma can be mitigated by glutathione. In other words, the inside of the infected erythrocyte glutathione is beneficial to the parasite; outside of the erythrocyte it reduces the negative effects of the malarial infection. High oxidative stress could actually be detrimental for the survival of young parasites (Gallo, 2009; Patzewitz and Müller, 2010).

Glutathione transferases (GSTs) are versatile enzymes involved in the intracellular detoxification of numerous substances. GSTs have been investigated in parasite protozoans, like those involved in malaria, with respect to their biochemistry and as targets in synthesis of new antiparasitic agents. *P. falciparum* possesses high quantity of these enzymes (PfGST) and their activity was found to be increased in chloroquine-resistant cells, and it has been shown to act as a ligand for parasitotoxic hemin. PfGST represents a promising target for antimalarial drug development. A PfGST isolated from *P. falciparum* has been associated with chloroquine resistance. Plant extracts have been found to act at different vulnerable metabolic sites of PfGST, disturbing GSH-dependent detoxification processes, increasing cytotoxic peroxides levels and possibly increasing the concentrations of toxic hemin in the parasites. In the case of *S. myrtoides* alkaloids, malagashanine was found to prevent chloroquine efflux from and stimulated chloroquine influx into drug resistant *P. falciparum*, suggesting that its effects are more on the plasma membrane than inside the parasite. Malagashanine (100 μM) reduced the activity of PfGST to 80%, but showed a time-dependent inactivation of PfGST, suggesting a role of malagashanine as a chemomodulator in cases of PfGST overexpression in chloroquine-resistant strains.

Future areas of research in malaria

The malaria cycle of a parasite is based on two cycles, one involving the host and the other affecting the vector. During the mosquito cycle, again there are metamorphoses and reproduction by the parasite.

In consideration of the resistance phenomenon, new transmission-blocking agents, able to interrupt malaria transmission, are required. These blocking drug components can be effective in reducing gametocyte density in the human host (gametocytocidal activity) or disrupting parasite development in the vector (sporontocidal activity), resulting in a reduced number of infective vectors and, as a consequence, decreased incidence of malaria cases. In other words, control malaria's parasites through the cure of the vectors infested by the disease.

In the sexual stages of *Plasmodium* parasites, gametocytes are critical for the transmission of the parasite to its vectors. *P. falciparum* gametocytes are also important in the disease diffusion, since being exceptionally long-lived, they cause clinically cured patients to be reservoirs of infection. The cycle of propagation of the malaria parasites starts when the female *Anopheles* feeds on blood from an infected vertebrate. Immediately, the first metamorphosis starts. By the ingestion, the mature male and female gametocytes, namely micro- and macrogametocytes, enter the mosquito host. Immediately after reaching the mosquito's midgut, the two types of gametocytes undergo dramatic metamorphoses. We must remember that such transformations are a response to environmental stimulation, like a decrease in temperature, an increase in pH, and an influence of xanthurenic acid. Within 10–20 min, the rounded macrogametes leave the erythrocytes and diffuse inside the blood, together with the flagellates microgametes. Now comes the last change. Within the next 24 h, the motile male gametes can fecundate the macrogametes, and round zygotes develop that mature to elongated motile ookinets and move to the outer midgut surface, completing early sporogonic development. These changes can be obtained by severe transformation inside the intrinsic cell organization, involving the cytoskeleton directly.

An equatorial position of chromosomes in the metaphase plate in the middle of the spindle is necessary for mitosis and symmetric cell divisions. A symmetric metaphase plate position is essential for symmetric cell divisions, explaining why it is conserved in all metazoans, plants, and many fungi. Control of this parameter is essential, since differences in cell size have been linked to cell fate and generate a class of anticancer drugs. Movements of chromosomes are in charge of microtubules, which are elements of the cytoskeleton. The cytoskeleton is a network of protein fibers forming the "infrastructure" of eukaryotic and prokaryotic cells. In eukaryotic cells, protein filaments and motor proteins form a complex mesh of protein filaments and motor proteins. The cytoskeleton aids the inside cell movement and transportation of subunits, like organelles and molecule groups, stabilizes and maintains cell shape, and gives support and order. The cytoskeleton is not a static structure but it is able to disassemble and reassemble its parts in order to enable internal and overall cell mobility. Intracellular movements include in particular manipulation of chromosomes during mitosis and meiosis from the equatorial plaque to the polar positions, in the formation of daughter cells, and also it is implicated in the immune cell response to pathogens. The cytoskeleton is composed of at least three different types of fibers: microtubules, microfilaments, and intermediate filaments. These fibers are distinguished by their size, with microtubules being the thickest and microfilaments being the thinnest. The assemblement of the proteins, tubulines a and b, makes microtubules, in form of long cave filaments. These hollow rods function primarily to help support and shape the cell and as "routes" along which organelles can move. Therefore, without the action of microtubules, the cell is unable to reproduce. The cell is blocked in a limb, with part of the mitosis already done and the final act in progress. The result is a polyploid cell, meaning a cell with double or more than the normal number of chromosomes. Because chromosomes cannot move alone, they must be dragged by the cytoskeleton. The mechanisms of action of several important antitumoral drugs derived from natural products are characterized by promotion of the assembly or disassembly of microtubules, meaning stabilization or destabilization of the tubules against depolymerization, resulting in mitotic arrest. Treated cells have defects in mitotic spindle assembly, chromosome segregation and movements, and consequently in cell division. The main problem of the utilization of these compounds in combination chemotherapy for sensitive tumor types concerns their selectivity against malignant cells. Cancer is basically a disease of uncontrolled cell division, including too-active mitosis, multiplying the cancerous mass. In most cases, these changes in activity are due to mutations in the genes that encode cell cycle regulator proteins. However, although cancer cells are a selected target, in consideration of their high level of mitosis, other tissues can be involved in the action of positive regulators of cell division. Molecular agents of plant origin are of primary importance in cancer treatment. Those acting on the cytoskeleton can be classified into two main groups: antimicrotubule agents like colchicine and the Vinca alkaloids, which induce depolymerization of microtubules, and taxol and taxotere, which induce tubulin polymerization and form extremely stable and nonfunctional microtubules (Rowinsky et al., 1990).

Neem products have been seriously explored in recent years, in several sectors, mainly in the fight against insect-borne diseases. However, it seems that so far the potentiality of neem has been only lightly touched on. NeemAzal is a marketed neem product consisting of a quantified alcoholic extract obtained from *Azadirachta indica* seeds, with a reported limonoid concentration of 57.7%, consisting of azadirachtin A 34%, azadirachtins B–K 17.7%, salanins 4%, and nimbins 2% (Dembo et al., 2015; Habluetzel et al., 2007). NeemAzal completely blocks transmission of the rodent malaria parasite *P. berghei* to *Anopheles stephensii* in vivo, when administered to gametocytemic mice at a corresponding azadirachtin A dose of 50 mg/kg. Other in vivo transmission blocking studies suggested that NA may have stronger transmission blocking activity than azadirachtin A alone, evaluating the activity of nonazadirachtin A constituents of NeemAzal. In an ex vivo assay, which exploits a major target process of azadirachtin A against *P. berghei*, microgamete formation inhibition of *Plasmodium* was used to estimate the pharmacodynamics of two varying doses of NeemAzal and azadirachtin A.

A team led by Prof. G. Chianese (University of Salerno, Italy) explored the possibility of influencing *Plasmodium* gametocytes by neem products, demonstrating the potential of blocking the reproduction stages of the parasite. NeemAzal is a marketed neem product consisting in a quantified alcoholic extract obtained from *Azadirachta indica* seeds, with a reported limonoid concentration of 57.7%, consisting in azadirachtin A 34%, azadirachtins B–K 17.7%, salanins 4%, and nimbins 2%. NeemAzal completely blocks transmission of the rodent malaria parasite *P. berghei* to *Anopheles stephensii* in vivo, when administered to gametocytemic mice at a corresponding azadirachtin A dose of 50 mg/kg. Other in vivo transmission blocking studies suggested that NA may have stronger transmission blocking activity than azadirachtin A alone, evaluating the activity of nonazadirachtin A constituents of NeemAzal.

Azadirachtins exert relevant effects on microtubules assembly and organization, interfering with the expression and/or function of adhesive proteins during the genesis of microgametocytes, through disruption of the organization of mitotic spindles and cytoskeleton formation and activity. These molecules can interfere with cytoplasmic microtubule organization and distribution, causing severe depletion of actin levels. In this action, NeemAzal proved to be more effective than azadirachtin A. In confirmation, another study showed that the product completely inhibits the growth of *P. falciparum* field isolates in *An. coluzzii* mosquitoes at a dose of 70 ppm in direct membrane feeding assays.

Microorganisms have not finished producing surprises and breaking the boundaries reported in books. Meanwhile researchers are investigating malaria parasites more and more deeply in search of their weak points, but their study is complicated by the parasite's metamorphosis, which involves not only the shape but also fundamental aspects of the metabolism (Becker and Kirk, 2004). Asexual stages of the parasite contain a single mitochondrion, whereas gametocytes can have several mitochondria. The energy production is very important. *Plasmodium falciparum*, as well as other similar Apocomplexa protozoans, possesses an intriguing nonphotosynthetic plastic, discovered in the 1970s. The surprise was that apicoplastides possess their own nucleic acid. Regarding their role, they were considered by Kilejian (1991) as "a source of some substrate essential for energy production of mitochondrion." In view of their other characteristics, they could be considered a possible bridge between organisms or the ancestral point of divergence from green algae and protozoans. In conclusion, apicoplastides could be part of the endosymbiosis pathway, wherein degenerated chloroplasts were useful to increase a mitochondrial efficiency still in evolution. Thus, endosymbiosis started with the inclusion of the two main bacterial forms, the hetero- and the autotrophic one, but later the ancestral (green or red) primordial alga degenerated the chloroplast in favor of a clear evolution toward the heterotrophic metabolism (Fig. 7.23).

Usually the shift to the eukaryotic cell is considered a consequence of environmental factors, such as the increase of oxygen in the oxidative atmosphere; however, it is possible that in some cases interactions between organisms could also have played an important role.

The study on apicoplastides allowed researchers to evidence similarities (Keeling, 2008; Kilejian, 1975; Köhler et al., 1997) between different arthropod-borne diseases, such as avian malaria, eimoriosis, and toxoplasmosis, confirming once more the occurrence of common survival strategies in different organisms. Other differences concern the enzymes network and the membrane transport mechanisms. The new knowledge about parasite-specific organelles could be of fundamental importance to the development of future antimalarial drugs, increasing efficiency and decreasing side effects, like resistance.

Another important research front full of possibilities is focused on the membrane mechanism of CQ's extrusion by permeability pathways induced by the parasite in the host red blood cells (Saliba et al., 1998). This is related to the CQ of interfering with the detoxification of toxic heme monomers.

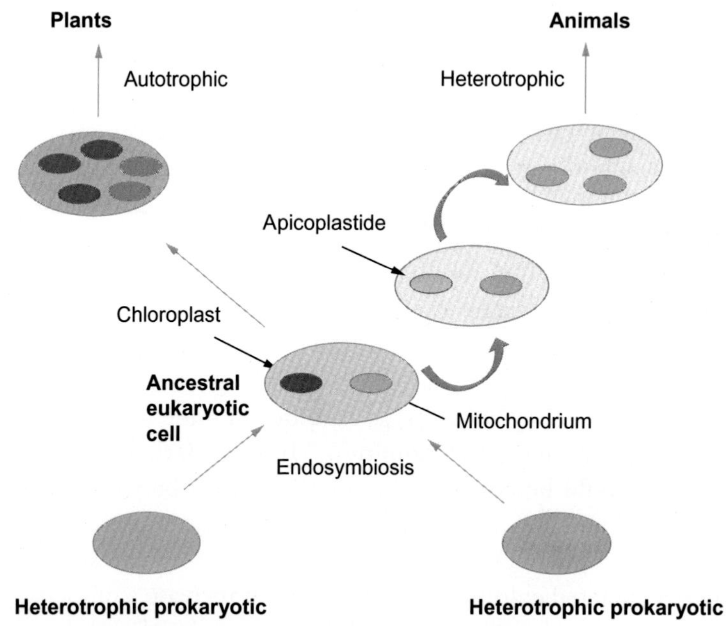

Fig. 7.23 The possible role of apicoplastide in the endosymbiosisi.

The studies showed that 12–16 h after the invasion by the parasite, the so-called new permeability pathways act on the interchanges, i.e., the entry of nutrients, as well as mediating the efflux of metabolic wastes. Several groups advanced the hypothesis of a number of channel types, activated by particular stress or stimuli (Ginsburg and Stein, 2004; Kirk et al., 1994; Duranton et al., 2004; Staines et al., 2004; Thomas and Lew, 2004).

All these references testify to and confirm the presence of a wide range of studies in search of an answer to the challenge of resistance. The front is still too large and undetermined, but every year the knowledge of host cell reaction is increasing and there is a high probability that the problem will be solved in the coming years.

Conclusions

During the development of the arguments contained in this book, it was necessary link the insect-borne diseases argument to several collateral items. The idea in particular was concentrated on a possible utilization of this particular topic as an epiphany, meaning an enlightening subject, which allows a revision of the problem from a new perspective. The interpretation of a new and key piece of information can allow the process of significant thought about a problem, until, in accordance with the original significance of the term in ancient Greek, the ἐπιφάνεια (*epiphanea*) appears like a manifestation, with a striking appearance.

This book started with considerations about gaps and books. Let us now return to these two points.

The lesson from Carson's book about the fundamental role of beneficial insects in the survival of mankind, has arguably not been understood. Throughout all warm terrestrial ecosystems, insects are a dominant component and they are part of the lives of any organism. The insect-plant relationship is a fundamental biotic interaction, and plants account for a large part of the planet's biomass, many times the biomass of all animals together (New, 2002; Jankielson, 2018; Dunn, 2005). The animal biodiversity is dominated by that of insects. They are a beautiful example of variability, in terms of both number of species (more than 1 million) and abundance (more than half of all living organisms), although at most only about 7%–10% of insects are scientifically described. This diversity, consisting of large numbers of individuals and great intra- and interspecific variety, is a consequence of the enormous functional significance of insects in habitats. Primitive insects appeared very early in the Silurian period, when plants and animals finally emerged from the sea and colonized dry land, and over the last 400 million years the number of insect families has been rising. They were able to colonize any part of the territory, including the sky. Today, the number of reported insect families is about 600 and they have survived various negative major impacts, including the mass extinction event at the end of the Cretaceous period. A review analysis, published in the journal *Biological Conservation* by Francisco Sánchez-Bayo, at the University of Sydney, Australia, and Kris Wyckhuys, at the China Academy of Agricultural Sciences, Beijing China, attests a current insect collapse. The decline's hypothesis is based on a study of 73 recent selected studies. The causes and significant factors include intensive agriculture, the heavy use of pesticides, urbanization, and climate change. The loss of insect population is calculated in an annual 2.5% rate over the last 20–25 years, and the future tendency is evaluated to 25% in the next 10 years and increasing continuously until only half left in 100 years. This scenario is already underway. In Puerto Rico, a recent study revealed a 98% fall in ground insects over 35 years. The catastrophic cascade effects on the planet's ecosystems include ants, aphids, shield bugs, and crickets, which are the food for many birds, reptiles, amphibians, and fish that eat insects. There are many indicators supporting the scenario (Sánchez-Bayo and

Wyckhuys, 2019; Diamond, 1989). In England, between 2000 and 2009, the number of widespread butterfly species fell by 58% on farmed land, suffering the biggest recorded insect falls overall—though that is probably a result of this area being more intensely studied than most places. A particular alert concerning bees being seriously affected has also been raised in Europe and the USA; for example, only half of the bumblebee species found in Oklahoma in the USA in 1949 were present in 2013 (Alburaki et al., 2015, 2018; Aizen, 2009). The number of honeybee colonies in the USA was 6 million in 1947, but 3.5 million have since been lost. In 2013, according to EU data, there were around 630,000 beekeepers and 16 million hives in the EU, producing 234,000 tons of honey per year, but the same source tells us that many insect pollinator populations are now in clear decline. There is similar news from Brazil, with half a million bees dead. On one side, this is considered the effect of the use of some pesticides, toxic to bees. On the other side, it is a classic example of rapid and intense environmental change to improve agricultural intensification and pasture, with the systematic elimination of all trees and shrubs that normally surround the fields, so there are plain, bare fields that are treated with synthetic fertilizers and pesticides.

Dr. Sanchez-Bayo said: "We are not alarmists, we are realists. We are experiencing the sixth mass extinction on Earth. If we destroy the basis of the ecosystem, which are the insects, then we destroy all the other animals that rely on them for a food source." He added, "It will collapse altogether and that's why we think it's not dramatic, it's a reality."

The situation comprehends micro- and macroepisodes, like the continuous devastation of equatorial tropical forests, in particular the Amazonia territory. The sequence is clear and well-known, and it always works: first, the fire destroys the vegetation, in particular the woody plants; second, the soil is cleaned, otherwise the plants could replace the habitat rapidly; and third, the territory is declared totally compromised and ready for further utilizations. However, as observed by Samways in *Biodiversity and Conservation* (1993, and later confirmed by this author in a series of further papers) in a paper titled "Insects in Biodiversity Conservation: Some Perspectives and Directives," the main concerns are the "Lack of human appreciation of importance, coupled with the general disregard and dislike of insects, is an enormous perception impediment to their conservation. This impediment coupled with the taxonomic impediment must be overcome for realistic biodiversity conservation management. As it is not possible to know all the species relative to the rate at which they are becoming extinct, it is essential to conserve as many biotopes and landscapes as possible." There is a sentiment of urgency for measures "essential to preserve species dynamo areas as an insurance for future biodiversity," such that "preserved areas must also be linked by movement and gene-flow corridors as much as possible." The last point of view is crucial. Preservation must be considered not only as an opportunity to maintain the presence of species in selected habitats against their disappearance, but it must be considered changes as opportunities to perform a positive future. In this regard, entomologists are asked to contribute in control of vectors affecting humans, crops, and livestock, but also to take an active part in the consideration due to the beneficial species.

The central task is the possibility to predict accurately the environmental effects of any intervention. Once the inherent risks connected with traditional control methods have been considered, the consequences of new introductions must be carefully predicted, including any synergist effect. The rate of insect species extinction is estimated as being eight times faster than those of mammals, birds, and reptiles (Barnosky et al., 2011; Dirzo et al., 2014).

Another important current gap concerns scientific information. Most ordinary people do not have access to data obtained by the scientific community, as well as opinions and models produced by experts and scientists. Information, when available, is usually distorted and adapted to the dominant axioms by a plethora of generalist supposed experts. The proposed idea is that these kinds of people are able to know and comment on everything. The distortion, sometimes voluntarily pursued and often a consequence of general confusion, generates progressive modification of the starting points and even the concealment of important facts. The recent phenomenon of fake news is clearly generated from the same situation. Although most research information is now easily accessible and can be obtained directly from the internet, its utilization remains restricted to dedicated people. In contrast, some scientific information is amplified far away from its real impact. How many times did you read about the discovery of a definitive cure to cancer? Or about the already obtained solution to any physiological problem using staminal cells? In our era of globalized knowledge, news are obtained and fluxed indirectly, without few possibility of checking the origin and the reliability. It is necessary to consider that more than 46% of the human population, consisting of 3.5 billion people, are connected via the internet, and 2.5 billion utilize social networks regularly. These numbers are likely to increase 10% every year. All these people have access to information only through selected channels and although they are in a condition to verify it, science and general information are on different and distant levels. The main problem is that the information is reduced to a few soundbites, and there is no place for elaboration or proposals of other possible interpretations or points of view. This is not a recent case, produced by digitalization of communication.

Beside the sources, the problem of the quality of scientific information was fully evidenced more than 30 years ago, in the "Public Understanding of Science." This is the title of a report requested in 1985 by the Royal Society and prepared by a group of experts, whose leader was the geneticist Sir Walter Fred Bodmer. The report evidenced the general lack of knowledge about scientific themes. On one side, most of the population, accounting for two-thirds of Europeans, was confident about science and technologies, considering that scientists were able to solve human problems and make human life "easier, healthier and comfortable." On the other side, the sequence "more communication = more knowledge = more social adherence to scientific arguments" appears largely inadequate. The dominant problem about scientific communication is that ordinary people need an alphabetization to understand and meet the complexity of the scientific items. The conversion of the original scientific information is usually distorted and changed, at best "adapted," but more often polluted by political, social, and cultural interests. The result is a reductive metamorphosis, in the best case, or complete revision to be adapted and useful to already-made opinions. Among the various examples of this operation we find never-ending debates, such as those concerning OGM, vaccines, or the consequences of climate changes, without considering abnormal and artificially created themes, such as the contraposition between vegans and meat-eaters. The manipulation is based on a presumed "democratic" interpretation of scientific data. No vote is necessary to assure the consistency of a scientific law based on adequate experimentation, but the aim is that reliability must be obtained by public consensus and even agreement. Independence has always been a necessary character of science, but manipulation was never pursued. History tells us that any political or social manipulation of science led to disaster. In contrast, priorities, when based on correct scientific information, as well as consequent implications and decisions, must be subject to the most ample democracy.

At the end of this little journey through macro-, micro-, and nanoworlds, it is undeniable how long the road still is to understand and discover the mysteries of insect-borne diseases. In the meantime, we await the next surprises. The COVID-19 pandemy dramatically evidenced all the current limits of science and technology to face this kind of challenges. The virus was faster and clever. Predictively and prevention were insufficient. Despite the potentiality, the debacle and medicine was evident and the consequent economic and social damages were enormous. Microorganisms will continue to play their role inside the habitats and next time their target could be the industrialized sources of our food. However, it is clear that is society will continue to ignore the alerts of researchers and scientists, the next pandemy will be the worst one.

References

Adovelande, J., Delèze, J., Schrével, J., 1998. Synergy between two calcium channel blockers, verapamil and fantofarone (SR33557), in reversing chloroquine resistance in Plasmodium falciparum. Biochem. Pharmacol. 55 (4), 433–440.

Aguilera, M., et al., 2003. Active and intelligent packaging: an introduction. In: Ahvenainen, R. (Ed.), Novel Food Packaging Techniques. Woodhead Publishing Ltd, Cambridge, UK, pp. 5–21.

Aizen, M.A., 2009. How much does agriculture depend on pollinators? Lessons from long-term trends in crop production. Ann. Bot. 103, 1579–1588.

Alburaki, M., et al., 2015. Neonicotinoid-coated *Zea mays* seeds indirectly affect honeybee performance and pathogen susceptibility in field trials. PLoS One. 10, e0125790.

Alburaki, M., et al., 2018. Honey bee survival and pathogen prevalence: from the perspective of landscape and exposure to pesticides. Insects 9, 65.

Amirthalingam, M., 1998. Sacred Trees of Tamilnadu. C.P.R. Environmental Education Centre, Chennai.

Anbu, P., et al., 2017. Green-synthesised nanoparticles from Melia azedarach seeds and the cyclopoid crustacean Cyclops vernalis: an eco-friendly route to control the malaria vector Anopheles stephensi? Nat. Prod. Res. 30 (18), 2077–2084.

Ansari, M.A., Razdan, R.K., 1996. Operational feasibility of malaria control by burning neem oil in kerosene lamp in Beel Akbarpur village, District Ghaziabad, India. Indian J. Malariol. 33, 81–87.

Appendini, P., Hotchkiss, J.H., 2002. Review of antimicrobial food packaging. Innov. Food Sci. Emerg. Technol. 3, 113–121.

Bagavan, A., Rahuman, A.A., 2010. Evaluation of larvicidal activity of medicinal plant extracts against three mosquito vectors. Asian Pac J Trop Med 8, 29–34.

Barnosky, A.D., et al., 2011. Has the Earth's sixth mass extinction already arrived? Nature 471, 51–57.

Baswa, M., Rath, C.C., Dash, S.K., Mishra, R.K., 2001. Antibacterial activity of karanj (*Pongamia pinnata*) and neem (*Azadirachta indica*) seed oil: a preliminary report. Microbios 105, 183–189.

Becker, K., Kirk, K., 2004. Of Malaria, metabolism and membrane transport. Trends Parasitol. 20 (12), 590–596.

Benelli, G., Madhiyazhagan, P., Conti, B., Nicoletti, M., 2015a. Old ingredients for a new recipe? Neem cake, a low-cost botanical by-product in the fight against mosquito-borne diseases. Parasitol. Res. 114 (2), 391–397.

Benelli, G., Bedini, S., Cosci, F., Toniolo, C., Conti, B., Nicoletti, M., 2015b. Larvicidal and ovideterrent properties of neem oil and fractions against the filariasis vector Aedes albopictus (Diptera: Culicidae): a bioactivity survey across production sites. Parasitol. Res. 114 (1), 227–236.

Benelli, G., Conti, B., Garreffa, R., Nicoletti, M., 2015c. Shedding light on bioactivity of botanical by-products: neem cake compounds deter oviposition of the arbovirus vector Aedes albopictus (Diptera: Culicidae) in the field. Parasitol. Res. 113 (3), 933–940.
Benelli, G., et al., 2016. Neem (*Azadirachta indica*): towards the ideal insecticide? Nat. Prod. Res. 31 (4), 369–386.
Benelli, G., et al., 2017a. Synergized mixtures of Apiaceae essential oils and related plant-borne compounds: larvicidal effectiveness on the filariasis vector Culex quinquefasciatus Say. Ind. Crop. Prod. 96, 186–195.
Benelli, G., et al., 2017b. Chemical composition and insecticidal activity of the essential oil from *Helichrysum faradifani* endemic to Madagascar. Nat. Prod. Res. 32 (14), 1690–1698.
Benelli, G., Maggi, F., Nicoletti, M., 2018. Ethnopharmacology in the fight against Plasmodium parasites and brain disorders: in memoriam of Philippe Rasoanaivo. J. Ethnopharmacol. 193, 726–728.
Benelli, G., et al., 2020. Insecticidal and mosquito repellent efficacy of the essential oils from stem bark and wood of *Hazomalania voyronii*. J. Ethnopharmacol. 248, 112333.
Bhowmik, D., Chiranjib, J., Yadav, K.K., Tripathi, S., Sampath, K.P., Kumar, S., 2010. Herbal remedies of *Azadirachta indica* and its medicinal application. J. Chem. Pharm. Res. 2, 62–72.
Biswas, K., Chattopadhyay, I., Banerjee, R.K., Bandyopadhyay, U., 2002. Biological activities and medicinal properties of neem (Azadirachta indica). Curr. Sci. 82, 1336–1345.
Blanchard, R., 1901. Le paludisme à Madagascar. Rev. Madagascar 3, 233–243.
Boa, E.R., 1995. A Guide to the Identification of Diseases and Pests of Neem (*Azadirachta indica*). RAP Publ. 1995/41, FAO, Bangkok.
Brahmachari, G., 2004. Neem-an omnipotent plant: a retrospection. Chembiochem 5 (4), 408–421.
Brockgreitens, J., Abbas, A., 2016. Responsive food packaging: recent progress and technological prospects. Compr. Rev. Food Sci. Food Saf. 5, 3–115.
Brown, A.W.A., 1986. Insecticide resistance in mosquitoes: a pragmatic review. J. Am. Mosq. Control Assoc. 2, 123–140.
Cantrel, C.L., Dayan, F.E., Duke, S.O., 2012. Natural products as sources for new pesticides. J. Nat. Prod. 75 (6), 1231–1242.
Chandramohan, B., Murugan, K., Panneerselvam, C., Madhiyazhagan, P., Nicoletti, M., 2016. Emergency and mosquitocidal potential of neem cake-synthesized silver nanoparticles: genotoxicity and impact on predation efficiency of mosquito natural enemies. Parasitol. Res. 115 (3), 1015–1025.
Chava, V.R., Manjunath, S.M., Rajanikanth, A.V., Sridevi, N., 2012. The efficacy of neem extract on four microorganisms responsible for causing dental caries viz *Streptococcus mutans*, *Streptococcus salivarius*, *Streptococcus mitis* and *Streptococcus sanguis*: an *in vitro* study. J. Contemp. Dent. Pract. 13, 769–772.
Chopra, I.C., Gupta, K.C., Nair, B.N., 1952. Biological activities and medicinal properties of neem (*Azadirachta indica*). Indian J. Med. Res. 40, 511–515.
Citti, C., 2019. A novel phytocannabinoid isolated from *Cannabis sativa* L. with an *in vivo* cannabimimetic activity higher than Δ^9-tetrahydrocannabinol: Δ^9-Tetrahydrocannabiphorol. Sci. Rep. 9, 20335.
Coma, V., 2008. Bioactive packaging technologies for extended shelf life of meat-based products. Meat Sci. 78, 90–103.
Cooksey, K., 2005. Effectiveness of antimicrobial food packaging materials. Food Addit. Contam. 22, 980–987.
Dahiya, N., et al., 2016. In vitro and ex vivo activity of an Azadirachta indica A. Juss seed kernel extract on early sporogonic development of Plasmidium in comparison with azadirachtin A, its most abundant constituent. Phytomedicine 23, 1743–1752.

De Kruijf, N., et al., 2002. Active and intelligent packaging: applications and regulatory aspects. Food Addit. Contam. 19, 144–162.
De Matteis, G., Domenico, R., Claps, S., Veneziano, V., Di Sotto, A., Nicoletti, M., Del Serrone, P., 2015. Assessment of neem oil effect on haematological profile and towards peripheral blood mononuclear cells of goat. Adv. Appl. Sci. Res. 6 (2), 46–54.
Del Serrone, P., Nicoletti, M., 2013. Antimicrobial activity of a neem cake extract in a broth model meat system. Int. J. Environ. Res. Public Health 10, 3282–3295.
Del Serrone, P., Nicoletti, M., 2014. Evaluation of a mono-component and a multi-component herbal extracts as candidates for antimicrobial packaging of fresh retail meat. In: Cimmino, S., Pezzuto, M., Silvestre, C. (Eds.), Proceedings: Eco-sustainable Food Intelligent and Smart Packaging. Packaging Based on Polymer nanomaterials. International Conference COST ACTION FA0904, 26–28 February. CNR Rome, Italy, p. 36.
Del Serrone, P., et al., 2006. Assessment of microbiological quality of retail fresh pork meat in central Italy. Ital. J. Food. Sci. 18, 397–407.
Del Serrone, P., Toniolo, C., Nicoletti, M., 2015. Neem (Azadirachta indica A. Juss) oil to tackle enteropathogenic Escherichia coli. Biomed. Res. Int. 343610. https://doi.org/10.1155/2015/343610.
Dembo, E., et al., 2015. Impact of repeated NeemAzal® treated blood meals on the fitness of Anopheles stephensi mosquitoes. Parasit. Vectors 8, 84.
Diamond, J.M., 1989. The present, past and future of human-caused extinctions. Philos. Trans. R. Soc. Lond. Ser. B Biol. Sci. 325, 469–477.
Dirzo, R., et al., 2014. Defaunation in the Anthropocene. Science 345, 401–406.
Dobrucka, R., Cierpiszewski, R., 2014. Active and intelligent packaging food—research and development—a review. Pol. J. Food Nutr. Sci. 64 (1), 7–15.
Dunn, R.R., 2005. Modern insect extinctions, the neglected majority. Conserv. Biol. 19, 1030–1036.
Duranton, C., et al., 2004. Organic osmolyte permeabilities of the malaria-induced anion conductances in human erythrocytes. J. Gen. Physiol. 123, 417–426. 46.
Elder, R.J., Smith, D., Bell, K.L., 1998. Successful parasitoid control of Aonidiella orientalis (Newstead) (Hemiptera: Diaspididae) on Carica papaya L. Aust. J. Entomol. 37, 74–79.
EPA (US Environmental Protection Agency), 2012. Biopesticide Registration Action Document. Office of Pesticide Programs. Cold Pressed Neem Oil. PC Code 025006. Margosa extract PT-18. Assessment Report. Standing Committee on Biocidal Products. CIRCABC Europe. 9/12/2011.
Forin, M.R., das Gracas Fernandes da Silva, M.F., da Silva, F., 2011. Secondary metabolism as a measurement of the efficacy of botanical extracts: the use of Azadirachta indica (Neem) as a model insecticide. In: Perveen, F. (Ed.), Advances in Integrated Pest Management, ISBN: 978-953-307-780-2, pp. 367–390.
Foster, P., Moser, G., 2000. Status Report on Global Neem Usage. Universun Verlagsalt, Weisbadan, Germany.
Frederich, M., Tits, M., Angenot, L., 2003. Indole alkaloids from *Strychnos* species and their antiplasmodial and cytotoxic activities. Chem. Nat. Compd. 39, 513519.
Gallo, V., 2009. Inherited glutathione reductase deficiency and Plasmodium falciparum malaria—a case study. PLoS One. 4(10), e7303.
Ghosh, A., Chowdhury, N., Chandra, G., 2012. Plant extracts as potential mosquito larvicides. Indian J. Med. Res. 135, 581–598.
Gibbons, S., 2008. Phytochemicals for bacterial resistance-strengths, weaknesses and opportunities. Planta Med. 74, 594–602.
Ginsburg, H., Stein, W.D., 2004. The new permeability pathways induced by the malaria parasite in the membrane of the infected erythrocyte: comparison of results using different experimental techniques. J. Membr. Biol. 197, 113–134. 48.
Girish, K., Bhat, S.S., 2008. Neem—a green treasure. J. Biol. 4, 102–111.

Govindachari, T.R., 1992. Chemical and biological investigations on *Azadirachta indica* (the neem tree). Curr. Sci. 63, 117–122.
Gupta, S.M., 2001. Plants Myths and Traditions in India. Munshi Manoharlal Publishers, New Delhi.
Habluetzel, A., et al., 2007. Impact of the botanical insecticide Neem Azal® on survival and reproduction of the biting louse *Damalinia limbata* on angora goats. Vet. Parasitol. 144, 328–337.
HLPE, 2014. Food Losses and Waste in the Context of Sustainable Food Systems. In: A Report by the High Level Panel of Experts on Food Security and Nutrition of the Committee on World Food Security. HLPE, Rome.
Isman, M.B., 1997. Neem pesticides. Pestic. Outlook 8, 32–38.
Jankielson, A., 2018. The importance of insect in agricultural ecosystems. Business. https://doi.org/10.4236/ae.2018.62006.
Jones, P.S., Ley, P., Morgan, E.D., Santafianos, D., 1989. Focus on Phytochemical Pesticides. The Neem Tree. CRC Press, Boca Ranton, FL, pp. 19–45.
Kader, A.A., 2005. Increasing food availability by reducing postharvest losses of fresh produce. Acta Hortic. 682, 2169–2175.
Karlson, P., 1956. Chemische Untersuchungen über die Metamorphosehormone der Insekten. Ann. Sci. Nat. Zool. Biol. Anim 18, 125–137. séries 11.
Kaushik, H.S., Lakshmi, M., Muralitharan, R., Hegde, A.K., 2014. NeeMDB: convenient database for neem secondary metabolites. Bioinformation 10 (5), 314–315.
Keeling, P.J., 2008. Bridge over troublesome plastids. Nature 451, 896–897.
Kerry, J.P., O'Grady, M.N., Hogan, S.A., 2006. Past, current and potential utilization of active and intelligent packaging systems for meat and muscle-based products: a review. Meat Sci. 74, 113–130.
Kilejian, A., 1975. Circular mitochondrial DNA from the avian malarial parasite Plasmodium lophurae. Biochim. Biophys. Acta 390, 276–284.
Kilejian, A., 1991. Spherical bodies. Parasitol. Today 7, 309.
Kirk, K., et al., 1994. Transport of diverse substrates into malariainfected erythrocytes via a pathway showing functional characteristics of a chloride channel. J. Biol. Chem. 269, 3339–3347.
Köhler, S., Delwiche, C.F., et al., 1997. A plastid of probable green algal origin in apicomplexan parasites. Science 275, 1485–1488.
Kong, A., et al., 2016. Malagashanine: a chloroquine potentiating indole alkaloid with unusual stereochemistry. Chem. Sci. 8 (1), 697–700.
Koul, O., Wahab, S. (Eds.), 2007. Neem: Today and in the New Millennium. Kluwer Academic Publishers, New York.
Kumar, V.S., Navaratnam, V., 2000. Neem (*Azadirachta indica*): prehistory to contemporary medicinal uses to humankind. Asian Pac. J. Trop. Biomed. 3, 505–514.
Lale, N., 1998. Neem in the conventional Lake Chad Basin area and the threat of oriental yellow scale insect (Aonidiella orientalis Newstead) (Homoptera: Diaspididae). J. Arid Environ. 40 (2), 191–197.
Lee, S.Y., Lee, S.J., Choi, D.S., Hur, S.J., 2015. Current topics in active and intelligent food packaging for preservation of fresh foods. J. Sci. Food Agric. 95 (14), 2799–2810.
Maggi, F., Petrelli, R., Canale, A., Nicoletti, M., Rakotosaona, R., Rasoanaivo, P., 2017. Not ordinary antimalarial drugs: Madagascar plant decoctions potentiating the chloroquine action against Plasmodium parasites. Ind. Crop. Prod. 103, 19–38.
Mangoyi, R., et al., 2010. Glutathione transferase from Plasmodium falciparum—interaction with malagashanine and selected plant natural products. J. Enzyme Inhib. Med. Chem. 25 (6), 854–862.
Malhotra, B., Keshwani, A., Kharkwa, H., 2015. Antimicrobial food packaging: potential and pitfalls. Front. Microbiol. 6, 611.

Manners, G.D., 2007. Citrus limonoids: analysis, bioactivity, and biomedical prospects. J. Agric. Food Chem. 55 (21), 8285–8294.

Mariani, S., Nicoletti, M., 2013. Antilarval activity of neem cake extracts against *Aedes* albopictus. Pharmacology 3, 137–140.

Mariani, S., Nicoletti, M., Serafini, M., 2013. Composizione biologica con proprietà fortemente biocide a basso contenuto di azadiractina e procedimento per la sua realizzazione. Patent No RM2013A000342 del 14.06.

Marino, G., Gaggia, F., Baffoni, L., Nicoletti, M., 2014. Antimicrobial activity of *Melia azadirachta* fruit extracts for control of bacteria in inoculated in vitro shoots of 'MRS-2/5' plum hybrid and calla lily and extract influence on the shoot cultures. Eur. J. Plant Physiol. 141 (3), 505–521.

Martin, S.K., Oduola, A.M., Milhous, W.K., 1987. Reversal of chloroquine resistance in Plasmodium falciparum by verapamil. Science 235 (4791), 899–901.

Martiney, J.A., Cerami, A., Slater, A.F., 1995. Verapamil reversal of chloroquine resistance in the malaria parasite Plasmodium falciparum is specific for resistant parasites and independent of the weak base effect. J. Biol. Chem. 270 (38), 22393–22398.

Maruchecka, A., Greis, N., Menac, C., Cai, L., 2011. Product safety and security in the global supply chain: issues, challenges and research opportunities. J. Oper. Manag. 29, 707–720.

Mossini, S.A.G., Arrotéia, C.C., Kemmelmeier, C., 2009. Effect of neem leaf extract and on *Penicillum* growth, sporulation, morphology and ochratoxin A production. Toxins 1, 3–13.

Murugan, K., et al., 2015. Mosquitocidal and antiplasmodial activity of Senna occidentalis (Cassiae) and Ocimum basilicum (Lamiaceae) from Maruthamalai hills against Anopheles stephensi and Plasmodium falciparum. Parasitol. Res. 114 (10), 3657–3664.

Murugan, K., et al., 2016. In vivo and in vitro effectiveness of Azadirachta indica-synthesized silver nanocrystals against Plasmodium berghei and Plasmodium falciparum, and their potential against Malaria Mosquitoes. Res. Vet. Sci. 106, 14–22.

NAS, 1992. Neem, A Tree for Solving Global Problems. National Academy of Science, Washington, DC.

National Research Council, 1992. Neem: A Tree for Solving Global Problems. Report of an Ad Hoc Panel of the Board on Science and Technology for International Development. Washington, DC, Vietmeyer, N. D. (Director) USA, National Academy Press, Washington, DC. 9168332.

New, T.R. (Ed.), 2002. Insect Conservation: Past, Present and Perspectives. Springer, NY.

Newman, D.J., Cragg, G.M., 2016. Natural products as sources of new drugs from 1981 to 2014. J. Nat. Prod. 79 (3), 629–661.

Nicoletti, M., 2011. HPTLC fingerprint: a modern approach for the analytical determination of botanicals. Rev. Bras. Farmacogn. 21, 818–823.

Nicoletti, M., 2013. Traceability in multi-ingredient botanicals by HPTLC fingerprint approach. J. Plan. Chromatogr. 26 (3), 243–247.

Nicoletti, M., 2014a. Advances in production of functional foods and nutraceuticals. Chapter 1, In: Brar Kaur, S.K., Kaur, S., Singh Dhillon, G. (Eds.), Nutraceuticals and Functional Foods. Natural Remedy. Food Science of Technology, NOVA Publisher, New York, USA.

Nicoletti, M., 2014b. Advanced in production of functional foods and nutraceuticals. Chapter 1, In: Brar, S.K., Kaur, S., Dhillon, G.S. (Eds.), Nutraceuticals and Functional Foods. Natural Remedy. Food Science and Technology, NOVA Publisher.

Nicoletti, M., Del Serrone, P., 2017. Intelligent and smart packaging. In: Mikkola, H. (Ed.), Future Foods. In Tech Open, London, UK.

Nicoletti, M., Murugan, K., 2013. Neem the tree of XXI century. Pharmacology 3, 115–121.

Nicoletti, M., Toniolo, C., 2012. HPTLC fingerprint analysis of plant staminal products. Comput. Sci. https://doi.org/10.4172/2157-7064.1000148.

Nicoletti, M., Toniolo, C., 2015. Analysis of multi-ingredient food supplements by fingerprint HPTLC approach. J. Chem. Chem. Eng. 9, 239–244.

Nicoletti, M., Serafini, M., Aliboni, A., D'Andrea, A., Marian, S., 2010. Toxic effects of neem cake extracts on Aedes albopictus (Skuse) larvae. Parasitol. Res. 107, 89–94.

Nicoletti, M., Mariani, S., Maccioni, O., Coccioletti, T., Murugan, K., 2012a. Neem cake: chemical composition and larvicidal activity on Asian tiger mosquito. Parasitol. Res. 111, 205–213.

Nicoletti, M., Petitto, V., Gallo, F.R., Multari, G., Federici, E., Palazzino, G., 2012b. The modern analytical determination of Botanicals and similar novel natural products by the HPTLC Fingerprint approach. In: Atta-ur-Rahman (Ed.), Studies in Natural Products Chemistry. Elsevier, Oxford, UK, pp. 217–258.

Nicoletti, M., Murugan, K., Del Serrone, P., 2014. Current mosquito-borne disease emergencies in Italy and climate changes. The neem opportunity. Trends Vector Res. Parasitol. 1, 2.

Nicoletti, M., Murugan, K., Canale, A., Benelli, G., 2017. Neem-borne molecules as eco-friendly control tools against mosquito vectors of economic importance. Curr. Org. Chem. 20 (25), 2681–2689.

Nicoletti, M., Serafini, M., Maggi, F., Benelli, G., 2018. Professor Philippe Rasoanaivo. Nat. Prod. Res. 30 (19), 2135–2136.

Nix, S., 2007. Neem Tree—"The Village Pharmacy", CBS Publishers and Distributors OVT LTD, Lucknow (UP), India.

Ofek, G., et al., 1998. The control of the oriental red scale, *Aonidiella orientalis* Newstead and the California red scale, *A. aurantii* (Maskell) (Homoptera: Diaspididae) in mango orchards in Hevel Habsor (Israel). Alon Hanotea 51 (5), 212–218.

Otoni, C.G., Espitia, P.J.P., Avena-Bustillos, R.J., McHugh, T.H., 2016. Trends in antimicrobial food packaging systems: emitting sachets and absorbent pads. Food Res. Int. 83, 60.

Ozdemir, M., Floros, J.D., 2004. Active food packaging technologies. Crit. Rev. Food Sci. Nutr. 44 (3), 185–193.

Palanappian, K., Holley, R.A., 2010. Use of natural antimicrobials to increase antibiotic susceptibility of drug resistant bacteria. Int. J. Food Microbiol. 140, 164–168.

Patzewitz, E.-M., Müller, S., 2010. Glutathione biosynthesis and metabolism in *Plasmodium falciparum*. Malar. J. 9, P37.

Pavela, R., et al., 2016a. Traditional herbal remedies and dietary spices from Cameroon as novel sources oflarvicides against filariasis mosquitoes? Parasitol. Res. 115 (12), 4617–4626.

Pavela, R., et al., 2016b. Chemical composition of Cinnamosma madagascariensis (Cannelaceae) essential oil and its larvicidal potential against the filariasis vector Culex quinquefasciatus Say. S. Afr. J. Bot. 108, 359–363.

Perisco, M., et al., 2017. The interaction ofheme with plakotin and a synthetic endoperoxide analogue: new insights into the heme-activated antimalarial mechamism. Sci. Rep. 7, 45485.

Puri, H.S., 1999. Neem: The Divine Tree Azadirachta indica. Harwood Academic Publishers, Australia. ISBN: 9057023482.

Rafatro, H., 2000. Reversal activity of the naturally occurring chemosensitizer malagashanine in Plasmodium malaria. Biochem. Pharmacol. 59 (9), 1053–1061.

Ragasa, C.Y., Nacpil, Z.D., Natividad, G.M., Tada, M., Coll, J.C., Rideout, J.A., 1997. Tetranortriterpenoids from *Azadirachta indica*. J. Phytochem. 46, 555–558.

Rakotosaona, R., et al., 2015. Effect of the Leaf Essential Oil from *Cinnamosma madagascariensis* Danguy on Pentylenetetrazol-induced Seizure in Rats. Chem. Biodivers. 14(10), e1700256.

Ramanitrahasimbola, D., et al., 2001. Biological activities of the plant-derived bisindole voacamine with reference to malaria. Phytother. Res. 15, 30–33.
Ramanitrahasimbola, D., Rasoanaivo, P., Ratsimamanga, S., Vial, H., 2006. Malagashanine potentiates chloroquine antimalarial activity in drug resistant plasmodium malaria by modifying both its efflux and influx. Mol. Biochem. Parasitol. 146 (1), 58–67.
Rasoanaivo, P., Galeffi, C., De Vicente, Y., Nicoletti, M., 1991. Malagashanine and malagashine, the alkaloids of Strychnos mostuoides. Rev. Latinoam. Quìm. 22 (1), 32–34.
Rasoanaivo, P., Petitjean, A., Ratsimamanga-Urverg, S., Rakoto-Ratsimamanga, A., 1992. Medicinal plants used to treat malaria in Madagascar. J. Ethnopharmacol. 37 (2), 117–127.
Rasoanaivo, P., Galeffi, C., Palazzino, G., Nicoletti, M., 1996. Revised Structure of malagashanine: a new Nb,C(21)-secocuran alkaloid. Gazz. Chim. Ital. 126 (8), 1517–1519.
Rasoanaivo, P., Ratsimamanga-Uveg, S., Frappier, F., 1996a. Reversing agents in treatment of drug-resistance malaria. Curr. Med. Chem. 3 (1), 1–10.
Rasoanaivo, P., Palazzino, G., Nicoletti, M., Galeffi, C., 2001. The co-occurrence of C(3) epimer Nb,C(21)-secocuran alkaloids in *Strychnos diplotricha* and *Strychnos myrtoides*. Phytochemistry 56 (8), 863–867.
Rasoanaivo, P., et al., 2004. Screening extracts of madagascan plants in search of antiplasmodial compounds. Phytother. Res. 18 (9), 742–747.
Rowinsky, E.K., Cazenave, L.A., Donehower, R.C., 1990. Taxol: a novel investigational antimicrotubule agent. J. Natl. Cancer Inst. 82 (15), 1247–1259.
Roy, A., Saraf, S., 2006. Limonoids: overview of significant bioactive triterpenes distributed in plants Kingdom. Biol. Pharm. Bull. 29 (2), 191–201.
Ruskin, F.R., 1992. Neem, A Tree for Solving Global Problems. National Academy Press, Washington, DC.
SaiRam, M., et al., 2002. Anti-microbial activity of a new vaginal contraceptive NIM-76 from neem oil (*Azadirachta indica*). J. Ethnopharmacol. 71, 377–382.
Saliba, K.J., et al., 1998. Transport and metabolism of the essential vitamin pantothenic acid in human erythrocytes infected with the malaria parasite Plasmodium falciparum. J. Biol. Chem. 273, 10190–10195.
Samways, M.J., 1993. Insects in biodiversity conservation: some perspectives and directives. Biodivers. Conserv. 2 (3), 258–282.
Sánchez-Bayo, F., Wyckhuys, K.A.G., 2019. Worldwide decline of the entomofauna: a review of its drivers. Biol. Conserv. 232, 8–27.
Sandanasamy, J.D.O., Nour, A.H., Nur, S.N.B., Tajuddin, A.H., 2013. Fatty acid composition and antibacterial activity of Neem (*Azadirachta indica*) seed oil. Open Conf. Proc. J. 4, 43–48.
Sara, S.B., Folorunso, O.A., 2002. Potentials of utilizing neem tree for desertification control in Nigeria. In: Ukwe, C.N., Folorunso, A.O., Ibe, A.C., Lale, N.E.S., Sieghart, L. (Eds.), Sustainable Industrial Utilization of Neem Tree (*Azadirachta indica*) in Nigeria. UNIDO Regional Dev Centre, Lagos, pp. 45–51.
Schmutterer, H., 1998. Some arthropod pests and a semi-parasitic plant attacking neem (*Azadirachta indica*) in Kenya. Anz. Schadlingskde. Pflanzenschutz Umweltschutz 71, 36.
Schumetter, H., Singh, R.P., 1995. List of insect pests susceptible to neem products. In: Schumetterer, H. (Ed.), The Neem Tree: *Azadirachta indica* A. Juss. and Other Meliaceous Plants, Sources of Unique Natural Products for Integrated Pest Management, Medicine, Industry and Other Purposes. VCH Weinheim, New York, pp. 325–326.

Schumutterer, H. (Ed.), 1995. The Neem Tree: Source of Unique Natural Products for Integrated Pest Management, Medicine, Industry and Other Purposes. VCH, Wenheim, Germany, pp. 1–696.
Schumutterer, H., 2002. The Neem Tree (Azadirachta indica A. Juss) and Other Meliaceous Plants: Sources of Unique Natural Products for Integrated Pest Management, Medicine, Industry and Purposes, first ed. Neem Foundation, Mumbai, India.
Shaalan, E.A.S., Canyonb, D., Younesc, M.W.F., AbdelWahaba, H., Mansoura, A.H., 2005. A review of botanical phytochemicals with mosquitocidal potential. Environ. Int. 3, 1149–1166.
Sharma, V.P., Ansari, M.A., Razdan, R.K., 1993. Mosquito repellent action of neem (*Azadirachta indica*) oil. J. Am. Mosq. Control Assoc. 9 (3), 359–360.
Singh, R.K., Singh, N., 2005. Quality of packaged foods. In: Han, J.H. (Ed.), Innovations in Food Packaging. Elsevier Academic Press, Amsterdam, pp. 22–24.
Staines, H.M., et al., 2004. Plasmodium falciparum-induced channels. Int. J. Parasitol. 34, 665–673.
Sujarwo, W., Keim, A.P., Caneva, G., Toniolo, C., Nicoletti, M., 2016. Ethnobotanical uses of neem (Azadirachta indica A. Juss.; Meliaceae) leaves in Bali (Indonesia) and the Indian subcontinent in relation with historical background and phytochemical properties. J. Ethnopharmacol. 189, 186–193.
Tehri, K., Singh, N., 2013. The role of botanicals as green pesticides in integrate mosquito management—review. Int. J. Mosq. Res. 2 (1), 18–23.
Tewari, D.N., 1992. Monograph on Neem (*Azadirachta indica* A. Juss.). International Book Distributors, Dehra Dun.
Thakur, M.S., Ragavan, R.V., 2013. Biosensors in food processing. J. Food Sci. Technol. 50 (4), 625–641.
Thomas, S.L., Lew, V.L., 2004. Plasmodium falciparum and the permeation pathway of the host red blood cell. Trends Parasitol. 20, 122–125.
Tikar, S.N., Mendki, M.J., Chandel, K., Parashar, B.D., Prakash, S., 2018. Susceptibility of immature stages of Aedes aegypti, the vector of dengue and chikungunya to insecticides from India. Parasitol. Res. 102, 907–913.
Tiwari, R., et al., 2014. Neem (*Azadirachta indica*) and its potential for safeguarding animals and humans. J. Biol. Sci. 14 (2), 110–123.
Toniolo, C., Nicoletti, M., Murugan, K., 2013. The HPTLC approach to metabolomic determination of neem products composition. Pharmacology 3, 122–127.
Toniolo, C., Nicoletti, M., Maggi, F., Venditti, A., 2014. Determination by HPTLC of chemical composition variability in raw material used in botanicals. Nat. Prod. Res. 28 (2), 119–126.
Toniolo, C., Di Sotto, A., DiGiacomo, S., Carsoli, E., Belluscio, A., Nicoletti, M., Ardizzone, G., 2018. Costa Concordia disaster: environmental impact from phytochemical point of view. Ecosphere. https://doi.org/10.1002/ecsz.2054.
Trapanelli, S., et al., 2016. Trasmission blocking effects of neem (Azadiracha indica) seed kernel limonoids on Plasmodium berghei sporogonic development. Fitoterapia 114, 122–126.
Valletta, A., Salvadori, E., Santamaria, A.R., Nicoletti, M., et al., 2015. Ecophysiological and phytochemical response to ozone of wine grape cultivars of Vitis vinifera. Nat. Prod. Res. 30 (22), 1–9.
Van der Nat, J.M., van der Sluis, W.G., de Silva, K.T., Labadie, R.P., 1991. Ethnopharmacognostical survey of *Azadirachta indica* A Juss. J. Ethnopharmacol. 35 (1), 1–24.
Vieira, I.J.C., Braz-Filho, R., 2006. Quassinoids: structural diversity, biological activity and synthetic studies. In: Studies in Natural Products Chemistry.vol. 33, pp. 433–492.

WHO/UNEP. Public health impact of pesticides used in agriculture: reportage of a World Health Organization and U.N. Environmental Programme. 1989. https://apps.who.int/iris/handle/10665/61414.

WHO, 2010. Malaria Treatment Guidelines. World Health Organization, Geneva.

Willcox, G., Bodeker, G., 2004. Traditional herbal medicines for malaria. BMJ 329, 1156–1159.

Index

Note: Page numbers followed by *f* indicate figures and *t* indicate tables.

B

C

D

E

F

G

I

O

P

Q

R

Printed in the United States
By Bookmasters